Earth and Environmental Sciences Library

Series Editors

Abdelazim M. Negm, Faculty of Engineering, Zagazig University, Zagazig, Egypt

Tatiana Chaplina, Antalya, Türkiye

Earth and Environmental Sciences Library (EESL) is a multidisciplinary book series focusing on innovative approaches and solid reviews to strengthen the role of the Earth and Environmental Sciences communities, while also providing sound guidance for stakeholders, decision-makers, policymakers, international organizations, and NGOs.

Topics of interest include oceanography, the marine environment, atmospheric sciences, hydrology and soil sciences, geophysics and geology, agriculture, environmental pollution, remote sensing, climate change, water resources, and natural resources management. In pursuit of these topics, the Earth Sciences and Environmental Sciences communities are invited to share their knowledge and expertise in the form of edited books, monographs, and conference proceedings.

Ayad M. Fadhil Al-Quraishi · Yaseen T. Mustafa
Editors

Natural Resources Deterioration in MENA Region

Land Degradation, Soil Erosion, and Desertification

Springer

Editors
Ayad M. Fadhil Al-Quraishi
Petroleum and Mining Engineering
Department
Tishk International University
Erbil, Kurdistan Region, Iraq

Yaseen T. Mustafa
Environmental Sciences Department
University of Zakho
Zakho, Kurdistan Region, Iraq

ISSN 2730-6674   ISSN 2730-6682 (electronic)
Earth and Environmental Sciences Library
ISBN 978-3-031-58314-8   ISBN 978-3-031-58315-5 (eBook)
https://doi.org/10.1007/978-3-031-58315-5

© The Editor(s) (if applicable) and The Author(s), under exclusive license to Springer Nature Switzerland AG 2024

This work is subject to copyright. All rights are solely and exclusively licensed by the Publisher, whether the whole or part of the material is concerned, specifically the rights of translation, reprinting, reuse of illustrations, recitation, broadcasting, reproduction on microfilms or in any other physical way, and transmission or information storage and retrieval, electronic adaptation, computer software, or by similar or dissimilar methodology now known or hereafter developed.
The use of general descriptive names, registered names, trademarks, service marks, etc. in this publication does not imply, even in the absence of a specific statement, that such names are exempt from the relevant protective laws and regulations and therefore free for general use.
The publisher, the authors and the editors are safe to assume that the advice and information in this book are believed to be true and accurate at the date of publication. Neither the publisher nor the authors or the editors give a warranty, expressed or implied, with respect to the material contained herein or for any errors or omissions that may have been made. The publisher remains neutral with regard to jurisdictional claims in published maps and institutional affiliations.

This Springer imprint is published by the registered company Springer Nature Switzerland AG
The registered company address is: Gewerbestrasse 11, 6330 Cham, Switzerland

Paper in this product is recyclable.

Preface

This comprehensive volume presents a vital multidisciplinary exploration of environmental challenges in the MENA region. Titled *Natural Resources Deterioration in MENA Region: Land Degradation, Soil Erosion, and Desertification*, the book is structured into three distinct parts, spread across fifteen chapters, and written by over fifty international experts. It delves into the complexities of land degradation, hydrological dynamics, soil erosion, drifting sands, sand/dust storms, and desertification. Part I, 'Land Degradation—Geological and Hydrological Status,' analyzes the interplay between geological, hydrological, and human factors in land degradation. The 'Soil Erosion' part offers an in-depth examination of soil erosion processes, while the final part on 'Drifting Sands, Sand/Dust Storms, and Desertification' highlights the critical impact of these phenomena on ecosystems and human livelihoods. This volume aims to align with Sustainable Development Goals related to environmental stewardship, offering scientific insights and practical solutions for effective environmental protection and sustainable resource management in the MENA region. It serves as an indispensable resource for policymakers, researchers, and stakeholders committed to fostering sustainable environmental practices.

Chapter titled "Remote Sensing Techniques for Investigating Natural Resources Deterioration: Application on Agricultural Degradation in Sultanate Oman," explores agricultural degradation in Oman, attributing it to urbanization, climate change, water scarcity, natural disasters, and agricultural challenges like pests, diseases, and soil salinity. The chapter critiques traditional ground-based monitoring methods for their labor-intensiveness and inefficiency and employs remote sensing techniques using satellite and aerial imagery instead. This method allows for an extensive analysis of agricultural land, highlighting the decline in arable land due to urban expansion, particularly in the Dhofar Governorate, and detailing the adverse effects of climatic change on agricultural productivity, including heat, droughts, and cyclones. It also addresses the decrease in per capita water availability in arid Oman and its impact on agriculture. The results show a marked decline in vegetation cover due to urbanization, climate change, water scarcity, and cyclones like Shaheen. This chapter further examines the effects of pests, diseases, and soil salinity on crop yields. In conclusion, this study underscores the escalating threat of environmental and climatic challenges to agricultural sustainability in Oman, highlighting the importance of the adopted remote sensing approach for effectively managing and preserving agricultural land productivity and sustainability.

Chapter "Spectral Angle Mapper Approach (SAM) for Land Degradation Mapping: A Case Study of the Oued Lahdar Watershed in the Pre-Rif Region (Morocco)," examines land degradation in the Oued Lahdar Watershed, Morocco. Utilizing the SAM approach with Sentinel-2 imagery, this study effectively mapped land use/land cover (LULC) to identify land degradation features. The methodology involves a detailed supervised classification process, including data preparation, training data collection, feature selection, SAM classification, and post-classification processing. The results revealed varied LULC classifications, showing that significant areas, especially bare soil (33%) and agricultural land (27%), are prone to land degradation. These insights highlight the need for immediate action to mitigate land degradation risks. The chapter concludes that the SAM approach, combined with remote sensing, is highly effective for mapping and understanding land degradation patterns. This information is crucial for environmental monitoring, ecosystem conservation, and informed land-use planning, providing valuable insights for policymakers, researchers, and stakeholders to support sustainable land management and combat land degradation.

Chapter titled "Remotely Sensed Data and GIS for Long-Term Monitoring of the Ghout Oases Degradation in the Region of Oued Souf (Northeastern Algerian Sahara)," delves into the degradation of Ghout oases in Algeria's Oued Souf region. The chapter utilized medium-resolution Landsat imagery and GIS, applying spectral indices and change detection methods from 1987 to 2018 to assess and monitor the dynamics of Ghout oases across 18 municipalities. The study revealed a significant decrease in the area occupied by Ghout, from 4462 hectares in 1987 to 1033 hectares in 2018, with a notable reduction, particularly in the northern areas. Concurrently, the built-up areas expanded, and new irrigated perimeters extended throughout the region. This transition from a traditional Oasian system to one driven by a free-market economy resulted in ecological imbalances, primarily manifested in groundwater level fluctuations, culminating in the gradual disappearance of Ghout oases. The chapter concluded that, to achieve sustainable agro-development in the Saharan region, adopting new economic strategies that ensure rational and controlled exploitation of soil and water resources is imperative, thereby preserving the delicate ecological balance of these unique agroecosystems.

Chapter entitled "Tectonic and Erosion in the Zagros Fold-and-Thrust Belt (ZFTB)," delves into the complex interplay among tectonics, climate, and geomorphological features within the ZFTB. This study utilized geomorphic indices, such as the hypsometric index and river profile steepness, to elucidate erosional processes and their connection to environmental land degradation. The chapter revealed a distinct uniformity among regions characterized by higher precipitation, steeper slopes, increased river steepness, and elevated hypsometric index values, particularly in the northeast of the Kirkuk Embayment and Bakhtyari Culmination areas. Areas with higher-strength lithologies were correlated with regions exhibiting greater hypsometric index steepness and precipitation levels. This finding suggests that the erodibility of the exposed lithology is not the sole determinant of the elevated geomorphic index values. The study highlighted a significant correlation between climatic conditions and tectonic uplift, which govern areas of heightened erosion marked by steep slopes, high hypsometric indices, and increased steepness values. The chapter underscored the importance of comprehending the synergy between climate, tectonics, and erosion to effectively assess land degradation and formulate strategic, sustainable land-use management practices in the region.

Chapter titled "Characterization of Post-uprising Impacts on Landcover and Land Use: Al Wasita-Satish—Area Northeast Libya Case Study," presents a detailed analysis of the land use and land cover (LULC) changes in Northeast Libya and the consequences of the civil war. Utilizing Landsat 5 Thematic Mapper (TM) and Landsat 8 OLI-TIRs imagery from 2010 and 2022, the study meticulously assessed alterations in LULC. It identified predominant LULC types, such as agriculture, arid urban areas, and dense and sparse vegetation. The results indicated notable shifts: a significant increase in barren land by 30.50%, and reductions in sparse and dense vegetation by 29.09% and 15.81%, respectively. In addition, the study observed a marginal increase in agricultural and urban areas. The findings concluded that the Libyan Civil War profoundly impacted the LULC, underscoring the period's marked ineffectiveness of land and environmental protection laws. The displacement of populations from conflict zones significantly contributed to these land-cover changes. The chapter underscores the necessity for alternative economic activities, such as tourism, to support local communities without adequate financial resources. This study offers vital insights, aiding policymakers and stakeholders in devising strategies for sustainable land management and counteracting the impacts of land degradation.

Chapter titled "Construction Industry Role in Natural Resources Depletion and How to Reduce It," critically examines the significant impact of the construction industry on natural resource depletion. This chapter underscored the construction sector's role as a major consumer of Earth's resources, contributing extensively to environmental pollution, greenhouse gas emissions, and substantial energy consumption. Industry accounts for a significant portion of national energy use, material consumption, and waste production, particularly in the United States. The chapter presents various statistics to depict the global extent of resource usage and pollution attributed to construction activities. Emphasizing the criticality of sustainable design and material recycling, this study highlights these strategies as pivotal for mitigating the industry's environmental footprint. In conclusion, the chapter advocated adopting more sustainable practices within the construction industry as a key measure to conserve natural resources and diminish environmental degradation.

Chapter titled "Assessing Soil Erosion Vulnerability in Semi-Arid Haouz Plain, Marrakech, Morocco: Land Cover, Socio-Spatial Mutations, and Climatic Variations," presents a comprehensive examination of soil erosion in the Haouz Plain, Morocco. Employing the Revised Universal Soil Loss Equation (RUSLE) within a Geographic Information System (GIS) framework, this study estimated soil erosion from 1992 to 2020. This study focuses on the interplay between land-cover dynamics, socio-spatial changes, and climate variations. This methodology entailed an in-depth analysis of precipitation, soil characteristics, land use, and digital terrain models to construct individual variables for the RUSLE model. The findings revealed significant fluctuations in soil erosion rates over the years, with a marked increase in areas with soil losses exceeding 26 tons per hectare per year. This study underscored the necessity of implementing effective and sustainable soil management practices to curb soil erosion, highlighting its critical importance for the sustainability of the plain. This chapter emphasized that precise soil loss estimation using RUSLE and GIS is indispensable for sustainable land management and effective soil erosion control strategies.

Chapter titled "GIS-Based Erosion Potential Method (EPM) for Soil Degradation Evaluation: A Case Study the Northeast of Morocco," focused on assessing soil erosion in the Oued Amter watershed in Northeast Morocco using the GIS-based Erosion Potential Method (EPM). This study developed a comprehensive erosion risk assessment model by integrating various spatial data, such as topography, rainfall, land cover, and soil properties. The results revealed that soil loss varied significantly across the watershed, ranging from 55.91 to 44,088.33 m^3/km^2/year. These findings underscore the region's urgent need for soil conservation and land-use planning. The study concluded that the EPM outcomes could assist policymakers, land managers, and researchers in making informed decisions and implement effective measures to safeguard the soil and promote sustainable land-use practices.

Chapter titled "Predicting Soil Erosion Using RUSLE Model in Duhok Governorate, Kurdistan Region of Iraq," focuses on analyzing soil erosion in the Duhok Governorate, Kurdistan region of Iraq. This study applied the Revised Universal Soil Loss Equation (RUSLE) model and Geographic Information System (GIS) tools to assess and forecast soil erosion rates, incorporating key factors such as rainfall erosivity, soil erodibility, slope gradient, land use/cover, and conservation practices. The findings indicated that approximately 78% of the Duhok Governorate experienced very low to low soil erosion levels, whereas areas with high to very high erosion rates constituted only a minor portion of the region (4% and 7%, respectively). The study also emphasized the significant role of slope and land-use interactions in accelerating soil erosion, especially in agricultural areas with steep slopes. In conclusion, this chapter stressed the importance of comprehending these dynamic interactions to devise effective soil erosion control strategies and implement suitable conservation measures.

Chapter titled "Spatiotemporal Variability of Aerosol Optical Depth Over the Arabian Peninsula Using MODIS Data" analyzes aerosol optical depth (AOD) variations in the Arabian Peninsula. The study used Moderate Resolution Imaging Spectroradiometer (MODIS) data from 2003 to 2019 to examine the seasonal, interannual, and spatial distributions of AOD. The results showed high AOD levels in desert and coastal regions during summer, with mountainous areas having the lowest AOD due to fewer sand and dust particles. The spring and summer had the highest AOD, which was attributed to intensive wind patterns, increased dust emissions, and high humidity. The chapter concluded that annual AOD concentration levels increased by approximately 0.5% during the study period. This research provides critical insights into the variability of aerosol concentrations, which is vital for understanding environmental impacts and planning mitigation strategies.

Chapter titled "The Carrying Loads Composition of Storms Over Iraq" delved into the composition of dust storms in Iraq, particularly from 2007 to 2010. This study aims to develop a monitoring methodology and investigate the inorganic and organic components of these storms. Dust samples were collected and analyzed for texture, mineral content, heavy metal concentrations, and organic components, including pollen and microorganisms. The findings revealed that the average dust texture was sandy silty clay, with light minerals primarily consisting of quartz and carbonate, and feldspar, gypsum, clay, and heavy minerals. Additionally, heavy metals such as Fe, Pb, Zn, Ni, Co, Cd, and Cu were identified with uranium activity below the critical dose level. Pollens, fungi, bacteria, and some algae are also common during these storms. The chapter concludes that aerosols in Iraq are a mix of natural and anthropogenic pollutants, and understanding their composition is crucial for environmental risk assessment in the region.

Chapter titled "Monitoring Drifting Sand Using Spectral Index and Landsat TM/OLI Datasets in Bahr An-Najaf Area, Iraq," focuses on the study of dune movement and its environmental impact in the Bahr An-Najaf area of Iraq. This study utilized Landsat TM/OLI images from 1988 to 2018 and the Normalized Difference Sand Dunes Index (NDSDI) to map and analyze the drift and expansion of sand dunes. The results showed significant changes in the size of the sand dunes, especially between 1988 and 2008, with notable expansion in certain areas. This expansion and migration substantially threatened human activities, including in Al-Rahima, Eitha, Khiribah, and Um-Tharawi. The creep rate of sand dunes was calculated every decade, revealing their approach toward populated areas, with some dunes moving as much as 3000 m. The chapter concluded that the rapid movement and expansion of sand dunes in this region requires immediate attention to mitigate their impact on local communities and infrastructure.

Chapter titled "Assessment of the Growth of Urban Heat Island in a Mediterranean Environment: A Pathway Toward a Sustainable City," delves into studying the Urban Heat Island (UHI) phenomenon in Meknes, Morocco. Employing Landsat imagery spanning from 1990 to 2020 and MODIS data from 2000 to 2021, the research focused on analyzing the changes in Land Surface Temperature (LST) in relation to land-use and land-cover changes (LULCC). The findings revealed a marked increase in LST throughout the study period, with the maximum LST in Meknes escalating by approximately 10 Â °C. This significant increase was primarily attributed to the transformation of vegetation cover into urban built-up areas. Additionally, the study documented growth in urban and arboricultural areas, alongside a decline in cereal cultivation and peri-urban landscapes. The chapter concludes by emphasizing the urgent need for sustainable urban planning measures to effectively counteract the intensifying UHI effect exacerbated by ongoing urban expansion and the diminution of vegetation cover.

Chapter titled "Environmental Challenges, The Impacts of Climate Change in North Africa Region: A Review," analyzes the environmental stresses caused by climate change in North Africa. The chapter emphasized that the region, being one of the most sensitive to climate change due to its location and dry environment, faces major environmental issues and international disputes. This study reviews the regional impacts of global climate change in North Africa, focusing on the most pressing environmental issues and potential remedies. It identified that climate change is largely responsible for the environmental problems in North Africa, threatening the area's long-term stability, especially in terms of poverty, inequality, and underdevelopment. However, efforts to address these issues have been hampered by the ignorance of their causes and potential solutions. The chapter concluded that North African countries must consider environmental conservation in the context of regional sustainable development and propose responsibilities for government and external aid.

Chapter entitled "Soil Salinization Impacts on Land Degradation and Desertification Phenomenon in An-Najaf Governorate, Iraq," explores the problem of soil salinization in Iraq's An-Najaf Governorate, and its role in land degradation and desertification. This study focused on the environmental, economic, and social impacts of desertification, exacerbated by climate change, low rainfall, and unsustainable agricultural practices. This chapter analyzes the increase in soil salinity, particularly in summer, due to high temperatures, which leads to enhanced evaporation and salt deposition, adversely affecting plant growth and productivity. This process contributes to soil erosion and is regarded as a critical indicator of desertification. This chapter emphasizes the potential of remote sensing data for assessing, monitoring, and controlling desertification criteria. This conclusion highlights the importance of understanding the causes of desertification for effective management and the need for governmental and international interventions for environmental conservation in the context of regional sustainable development.

Erbil, Kurdistan Region, Iraq	Ayad M. Fadhil Al-Quraishi
Zakho, Kurdistan Region, Iraq	Yaseen T. Mustafa
November 2023	

Contents

Land Degradation—Geological and Hydrological Status

Remote Sensing Techniques for Investigating Natural Resources
Deterioration: Application on Agricultural Degradation
in Sultanate Oman 3
Yaseen A. Al-Mulla, Ahsan Ali, Mezna Alalawi,
and Mohammed Bait-Suwailam

Spectral Angle Mapper Approach (SAM) for Land Degradation
Mapping: A Case Study of the Oued Lahdar Watershed
in the Pre-Rif Region (Morocco) 15
Brahim Benzougagh, Ayad M. Fadhil Al-Quraishi, Youssef Bammou,
Shuraik Kader, Mohammed El Brahimi, Driss Sadkaoui, and Latifa Ladel

Remotely Sensed Data and GIS for Long-Term Monitoring
of the Ghout Oases Degradation in the Region of Oued Souf
(Northeastern Algerian Sahara) 37
Nouar Boulghobra

Tectonic and Erosion in the Zagros Fold-and-Thrust Belt (ZFTB) 55
Ahmed K. Obaid, Arsalan A. Othman, Sarkawt G. Salar,
Varoujan K. Sissakian, and Salahalddin S. Ali

Characterization of Post-uprising Impacts on Landcover and Land
Use: Al Wasita-Satish—Area Northeast Libya Case Study 73
Salah Hamad and Attia Alsanousi

Construction Industry Role in Natural Resources Depletion
and How to Reduce It 93
Bayan Salim Obaid Al-Numan

Soil Erosion

Assessing Soil Erosion Vulnerability in Semi-Arid Haouz Plain, Marrakech, Morocco: Land Cover, Socio-Spatial Mutations, and Climatic Variations 113
Youssef Bammou, Brahim Benzougagh, Brahim Igmoullan,
Ayad M. Fadhil Al-Quraishi, Fadhil Ali Ghaib, and Shuraik Kader

GIS-Based Erosion Potential Method (EPM) for Soil Degradation Evaluation: A Case Study the Northeast of Morocco 135
Mohammed El Brahimi, Brahim Benzougagh, Mohamed Mastere, Bouchta El Fellah, Ayad M. Fadhil Al-Quraishi, Najia Fartas, and Khaled Mohamed Khedhe

Predicting Soil Erosion Using RUSLE Model in Duhok Governorate, Kurdistan Region of Iraq 171
Azade Mehri, Hazhir Karimi, Yaseen T. Mustafa,
Ayad M. Fadhil Al-Quraishi, and Saman Galalizadeh

Drifting Sands, Sand/Dust Storms, and Desertification

Spatiotemporal Variability of Aerosol Optical Depth Over the Arabian Peninsula Using MODIS Data 191
Abdelgadir Abuelgasim and Ashraf Farahat

The Carrying Loads Composition of Storms Over Iraq 209
Moutaz A. Al-Dabbas

Monitoring Drifting Sand Using Spectral Index and Landsat TM/ OLI Datasets in Bahr An-Najaf Area, Iraq 241
Ghadeer F. Al-Kasoob, Ahmed H. Al-Sulttani,
Ayad M. Fadhil Al-Quraishi, and Ragad N. Hussein

Assessment of the Growth of Urban Heat Island in a Mediterranean Environment: A Pathway Toward a Sustainable City 261
Mohammed El Hafyani, Narjisse Essahlaoui, Ali Essahlaoui,
Meriame Mohajane, Abdelali Khrabcha, and Anton Van Rompaey

Environmental Challenges, The Impacts of Climate Change in North Africa Region: A Review 281
Afeez Alabi Salami and Olushola Razak Babatunde

Soil Salinization Impacts on Land Degradation and Desertification Phenomenon in An-Najaf Governorate, Iraq 295
Sa'ad R. Yousif

Land Degradation—Geological and Hydrological Status

Remote Sensing Techniques for Investigating Natural Resources Deterioration: Application on Agricultural Degradation in Sultanate Oman

Yaseen A. Al-Mulla, Ahsan Ali, Mezna Alalawi, and Mohammed Bait-Suwailam

Abstract Agriculture is an important economic sector that provides food and other resources for the global population. However, various environmental and climatic issues, such as declining arable areas, erosion, climate change, water scarcity, drought, cyclones, crop and livestock pests, disease and infection, and soil salinity, pose growing challenges to agricultural sustainability. These difficulties can lead to agricultural land degradation, affecting crop productivity and food security considerably. Remote sensing techniques involve the use of satellite or aerial pictures to analyze and monitor the status of agricultural land. These strategies can be used to identify and quantify agricultural degradation indicators, such as soil erosion, crop stress, and insect infestations. One of the most important advantages of remote sensing in this application is its capacity to cover enormous expanses of land rapidly and effectively without requiring ground-based measurements. This enables a more frequent and thorough investigation of agricultural conditions, which may be used to

Y. A. Al-Mulla (✉) · A. Ali · M. Alalawi · M. Bait-Suwailam
Remote Sensing and GIS Research Center, Sultan Qaboos University, Muscat 123, Oman
e-mail: yalmula@squ.edu.om

A. Ali
e-mail: basra240@gmail.com

M. Alalawi
e-mail: m.alawi@squ.edu.om

M. Bait-Suwailam
e-mail: msuwailem@squ.edu.om

Y. A. Al-Mulla
Department of Soils, Water and Agricultural Engineering, Sultan Qaboos University, Muscat 123, Oman

M. Bait-Suwailam
Department of Electrical and Computer Engineering, College of Engineering, Sultan Qaboos University, Muscat, Oman

© The Author(s), under exclusive license to Springer Nature Switzerland AG 2024
A. M. F. Al-Quraishi and Y. T. Mustafa (eds.), *Natural Resources Deterioration in MENA Region*, Earth and Environmental Sciences Library,
https://doi.org/10.1007/978-3-031-58315-5_1

influence management decisions and to prevent the negative consequences of degradation. These strategies can provide valuable insights into the health and productivity of agricultural fields and assist in identifying areas that may require further improvement.

Keywords Natural resources · Deterioration · Remote sensing · Satellite · UAV · Agriculture

1 Introduction

Sustainable agriculture has become a pillar of food security and natural resource management. Traditional methods for monitoring these resources are laborious, time-consuming, and expensive. In the last three decades, technologies for accessing data without physical contact have improved dramatically, helping the industry monitor large amounts of data efficiently. Remote sensing (RS) platforms, such as sensors and satellite imagery resolutions, have been improved (Retallack et al. 2023). Moreover, short-flight and long-flight unmanned aerial vehicles (UAV) and drones can perform real-time monitoring of the study area. Hence, remote sensing techniques can detect and map agricultural production and deterioration with spatial and temporal resolutions. The Main factors causing Oman's agricultural deterioration are explained in the following paragraphs.

2 Shrinking Arable Lands Erosions

Oman is highly diverse in terms of geographic and climatic conditions, and 75% of the total area is desert. In the early 1960s, Gulf Cooperation Council (GCC) countries experienced a sudden urban transformation after the discovery of oil and commercial activities in the region. In 1950, approximately 2.5% of the Omanis lived in urban areas, which increased to 11% and then to 62.1% in 1970 and 2000. Urbanization has reached 80% and is expected to reach 95% by 2050 (Benkari 2017; UN 2018). Urbanization is a common problem for arable-land shrinkage, especially in third-world countries (Al-Awadhi 2007).

The capital absorbs most of the construction and development budget in the Arabian Gulf because they are considered their countries' cultural, economic, and administrative centers because of the difference in development factors such as family income, provision of infrastructure, and access to social facilities. The migration of families from rural to urban areas has increased, leading to rapid urbanization in Oman.

Different pressure indicators were analyzed by Choudri et al. (2015) to monitor the development near the coastal areas of Oman. Their study conducted a detailed literature survey focusing on the influence of different drivers in Omani coastal areas.

Remote Sensing Techniques for Investigating Natural Resources ...

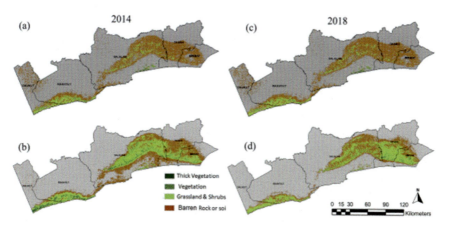

Fig. 1 Vegetation cover before (**a, c**) and after (**b, d**) the monsoon season for 2014 and 2018 in the Dhofar Governorate

They also observed the severity of urbanization pressure on the natural resources in coastal ecosystems. Al-Hatrushi (2013) monitored seashore changes using IRS and IKONOS satellite imagery in the Al-Hwasnah tidal inlet in the Al-Batinah coastal plains, north of Oman. Satellite imagery detected many changes in the study area over a period of six years. The study decreased the number of tidal inlets with an average length of 5.2% and a change rate of 12.4% annually.

Al-Mulla et al. (2022a) used satellite imagery to investigate the 40-year impact of urban expansion on vegetation patterns in Dhofar governorate, south of Oman (Fig. 1). Their study reported that the Dhofar governorate had seen a 173.3% increase in urbanization, which led to a 36.4% decline in vegetation cover. Their study concluded that 23.2% of the total plant species in coastal areas and 18.5% in mountain areas were reduced due to urban expansion.

The Al-Mulla et al. (2022a) study also discovered that the vegetation cover in 1978, with an area of 61,046 ha comprising forest, vegetation, grassland, and shrub, was more condensed in the southwest region of Dhofar (Dalkut and Rakhyut). Conversely, vegetation cover was more dispersed in Dhofar's northeast regions (Mirbat and Taqah). According to the Normalized Difference Vegetation Index (NDVI) analysis of 1978 and 1984 satellite images, the vegetation cover patterns in the southwest and northeast of Dhofar in 1984 were comparable to the pattern reported in 1978. The area covered by vegetation in 1984 was 48,882 ha. In 1988, a similar vegetation cover pattern was observed. However, there was no discernible change in the vegetation area in the study region compared to that in 1984. This year, the vegetation-covered area was 45,048 ha.

Although the vegetation cover pattern in 1993 was comparable to that in 1988, with an area of 45,529 ha, the vegetation coverage in Dhofar's northeast region was reduced compared to 1988. The 1998 vegetation cover pattern remained comparable to that of 1988. However, the northeast region of Dhofar (Mirabt and Taqah) had significantly less vegetation cover than in the previous years. At the same time, plant cover in the study region has remained unchanged since 1993. The vegetation cover in 1998 was 44,827 ha. In 2003, the northeastern part of Dhofar (Mirabt and Taqah) had more vegetation than in the previous years. In the northeastern area of Dhofar, the vegetation cover pattern in 2014 remained similar to that of the previous year. Simultaneously, vegetation coverage in the research region has increased over the years. In 2018, the vegetation-covered area in Dhofar was 40,768 ha, with a decreasing pattern in the northeast and a modest increase in the southwest compared to 2014.

3 Climate Changes

Earth observation data can help monitor climatic changes and better understand climate, biodiversity, and land cover changes. Oman is an arid region with adverse climatic conditions, including heat and drought. Since 1977, many cyclones and storms have affected the coastal areas of Oman. Over the last two decades, many severe cyclones have directly affected Oman's social and economic aspects. In addition, agricultural production is affected by cold and heat waves in Oman (Al-Yahyai et al. 2022). Caspari et al. (2019) used a combined approach of remote sensing and inhabitant science to help understand past local ecological niches and recent land-use shifts. Their study showed that only 1.5 km^2 of the entire peninsula was used effectively for agricultural practices. They concluded that very high-resolution satellite imagery can help decision-makers and policymakers locate suitable locations for man-made water reservoirs for better water management.

Coastal shoreline erosion due to climatic extremes was investigated by Al Ruheili and Boluwade (2021) using Sentinel-2 satellite imagery of the Sultanate of Oman caused by Cyclone Kyarr's impacts. Their study concluded that the Al-Batinah shoreline eroded with an average annual speed of 9 m during the 4-years study period between 2017 and 2020. Al-Mulla et al. (2022a) also reported that climate change has impacted the decline in vegetation cover in the Dhofar governorate. They examined the climate parameters and observed an approximately 1 °C increase in ambient temperature, while the rainfall decreased by 104 mm over the last 40 years. Rajendran et al. (2014) used the Earth observation system of the ASTER sensor on the TERRA platform to map CO_2 sequestration in Samail, Oman. Their study emphasized remote sensing as an effective and inexpensive tool for environmental safety monitoring. Their study detected and mapped the abundance of CO_2 sequestration in the study area.

4 Water Scarcity (Drought)

Water scarcity is also present in Oman, as water availability per capita decreased from 1330 m^3 per person per year in 1990 to only 281.4 m^3 in 2019 (FAO 2021), while the annual rainfall (a key source of freshwater) was less than 100 mm. Figure 2 shows the amount of rainfall in Muscat, Oman, in 2015 and 2016. Therefore, most domestic and agricultural water use depends on desalination plants (Moossa et al. 2022). Therefore, Oman, as a hyper-arid country, needs to monitor its water resources prudently. Approximately 90% of the water-scarce in Oman is consumed by agriculture, and the leading crop in consuming water in Oman is date palm trees (Fig. 3) (Al-Mulla and Al-Gheilani 2017).

Remote sensing techniques have been introduced to monitor and evaluate the water resources in Oman. Ali et al. (2021) used Landsat-8 satellite imagery to estimate actual evapotranspiration for the date palm trees in Oman. Their study used surface energy balance algorithms to estimate actual evapotranspiration by estimating energy residues in date palm trees. Thermal and multispectral satellite imagery bands are used in this study. Their study is unique in estimating actual evapotranspiration in

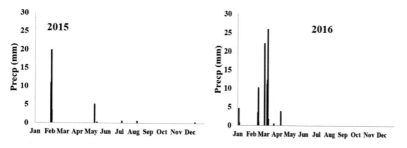

Fig. 2 Rainfall recorded for 2015 and 2016 at the Agriculture Experiment Station, Sultan Qaboos University, Muscat, Oman

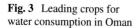

Fig. 3 Leading crops for water consumption in Oman

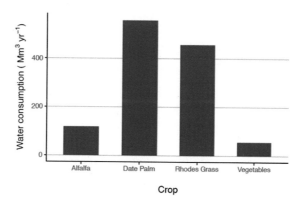

the Sultanate of Oman, as traditional techniques required in field estimation of actual evapotranspiration are time-consuming and laborious.

Akhtar et al. (2022) evaluated the potential groundwater storage in Wadi Al-Jizi, Oman, by integrating Remote sensing techniques with the infield data. Visual analysis of thematic maps and satellite imagery integrated different area characteristics, such as slope, geology, soil type, and geomorphology. Their study concluded that integrating remote sensing and GIS techniques with area characteristics was less time-consuming and cost-effective in evaluating potential groundwater resources in Oman.

5 Cyclones

Owing to climate change, cyclones have become a new norm in our cities. Scovell (2020) defined cyclones as a natural phenomenon created in a low-pressure area surrounded by circulating winds. Oman was struck directly and indirectly by five tropical cyclones. Gonu, in 2007, was the most robust cyclone to hit the Arabian Peninsula in 1945 (AlRuheili 2022). Furthermore, the Shaheen cyclone, which hit the northern part of Oman named Al-Batinah in 2021, has been considered the second-largest cyclone that has hit the Al-Batinah region since 1890 (CAA 2021). AlRuheili (2022) studied the consequences of the Shaheen cyclone along the shoreline of Al-Khaboura using remote sensing and GIS techniques. The results from the study reported that the shoreline witnessed 7.82% erosion and 92% sediment deposition. Their study also reported the downstream sediment deposition caused by flash floods. The values of the end-point rate (EPR), linear regression rate (LRR), and weighted linear regression rate (WLR) suggest similar outcomes for the Al Khaboura shoreline. These statistical metrics show that a massive amount of sediment deposition occurred during the short time frame of the Shaheen event.

Al-Mulla et al. (2022b) conducted the before and after impact assessment of the Shaheen cyclone at vegetation cover and urbanization in the Al-Batinah governorates of Oman using remote sensing and deep learning techniques. Their study integrated satellite imagery with deep learning technology for pre- and post-impact analyses. Their study showed that the cyclone impacted vegetation cover by reducing its area by 17.6% (Fig. 4).

6 Crop and Livestock Pests, Diseases, and Infections

Agriculture production in Oman peaked in the 1990s but declined (MAF 2012). Currently, fruit trees dominate only 31% of the total cropped area of Oman, but fruit crop production accounts for only 17% of the total crop production; date palms dominate 90% of the total fruit harvested area (Al-Yahyai et al. 2022). Bananas are

Fig. 4 Impact of the Shaheen cyclone on the Al-Batinah region. Top: before, bottom: after the cyclone. Red: farms degraded by cyclones. Blue: water inundation intensity

the second crop in Oman's production and export sector, while mangoes are also one of the major crops in Oman (MAF 2017).

Residues from the industry may have adverse effects on crops and trees. Al-Shali et al. (2019) conducted a study to check the possible effect of heavy metals concentration on Oman's date palm tree farms. Heavy metal concentrations were also examined in soil, water, and date fruits. The study showed a significant increase in heavy metal concentrations in date palm farms close to industrial sites. In GCC countries, especially Oman, the date palm has also been adversely affected by different bugs and pests, dominated by Dubas and Red Palm Weevil bugs. Many mango orchards have also been eradicated in many regions of northern Oman due to high infestation levels, in addition to other factors such as soil salinity, which significantly affects mango yield.

Computer neural networks were mostly used for leaf disease detection and were later improved by introducing object detection models. David et al. (2021) reported that the traditional methods of detecting diseases in tomatoes were slow and failed at the early stages of the disease. They proposed an early disease detection system using a hybrid deep learning algorithm.

7 Soil Salinity

Previous and present interactions between humans and natural resources have created unexpected consequences. Soil is an essential natural resource for lands, living organisms, and numerous ecosystems. Agriculture depends on major natural resources such as soil, water, plants, and air. Soil consists of organic materials and minerals with different proportions of water and air (Shahid and Omar 2022). Minerals and organic materials are important for all living organisms. Millions of tons of soil are degraded yearly worldwide, as 2000 ha of agricultural land is lost due to soil salinity (Shahid et al. 2018). Soil degradation is one of the biggest threats to sustainable agriculture, as ever-increasing soil salinity and soil erosion limit the functional activity of different ecosystems and the capacity of sustainable agriculture (Lal 2012; Alam 2014).

Soil salinity must be monitored, controlled, and preserved for better agricultural applications, especially in hyper-arid regions and countries like Oman. Shahid et al. (2018) reported an expected increase in soil salinity, especially in coastal areas, due to rising sea levels and temperatures. Oldeman et al. (1990) attempted to map the human-induced salinity. However, their study did not quantify the global extent owing to the improper scale of the projections. This issue was later resolved by Oldeman (1992) by digitizing map units with a geographic information system (GIS). A later study globally characterized and mapped man-made soil degradation at four salinity intensity levels.

The overexploitation of coastal groundwater has led to salt accumulation in the soil, causing the groundwater salinity level to reach 30,000 ppm along the coastal strip since the mid-first decade of this century (Al Barwani and Helmi 2006). Hence, the impact of salinity on Oman is evident in the agricultural sector, especially in the Al-Batinah coastal region (Abulibdeh et al. 2021). Moreover, owing to the limited resources of fresh water in Oman, there is an increasing trend of using reverse osmosis-based water desalination plants by farmers in their farms to sustain their agricultural production (Al Jabri et al. 2019). This solution creates a problem of brine accumulation. Al-Rahbi et al. (2019) used aerial imagery from unmanned aerial vehicles (UAV) to evaluate how northern Oman's soil and water salinity have impacted date palm trees. In this study, vegetation affected by soil and water salinity was accurately detected with $R^2 = 0.89$ and 0.86, respectively.

Soil salinity has detrimental effects on the physicochemical properties of the soil and on plant growth and production (Brady and Weil 2002). In GCC, especially in Oman, soil salinity is not limited only to the environment but also to the economy. Naifer et al. (2011) reported a loss of USD 1,604 per hectare on low to medium levels of secondary salinization and USD 4,352 per hectare on low to high levels of secondary salinization in the Al-Batinah region in Oman.

Disasters are not new and have been striking mankind since the dark ages. Nevertheless, with technological advancements, we can foresee them before a strike. In this sense, remote sensing can help mitigate the effects of these disasters because of its cost-effectiveness and on-the-spot access to the affected areas. Remote sensing platforms, such as sensors and satellite imagery resolution, have been improved. Moreover, the introduction of short- and long-flight unmanned aerial vehicles (UAV) and drones can enable real-time monitoring of the study area. Hence, remote sensing techniques can detect and map agricultural production and deterioration with good spatial and temporal resolution. There is a dire need to develop international policies and national strategies for detection, risk reduction, climate change, and resource monitoring based on our best understanding, especially its effect on agriculture and soil. This can be achieved by identifying the intensity of the damage and loss patterns in yield and production. A more detailed breakdown of impacts on livestock, crops, fisheries, aquaculture, and forestry is also necessary to create a profile for different possible disasters that can happen slowly (drought) and fast (cyclones). This can also be achieved by creating a nexus between disaster management, risk analysis, and change detection in the climate and the long-term effects of climate change.

8 Conclusions

Finally, various environmental and climatic issues, such as declining arable areas, erosion, climate change, water shortages, droughts, cyclones, crop and livestock pests, diseases and infections, and soil salinity, are progressively threatening agricultural sustainability. These difficulties can lead to the degradation of agricultural land, which can considerably affect crop productivity and food security. It is feasible to identify regions at risk of degradation and to implement appropriate management actions to offset these consequences by employing remote sensing to monitor the influence of these variables on agriculture. This can assist in preserving agricultural land productivity and sustainability as well as the long-term viability of this crucial economic sector. Overall, remote sensing methods provide an effective means of exploring and monitoring the state of agricultural fields and identifying and mitigating the detrimental effects of environmental and climatic variables on agriculture. Hence, it is recommended to initiate more efforts and funds for further research in the applications of remotes sensing technologies in the fields of environmental and climatic issues using high-resolution datasets.

References

Abulibdeh A, Al-Awadhi T, Al Nasiri N, Al-Buloshi A, Abdelghani M (2021) Spatiotemporal mapping of groundwater salinity in Al-Batinah, Oman. Groundw Sustain Dev 12:100551. https://doi.org/10.1016/j.gsd.2021.100551

Al-Awadhi T (2007) Monitoring and modeling urban expansion using GIS & RS: case study from Muscat, Oman. In; 2007 urban remote sensing joint event. IEEE pp 1–5

Al Barwani A, Helmi T (2006) Sea water intrusion in a coastal aquifer: a case study for the area between Seeb and Suwaiq, Sultanate of Oman. J Agricul Marine Sci [JAMS] 11:55–69

Akhtar J, Sana A, Tauseef SM, Chellaiah G, Kaliyaperumal P, Sarkar H, Ayyamperumal R (2022) Evaluating the groundwater potential of Wadi Al-Jizi, Sultanate of Oman, by integrating remote sensing and GIS techniques. Environ Sci Pollut Res 29:72332–72343

Al-Hatrushi SM (2013) Monitoring of the shoreline change using remote sensing and GIS: a case study of Al Hawasnah tidal inlet, Al Batinah coast, Sultanate of Oman. Arab J Geosci 6(5):1479–1484. https://doi.org/10.1007/s12517-011-0424-2

Alam F (2014) Evaluation of hydrogeochemical parameters of groundwater for suitability of domestic and irrigational purposes: a case study from central Ganga Plain, India. Arab J Geosci 7:4121–4131

Ali A, Al-Mulla YA, Charabi Y, Al-Wardy M, Al-Rawas G (2021) Use of multispectral and thermal satellite imagery to determine crop water requirements using SEBAL, METRIC, and SWAP models in hot and hyper-arid Oman. Arab J Geosci 14(7):1–21. https://doi.org/10.1007/s12517-021-06948-0

Al Jabri SA, Zekri S, Zarzo D, Ahmed M (2019) Comparative analysis of economic and institutional aspects of desalination for agriculture in the Sultanate of Oman and Spain. Desalin Water Treat 156:1–6

Al-Mulla Y, Al-Gheilani HM (2017) Increasing water productivity enhances water saving for date palm cultivation in Oman. J Agric Mar Sci [JAMS] 22:87–91

Al-Mulla Y, Al-Ruheili A, Al-Lawati A, Parimi K, Ali A, Al-Sadi N, Al-Harrasi F (2022a) Assessment of urban expansion's impact on changes in vegetation patterns in Dhofar, Oman, using remote sensing and GIS techniques. IEEE Access 10:86782–86792. https://doi.org/10.1109/ACCESS.2022.3198942

Al-Mulla Y, Parimi K, Bait-Suwailam M (2022b) Remote sensing and deep learning techniques for impact assessment of Shaheen cyclone on vegetation cover at Al Batinah governorate of Oman. In: Remote sensing for agriculture, ecosystems, and hydrology XXIV, vol 12262. SPIE, pp 117–121

Al-Rahbi SYA, Al-Mulla H, Jayasuriya BC (2019) Analysis of true-color images from unmanned aerial vehicle to assess salinity stress on date palm. J Appl Remote Sens 13(3):034514. https://doi.org/10.1117/1.JRS.13.034514

AlRuheili AM (2022) A tale of Shaheen's cyclone consequences in Al Khaboura city, Oman. Water 14(3):340. https://doi.org/10.3390/w14030340

Al Ruheili AM, Boluwade A (2021) Quantifying coastal shoreline erosion due to climatic extremes using remote-sensed estimates from sentinel-2a data. Environ Process 8:1121–1140. https://doi.org/10.1007/s40710-021-00522-2

Al-Shali MAM, Kalyani A, Ndaginna AI, Yardi K (2019) Can industrialization affect heavy metals bioconcentration in date palm tree farms in the Sultanate of Oman? Iraqi J Agric Sci 50:152–172

Al-Yahyai RA, Al-Kharusi LM, Khan MM, Al-Adawi AO, Al-Subhi AM, AL-Kalbani BS, Al-Sadi AM (2022) Biotic and abiotic stresses of major fruit crops in Oman: a review. J Agric Mar Sci [JAMS] 27(1):16–37. https://doi.org/10.53541/jams.vol27iss1pp16-37

Benkari N (2017) Urban development in Oman: an overview. WIT Trans Ecol Environ 226:143–156

Brady NC, Weil RR (2002) The nature and properties of soils, 13th edn. Prentice Hall, New Jersey, USA, p 249

Caspari G, Donato S, Jendryke M (2019) Remote sensing and citizen science for assessing land use change in the Musandam (Oman). J Arid Environ 171:104003. https://doi.org/10.1016/j.jaridenv.2019.104003

Choudri BS, Baawain M, Ahmed M, Al-Sidairi A, Al-Nadabi H (2015) Relative vulnerability of coastal Wilayats to development: a study of Al-Batinah North, Oman. J Coast Conserv 19:51–57. https://doi.org/10.1007/s11852-014-0355-7

Civil Aviation Authority (CAA) (2021) Directorate General of Meteorology; Preliminary report on tropical cyclone 1–4 October 2021. Department of Research and Development, Muscat, Oman

David HE, Ramalakshmi K, Gunasekaran H, Venkatesan R (2021) Literature review of disease detection in tomato leaf using deep learning techniques. In: 2021 7th international conference on advanced computing and communication systems (ICACCS), vol 1, March. IEEE, pp 274–278. https://doi.org/10.1109/ICACCS51430.2021.9441714

FAO (2021) AQUASTAT country profile—Oman. Revised version. Rome

Lal R (2012) Climate change and soil degradation mitigation by sustainable management of soils and other natural resources. Agric Res 1(3):199–212. https://doi.org/10.1007/s40003-012-0031-9

Ministry of Agriculture and Fisheries (2017) Annual statistical yearbook, directorate general of planning and investment promotion, ministry of agriculture & fisheries, Muscat, Sultanate of Oman

Ministry of Agriculture and Fisheries (2012) Oman salinity strategy report, main report, Muscat, Sultanate of Oman. http://maf.gov.om/Download.ashx?File=FCKupload/File/books/main.pdf

Moossa B, Trivedi P, Saleem H, Zaidi SJ (2022) Desalination in the GCC countries—a review. J Clean Prod 131717

Naifer A, Al-Rawahy SA, Zekri S (2011) Economic impact of salinity: the case of Al-Batinah in Oman. Int J Agric Res 6(2):134–142. https://doi.org/10.3923/ijar.2011.134.142

Oldeman LR (1992) Global extent of soil degradation. In: Bi-annual report 1991–1992/ISRIC. ISRIC, pp 19–36

Oldeman LR, Hakkeling RTA, Sombroek WG (1990) World map of the status of human-induced soil degradation: an explanatory note. International Soil Reference and Information Centre

Rajendran S, Nasir S, Kusky TM, Al-Khirbash S (2014) Remote sensing based approach for mapping of CO_2 sequestered regions in Samail ophiolite massifs of the Sultanate of Oman. Earth-Sci Rev 135:122–140

Retallack A, Finlayson G, Ostendorf B, Clarke K, Lewis M (2023) Remote sensing for monitoring rangeland condition: current status and development of methods. Environ Sustain Indic 100285

Scovell MD (2020) Investigating the psychological factors that influence cyclone mitigation behaviour. Doctoral dissertation, James Cook University

Shahid SA, Omar SA (2022) The soil that we attempt to classify. In: Kuwait soil taxonomy. Springer, Cham, pp 13–18

Shahid SA, Zaman M, Heng L (2018) Soil salinity: historical perspectives and a world overview of the problem. In: Guideline for salinity assessment, mitigation and adaptation using nuclear and related techniques. Springer, Cham, pp 43–53. https://doi.org/10.1007/978-3-319-96190-3_2

UN (2018) United Nations, Department of Economic and Social Affairs, Population Division (2018) World urbanization prospects: the 2018 revision

United Nation (2003) World urbanization prospects: the 2003 revision. http://esa.un.org/unup/. Last visit 13 Jan 2007

Spectral Angle Mapper Approach (SAM) for Land Degradation Mapping: A Case Study of the Oued Lahdar Watershed in the Pre-Rif Region (Morocco)

Brahim Benzougagh, Ayad M. Fadhil Al-Quraishi, Youssef Bammou, Shuraik Kader, Mohammed El Brahimi, Driss Sadkaoui, and Latifa Ladel

Abstract Land degradation is a complex and widespread environmental issue with significant implications for global sustainability. It encompasses various processes that negatively impact soil health and support productive ecosystems and human livelihoods. This chapter focused on mapping land use/land cover (LULC) and

---

B. Benzougagh (✉) · M. El Brahimi · L. Ladel
Geophysics and Natural Hazards Laboratory, Department of Geomorphology and Geomatics (D2G), Scientific Institute, Mohammed V University in Rabat, Avenue Ibn Batouta, Agdal, P.O. Box 703, 10106 Rabat City, Morocco
e-mail: brahim.benzougagh@is.um5.ac.ma

M. El Brahimi
e-mail: mohammed_elbrahimi2@um5.ac.ma

L. Ladel
e-mail: latifa_ladel@um5.ac.ma

A. M. F. Al-Quraishi
Petroleum and Mining Engineering Department, Tishk International University, Erbil 44001, Iraq
e-mail: ayad.alquraishi@tiu.edu.iq; ayad.alquraishi@gmail.com

Y. Bammou
Department of Geology, Faculty of Science and Technology, Laboratory of Geo-Resources, Geo-Environment and Civil Engineering (L3G), Cadi Ayad University, Marrakech, Morocco
e-mail: youssef.bammou@ced.uca.ma

S. Kader
School of Engineering and Built Environment, Griffith University, Nathan, QLD 4111, Australia

Green Infrastructure Research Labs (GIRLS), Cities Research Institute, Griffith University, Gold Coast, QLD 4215, Australia

S. Kader
e-mail: shuraik.mohamedabdulkader@griffithuni.edu.au

D. Sadkaoui
Laboratory for Applied and Marine Geosciences, Geotechnics and Geohazards (LR3G), Department of Geology, Faculty of Sciences, University Abdelmalek Essaâdi, Tetouan, Morocco
e-mail: d.sadkaoui@uae.ac.ma

© The Author(s), under exclusive license to Springer Nature Switzerland AG 2024
A. M. F. Al-Quraishi and Y. T. Mustafa (eds.), *Natural Resources Deterioration in MENA Region*, Earth and Environmental Sciences Library,
https://doi.org/10.1007/978-3-031-58315-5_2

identifying land degradation features using Sentinel-2 imagery and the Spectral Angle Mapper (SAM) approach. The primary objective was to gain insight into the spatial distribution of land degradation within the study area. The step-by-step supervised classification process involved data preparation, training data collection, feature selection, SAM classification, and post-classification processing. An accuracy assessment was conducted to validate the results and ensure the reliability of the land cover map. As a result of the different LULC classifications in the research area, the arboriculture class represented 18% of the study area. In contrast, the agriculture class showed coverage of 27%, followed by the forest class occupying 22% of the catchment area, and the bare soil class representing 33% of the total study area. The combination of substantial proportions of bare soil (33%) and agriculture (27%) suggested that a large portion of the landscape may be susceptible to land degradation if appropriate measures are not implemented. The derived LULC map is a valuable resource for environmental monitoring, ecosystem conservation, and land use planning. Policymakers, researchers, and stakeholders can use this information to make informed decisions for sustainable land management, protect natural resources, and mitigate the impact of land degradation.

Keywords Land degradation · Spectral angle mapper · Pre-Rif Morocco · LULC · Lahdar Watershed

1 Introduction

The earth's land is a necessary resource for humans because it enables more than 7.5 billion people living there to eat every day, but it is also a limited resource because there are currently only 33 million km^2 (or 6.4% of the earth's surface) that can be used for agriculture (Farah et al. 2021; Bammou et al. 2023). Land degradation is one of the most important and persistent environmental issues the Earth has long faced (Fadhil 2009; Ganasri and Ramesh 2016; Rawat et al. 2016; Benzougagh and Fellah 2023). Land degradation is a complex and widespread environmental issue with significant implications for global sustainability. It encompasses various processes that negatively impact soil health and its ability to support productive ecosystems and human livelihoods (Hossain et al. 2020; Benzougagh et al. 2022). Land degradation affects the vast expanses of the Earth's land surface, resulting in reduced crop yields, increased vulnerability to erosion, loss of biodiversity, and diminished ecosystem services (Eswaran et al. 2001; UNCCD 2017; Hossini et al. 2022). The Food and Agriculture Organization (FAO) has estimated that approximately 24% of the global land area is already somewhat degraded, and an additional 8% is at risk of degradation (FAO 2015). Moreover, it is estimated that land degradation costs the global economy by approximately 10% of the annual gross product (GDP) through the loss of ecosystem services and reduced agricultural productivity (Costanza et al. 2014; Sestras et al. 2023).

Both natural and anthropogenic variables affect the complicated processes of land degradation. Land deterioration can be accelerated by natural factors such as geological processes, extreme weather, and climate variability (Cavicchioli et al. 2019; Wong et al. 2021). For example, protracted droughts and strong rains can worsen agricultural areas by causing soil erosion, vegetation loss, and other environmental problems. Geological aspects, such as delicate soils, can further accelerate land degradation in some areas (Al-Quraishi and Negm 2020; Benzougagh et al. 2020a, b; Meshram et al. 2022). Conversely, human activity is a major contributor to land degradation (Al-Quraishi 2004; Gong et al. 2022; Kader et al. 2023a). Deforestation, excessive grazing, and poor irrigation are examples of unsustainable land use practices that can deplete soil nutrients and lower land production.

In addition to causing habitat loss and fragmentation, the expansion of agriculture and urbanization can transform natural ecosystems into degraded land (Zheng et al. 2021; Kader et al. 2023b). Additionally, socioeconomic factors, such as population increase, poverty, and lack of access to resources and technology, can exacerbate the pressures associated with land degradation (Seifollahi-Aghmiuni et al. 2022). Poor land management methods and ineffective land tenure structures may deter investments in sustainable land use strategies and hasten further deterioration (Searchinger et al. 2014; Asaaga et al. 2020). Overall, the interaction between natural and human-induced factors creates a complex web of drivers that influence the rate and extent of land degradation. Understanding these factors is crucial for developing effective land management strategies and policies to combat land degradation and promote sustainable land use practices for the well-being of the environment and society (Benzougagh et al. 2016, 2017).

The consequences of land degradation extend beyond ecological degradation and pose serious socio-economic challenges, particularly for communities reliant on agriculture and natural resources (Montanarella and Panagos 2021; Karimi et al. 2022a, b; Brandolini et al. 2023). As land degradation continues to accelerate owing to human activities and climate change, there is an urgent need for effective monitoring and mapping techniques to inform conservation efforts and promote sustainable land management practices.

Land use and land cover (LULC) information is paramount, supporting various environmental assessments and effective soil and water resource management. Land cover mapping serves as a valuable tool for landscape management, offering valuable insights into the patterns of environmental monitoring, natural resources, and human activities that influence the diverse surface areas of Earth. By accurately identifying and categorizing different types of LULC, we understand how human actions and natural processes interact with the environment. This knowledge is vital for making informed decisions and implementing sustainable practices to preserve and protect precious natural resources (Shi and Yang 2017; Kathwas and Saur 2022; Benzougagh and Fellah 2023).

The Oued Lahdar catchment is part of the Oued Inaouene watershed located in northeast Morocco and is also a Sebou sub-basin. The Lahdar watershed is also exposed to heavy soil erosion owing to its favorable geographical, climatic, geological, and geomorphological characteristics. Examining the relative significance of climate, LULC, and conservation strategies helps to better understand how soil erosion changes over time due to land degradation, endangering agricultural fertility, and the ecosystem. Watershed management entails monitoring spatial and temporal changes in basins, with LULC mapping being a crucial method for territorial management, planning, and studying the environment over time. This mapping technique plays a vital role in understanding and interpreting the evolving landscape, enabling effective decision-making to preserve and sustainably utilize natural resources within the watershed.

The outcomes of LULC mapping hold significant value for planners, engineers, and decision-makers when implementing conservation measures and managing natural resource development. Remote sensing is a convenient and effective data source for generating thematic maps of LULC. This data-driven approach aids in understanding the spatial distribution of land-use types and land-cover classes, enabling informed and well-targeted decisions to promote sustainable land management practices and protect the environment (Benzougagh and Fellah 2023). Many techniques have been developed to achieve this goal, with image classification being the most popular (Mathur and Foody 2008; Perez and Wang 2017; Gulzar 2023). Accurate and timely information on spatial distribution and severity is crucial. Remote sensing technologies have become indispensable tools for assessing and mapping land degradation over large areas (Zhang et al. 2022; Shahfahad et al. 2023). Remote sensing platforms, including satellites and airborne sensors, provide multispectral and hyperspectral imagery capable of capturing detailed information about the Earth's surface (Adão et al. 2017). These data enable monitoring changes in vegetation cover, soil moisture, and land use patterns, which are essential for identifying areas susceptible to degradation.

The Spectral Angle Mapper (SAM) approach is a prominent remote sensing technique for land degradation mapping. SAM is a spectral classification algorithm that quantifies the similarity between the spectral signatures of reference and target areas by calculating their spectral angle (Kruse et al. 1993). Two n-dimensional vectors form the spectral angle, each representing a pixel's spectral values in a multiband image. The smaller the angle between the two spectra, the more similar are their spectral characteristics, and vice versa. SAM has been widely used in various remote sensing applications, including land cover classification, mineral mapping, and vegetation analysis (Christovam et al. 2019; Chakravarty et al. 2021; Benzougagh and Fellah 2023).

This chapter explores the potential of the SAM approach for mapping land degradation and improving the understanding of the relative importance of dynamic parameters, particularly LULC, on land degradation in the Lahdar watershed in Northeast Morocco. In addition, erosion risk zones are monitored using cutting-edge supervised machine-learning algorithms, such as SAM, with data from Sentinel-2 images to identify areas susceptible to soil degradation. Integrating SAM-derived data with ground truth observations and soil samples will facilitate a comprehensive assessment of land degradation patterns and their implications for regional land management. These findings will aid managers and decision-makers in formulating appropriate conservation plans for natural resources in the research area.

2 Materials and Methods

2.1 *Study Area*

The Oued Lahdar catchment is part of the Oued Inaouene watershed, which covers 3680 km^2 and is also a Sebou subbasin (Fig. 1). After the Oued Ouergha, the Oued Inaouene is Sebou's second most important tributary; it travels east–west via the southern Rif corridor, eventually reaching the Idriss I dam. It encompasses a portion of the External Rif on the left, a portion of the Middle Atlas on the right, and southern Rif Corridor Fez-Taza. The Lahdar Basin, with a total area of 610 km^2, is located northwest of Taza in Taza Province, in the Fez-Meknes region (Morocco). The study area is located in the Eastern Pre-Rif in the northeast of Morocco between Cartesian coordinates X = 604.000 m W and 63.0000 m E) and Y = 446.000 m N and 404.000 S, respectively (Fig. 1).

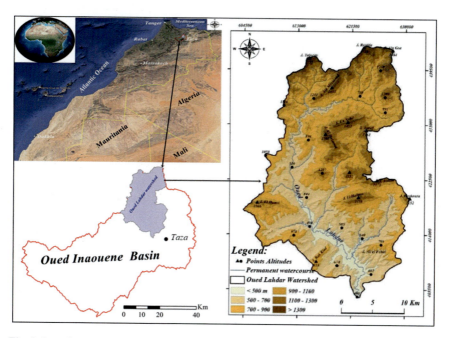

Fig. 1 Location map of the study area in Morocco

2.2 Erosion Susceptibility Assessment of the Lahdar Watershed

The Lahdar watershed is shaped in the Eastern Pre-Rif region in northeastern Morocco, covering an area of approximately 610 km². Geomorphologically, the reliefs are strong and uneven, with varied geological terrains. It is characterized by the dominance of soft rocks (marl, marly limestone, sandstone marl, schists, etc.). Morphologically, the upstream parts of the study area showed mountainous ridges at an altitude of 1700 m, and the slopes were strong and exceeded 25% (Fig. 2). Towards the south, low mountains and hills dominate mountains and hills, offering aerated relief and less pronounced slopes (Amhani and Tribak 2021).

Geologically, the Rifan domain is the only Moroccan massif that has formed as a result of alpine orogeny, and most of its facies are more similar to those of Andalusia than to those of the rest of Morocco (Khalis et al. 2021). The basin of Lahdar, which is a geological component of the Rifan chain, is unique in that it shares two structural domains with a wide variety of units (Fig. 3). The watershed of the Oued Lahdar ranges over the Outer Rif, which is characterized by fragile terrain and high altitudes. In the North, the internal domain of the Rif is constituted in the majority by allochthonous grounds with resistant material with some autochthonous grounds marly or schisto-sandstone (Fig. 3). To the south, the autochthonous marly terrain of the Perifaine nappe dominates, and is partly covered by allochthonous terrain belonging to the Ouezzane system (Leblanc 1977). A lithological mosaic seen as the outcrop of several units dominated by soft materials, mainly marl-limestone and marl-sandstone tertiary, within a tormented structure, is an important factor that conditions a rapid and disorderly evolution of the slopes (Tribak et al. 2017; Benzougagh et al. 2020a, b).

Tribak et al. (2017) state that this region is characterized by superficial and skeletal mineral soils. Of the soil classes, little evolution of erosion occurred on the strong slopes, vertisols, and calcimagnesic soils in the less hilly and less hilly areas, and soils of alluvial contribution developed on the terraces bordering the rivers. Consequently, the different types of soils in the study area show, globally, favorable behaviors to erosive processes, although the latter vary in nature, intensity, and degree of activity from one soil to another. They are characterized by low organic matter content and the predominance of silty or silty-clay textures, making them more unstable and highly sensitive to rainfall-aggressive rainfall phenomena.

The Mediterranean climate dominated the region with continental oceanic influence. Seasonal solid contrasts and clear irregularities in rainfall mark it. The research area is part of a semi-arid Mediterranean environment characterized by erratic annual rainfall and violent thunderstorms that cause serious water erosion. The necessary rainfall was measured between September and May (wet season). In contrast, the least amount was measured between June and August (dry season) (Fig. 4). A common method for investigating the hydrological interactions of rainwater with soils and simulating rainfall that causes runoff and soil erosion is the use of rainstorm simulators (Isa et al. 2018; Benzougagh et al. 2020a, b).

Spectral Angle Mapper Approach (SAM) for Land Degradation ...

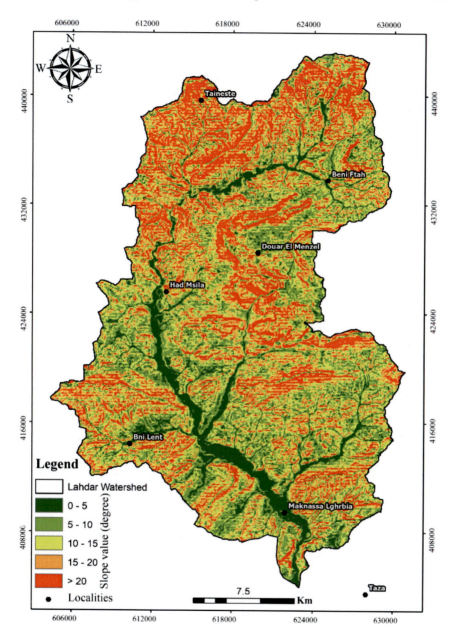

Fig. 2 Slope map of the Lahdar watershed

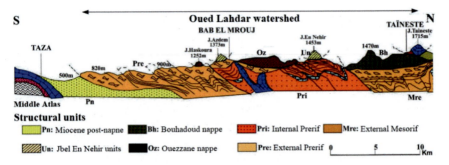

Fig. 3 Geological section of structural units in the Eastern Prerif (study area)

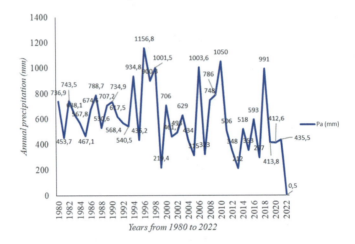

Fig. 4 Precipitation in the study area for the period 1980–2022

3 Methods

Using the SAM approach as a supervised classification technique, the current research demonstrated the capacity to use hyperspectral data to distinguish between areas prone to soil deterioration and those shielded from it. We employed several materials for this purpose, as shown in the workflow depicted in Fig. 5.

3.1 Image Datasets

The first step in developing a land-use map was to collect Sentinel-2 satellite imagery from the Copernicus Earth Observation Program distributed by the US Geological

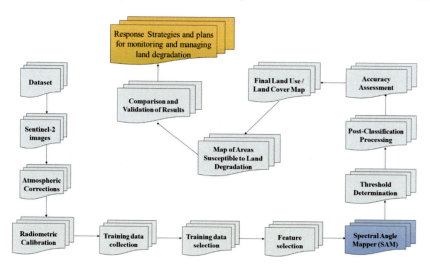

Fig. 5 Flowchart of the work methodology for the spectral angle mapper

Survey (https://earthexplorer.usgs.gov/). The Sentinel-2 satellite, which the European Space Agency (ESA) launched on June 23, 2015, has 13 bands that can record multispectral images (Table 1). Sentinel-2 data are used in the Global Monitoring for Environment and Security (GMES) program, which the European Commission and the European Space Agency jointly run. Services provided by this program include monitoring natural disasters, humanitarian efforts, and services related to land management, agricultural production, and forestry. Sentinel-2 imagery was acquired for the study area, ensuring that it covered the region of interest and was cloud-free or with minimal cloud cover. Ensure that the imagery is preprocessed, including atmospheric correction and radiometric calibration, to normalize reflectance values. All data were projected onto the UTM (zone 30) and WGS 84 data. The study area lies within 127 paths and 57 rows of the WRS2 reference system. The acquired data were used for the LULC mode classification.

3.2 Images Classifications

The European Space Agency's (ESA) Sentinel Application Platform (SNAP) software from the European Space Agency was used to classify Sentinel-2 images. We also used ENVI software, and for the cartography of the study results, this study employed QGIS software. We performed a detection analysis using a supervised classification approach based on the definition of training sites. No automatic classification processes exist. By manually entering the training polygons, the detector establishes the classes seen on the image using an in-depth understanding of the field. These polygons are then classified in accordance with the corresponding class

Table 1 Sentinel-2 image characteristics

Band name	Sensor	Band number	Sentinel-2A W (nm)	Sentinel-2A B (nm)	Sentinel-2B W (nm)	Sentinel-2B B (nm)	Resolution (m)
Coastal aerosol	MSI	B1	443.9	20	442.3	20	60
Blue	MSI	B2	496.6	65	492.1	65	10
Green	MSI	B3	560.0	35	559	35	10
Red	MSI	B4	664.5	30	665	30	10
Vegetation red edge	MSI	B5	703.9	15	703.8	15	20
	MSI	B6	740.2	15	739.1	15	20
	MSI	B7	782.5	20	779.7	20	20
NIR	MSI	B8	835.1	115	833	115	10
Narrow NIR	MSI	B9	864.8	20	864	20	20
Water vapour	MSI	B10	945.0	20	943.2	20	60
SWIR-cirrus	MSI	B11	1373.5	30	1376.9	30	60
SWIR	MSI	B12	1613.7	90	1610.4	90	20
SWIR	MSI	B13	2202.4	180	2185.7	180	20

NB: MSI: Multispectral Instrument; W: Central wavelength; B: Bandwidth

after the classification algorithm has finished analyzing the spectral signatures of each pixel contained in each class (Benzougagh and Fellah 2023). Representative samples of different land cover classes were chosen for training data collection, and ground-truth data were gathered through field surveys or existing references. Subsequently, we separated the ground truth data into training and validation datasets, ensuring that the training set adequately represented each land cover class.

Feature selection involves identifying the relevant spectral bands and indices from Sentinel-2 imagery, emphasizing the Red, Green, Blue, Near-Infrared, and Shortwave Infrared bands used in the SAM approach. We applied the SAM classifier with the training data and selected features, which calculates the spectral angle between each pixel's spectral signature and those of land cover classes in the training dataset. Threshold determination is important when assigning each pixel to a class. Pixels with spectral angles below the threshold were assigned to the nearest spectral signature, whereas the others were either unclassified or formed a separate class. Post-classification processing refines the results and removes misclassifications through spatial filtering or object-based refinement for coherent land cover polygons. We used the validation dataset to assess the accuracy and compare the classified map with the ground truth data, calculating classification metrics such as the overall accuracy and kappa coefficient. Finally, we incorporated post-classification refinements and accuracy assessment results into the final land-cover map. Visualization and interpretation ensured well-defined classes that were representative of the actual land cover distribution. Following these steps, the SAM approach with Sentinel-2 imagery yielded a reliable LULC map, effectively distinguishing land cover classes with subtle spectral differences.

3.3 SAM

The primary objective of this study is to identify and map the spatial distribution of land degradation features using hyperspectral image classification. We used the SAM classification algorithm to achieve this goal. SAM is a powerful technique employed for hyperspectral image classification, which allows us to accurately categorize different land cover and land degradation classes based on their spectral signatures. The SAM technique is based on the definition of the spectral similarities in the domain of (n-dimensional) structure. The angle (α) determines the difference in the spectra. By measuring the angle between the pixel and the reference spectrum (Fig. 6), an image can be separated into any number of classes (Kruse et al. 1993). Smaller angles correspond closely to the reference spectrum (Kruse et al. 1993; Abdolmaleki et al. 2022; Benzougagh and Fellah 2023).

Based on the trigonometric function of the cosine arc, all angles between the total of all target pixels (t) and all reference pixels (r) were calculated for all the band numbers (nb). The formula for Eq. (1) is as follows:

$$\alpha = \text{Cos}^{-1}\left(\frac{\left(\sum_{i=1}^{nb} t_i r_i\right)}{\left(\sum_{I=1}^{nb} t_i^2\right)1/2\left(\sum_{i=1}^{nb} r_i^2\right)1/2}\right) \quad (1)$$

where:

- α: spectral angle between vectors; nb: number of spectral bands; t: target pixel; r: reference pixel.

Fig. 6 Spectral angle in two-dimensional space between the reference and image spectra

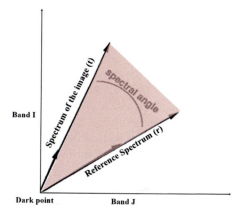

The group with the lowest pixel angle received a pixel on its own. Instead of measuring the vector size, the angle size of the radiant and angle is used.

4 Results and Discussion

4.1 Accuracy Assessment

The classification LULC map produced by supervised classification algorithms must first undergo an accuracy assessment to determine its dependability and quality. Comparing the categorized results with reference data, which are typically derived from field research or independently validated data sources, is what this process entails.

4.2 Final LULC Map

During the application of supervised classification using the SAM algorithm and Sentinel-2 imagery in the watershed of Oued Lahdar, four classes were defined as areas of interest to produce an LULC map: Arboriculture, Agriculture, Forests, and Bare Soil (Fig. 7). The table below shows the percentages and areas covered in the study area for each class (Table 2).

The distribution of the different LULC classifications in the research area is shown in Table 2, and we examine how this relates to the risk of land degradation. Land degradation refers to a number of factors, some of which are natural and others that are frequently brought on by human activity or natural processes. We analyze the table by concentrating on its implications for the risk of soil degradation.

- Arboriculture: This area accounts for 109.8 km^2, or 18% of the entire area. Growing trees, such as fruit or ornamental trees, are known as arbocultures. The risk of soil degradation in these areas may change depending on the management techniques and soil conservation measures used. Effective management and soil conservation practices in arboriculture settings can reduce the risks of erosion and deterioration.
- Agriculture: The area used for agriculture totals 164.7 km^2, or 27% of the total area. Soil deterioration, including erosion, nutrient depletion, and loss of soil structure, can be caused by intensive agricultural methods, such as monoculture, excessive use of fertilizers and pesticides, and inadequate soil management. Sustainable agricultural methods and measures for soil conservation are crucial for reducing the risk of soil degradation in these locations.
- Forests: 134.2 km^2, or 22% of the total area, is covered by forests. Forests are essential for the preservation and protection of soil. Tree roots contribute to soil stabilization and erosion reduction. These forests can lessen the danger of soil

Spectral Angle Mapper Approach (SAM) for Land Degradation ...

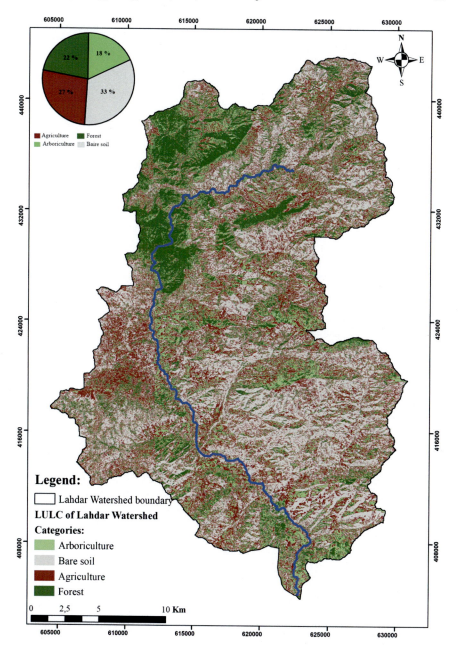

Fig. 7 The Lahdar watershed map of LULC

Table 2 Areas and percentages of each LULC class according to the SAM approach

Class	SAM (km^2)	SAM (%)
Arboriculture	109.8	18
Agriculture	164.7	27
Forests	134.2	22
Bare soil	201.3	33
Total	610	100

degradation if properly managed and not subjected to deforestation or destructive logging.
- Bare soil: A total of 201.3 km^2, approximately 33% of the total area, falls under the bare soil category. In particular, bare soil regions are more susceptible to land degradation if exposed to erosive influences such as water and wind. Without vegetation, there is a large increase in the risk of soil erosion and loss of fertile topsoil. Implementing soil conservation activities such as reforestation or ground cover planting is crucial for lowering the risk of soil degradation in bare soil areas.

The data presented in Table 2 and Fig. 7 indicate a significant concern for the overall land degradation risk in the Lahdar Watershed. The combination of substantial proportions of bare soil (33%) and agriculture (27%) suggested that a large portion of the landscape may be susceptible to land degradation if appropriate measures are not implemented (Fig. 8). Bare Soil areas are particularly vulnerable to erosion and loss of fertile topsoil, especially when exposed to erosive forces, such as water runoff and wind (Chamizo et al. 2017; Weeraratna 2022). The absence of vegetation cover leaves soil unprotected, making it more susceptible to degradation (Wassie 2020; Benzougagh and Fellah 2023). To mitigate the risk of soil degradation in these areas, urgent intervention is necessary, such as implementing soil conservation practices such as afforestation, reforestation, or ground cover planting. Agricultural lands covering a significant portion of the study area are also at considerable risk of soil degradation. Intensive agricultural practices, including monoculture, excessive use of fertilizers and pesticides, and improper soil management, can lead to soil erosion, nutrient depletion, and a decline in soil quality (Singh 2000; Nair and Nair 2019).

Sustainable agricultural practices such as crop rotation, cover cropping, reduced tillage, and organic farming techniques are vital for preserving soil health and mitigating degradation risks (Peigné et al. 2007; Crystal-Ornelas et al. 2021; Kader et al. 2022). Although Forested areas provide valuable protection against soil erosion owing to their stabilizing effect on the soil through tree root systems (Zuazo and Pleguezuelo 2009; Karimi et al. 2022a, b), it is essential to ensure that these forests are well managed and protected from deforestation or destructive logging practices. Sustaining healthy forests can significantly reduce the risk of soil degradation and maintain the integrity of the ecosystem.

Addressing the overall land degradation risk requires comprehensive LULC planning and targeted soil conservation strategies. To devise effective and sustainable management approaches tailored to each land cover class, it is crucial to consider

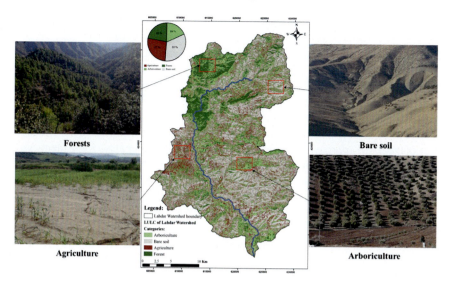

Fig. 8 Different classes of LULC and forms of soil degradation in the Lahdar watershed

site-specific factors such as soil properties, climatic context, topography, geomorphology, geology, and LULC history. Additional evaluations, including soil property analysis, erosion studies, and monitoring of land use practices, are essential for acquiring more precise knowledge of the risk of soil degradation. The findings will inform policymakers, land managers, and other stakeholders of such evaluations to take necessary precautions to protect the study area's soil health and overall environmental sustainability.

5 Conclusions

Remote sensing is vital in studying various earth science applications, particularly in addressing environmental issues such as land degradation and soil erosion mapping. Its significance lies in its ability to reduce expenses associated with field mapping, overcome challenges in accessing remote regions, and identify areas prone to land degradation. This study utilized Sentinel-2 imagery and the SAM approach to perform supervised classification for land cover mapping and identification of land degradation features. The application of SAM allowed us to accurately categorize different land-cover classes based on their spectral signatures, providing valuable insights into the spatial distribution of land degradation within the study area. We systematically acquired and preprocessed Sentinel-2 imagery, collected ground-truth data, selected relevant features, and applied the SAM classifier. Post-classification processing and accuracy assessment were conducted to refine the results and evaluate the reliability of the land cover map.

As a result of the different LULC classifications in the research area, the arboriculture class represented 18% of the study area, while the agriculture class showed coverage of 27%, followed by the forest class occupying 22% of the catchment area, and finally, the bare soil class representing 33% of the total study area. The combination of substantial proportions of bare soil (33%) and agriculture (27%) suggested that a large portion of the landscape may be susceptible to land degradation if appropriate measures are not implemented.

The accuracy assessment revealed promising results, demonstrating the effectiveness of the SAM approach in accurately classifying land cover classes and identifying areas prone to land degradation. This information is vital for informed land management and conservation strategies as it aids in identifying vulnerable areas and guiding targeted conservation measures for sustainable development. It is worth noting that the study's success is attributed to the comprehensive validation of results, ensuring the accuracy and credibility of land cover mapping through field missions and the use of Google Earth. The validation process allowed us to understand the strengths and limitations of the classification and enabled iterative improvements to enhance the classification methodology. The derived LULC map is valuable for environmental monitoring, ecosystem conservation, and land use planning. Policymakers, researchers, and stakeholders can use this information to make informed decisions for sustainable land management, protect natural resources, and mitigate the impact of land degradation.

6 Recommendations

At the end of this chapter, a number of recommendations have been made:

- Promote the use of remote sensing, specifically Sentinel-2 imagery, in addressing environmental issues such as land degradation and soil erosion mapping. Highlight the cost-effectiveness and efficiency of remote sensing compared to traditional field mapping methods.
- Encourage the application of the Spectral Angle Mapper (SAM) approach for supervised classification of land cover mapping and identification of land degradation features. Emphasize the accuracy and reliability of SAM in categorizing different land cover classes based on their spectral signatures.
- Advocate for the implementation of appropriate measures to prevent land degradation in areas with substantial proportions of bare soil and agriculture. Raise awareness of the vulnerability of these areas and the potential consequences if preventive actions are not taken.
- Highlight the importance of accuracy assessment and validation in remote sensing studies. Emphasize the need for ground-truth data collection, field missions, and the use of tools like Google Earth to validate and refine classification results.

- Stress the value of the derived land cover map for environmental monitoring, ecosystem conservation, and land use planning. Encourage policymakers, researchers, and stakeholders to utilize this information for informed decision-making in sustainable land management and resource protection.
- Promote collaboration between researchers, policymakers, and stakeholders to develop targeted conservation measures based on the identified vulnerable areas. Emphasize the role of collective efforts in mitigating the impact of land degradation and ensuring sustainable development.
- Encourage further research and studies in the field of remote sensing and its applications in addressing environmental issues. This can include exploring advanced classification algorithms, integrating multi-sensor data, and investigating other remote sensing platforms for improved accuracy and coverage.
- Promote capacity building and training programs for professionals and stakeholders involved in land management and conservation. Provide education and resources on remote sensing techniques, data analysis, and interpretation to enhance their understanding and utilization of these tools in decision-making processes.
- Foster collaboration between academia, government agencies, non-governmental organizations, and local communities to develop integrated land management strategies. Encourage the sharing of data, knowledge, and best practices to facilitate evidence-based decision-making and promote sustainable land use practices.

Funding This research received no external funding.

Data Availability Statement The data presented in this study are available on request from the corresponding author.

Conflicts of Interest The authors declare no conflict of interest.

References

Abdolmaleki M, Consens M, Esmaeili K (2022) Ore-waste discrimination using supervised and unsupervised classification of hyperspectral images. Remote Sens 14:6386. https://doi.org/10.3390/rs14246386

Adão T, Hruška J, Pádua L, Bessa J, Peres E, Morais R, Sousa JJ (2017) Hyperspectral imaging: a review on UAV-based sensors, data processing and applications for agriculture and forestry. Remote Sens 9(11):1110. https://doi.org/10.3390/rs9111110

Al-Quraishi AMF (2004) Design a dynamic monitoring system of land degradation using geoinformation technology for the northern part of Shaanxi Province, China. J Appl Sci 4(4):669–674

Al-Quraishi AMF, Negm AM (2020) Environmental remote sensing and GIS in Iraq. Springer Water, Springer Nature, Cham. https://doi.org/10.1007/978-3-030-21344-2

Amhani Z, Tribak A (2021) Mapping land use and water erosion in the Oued Lahdar watershed (Prérif oriental-Morocco): using remote sensing data and the RUSLE model. Bollettino dell'Associazione Italiana di Cartografia. https://doi.org/10.13137/2282-572X/33429

Asaaga FA, Hirons MA, Malhi Y (2020) Questioning the link between tenure security and sustainable land management in cocoa landscapes in Ghana. World Dev 130:104913. https://doi.org/10.1016/j.worlddev.2020.104913

Bammou Y, Benzougagh B, Bensaid A, Igmoullan B, Al-Quraishi AMF (2023) Mapping of current and future soil erosion risk in a semi-arid context (haouz plain-Marrakech) based on CMIP6 climate models, the analytical hierarchy process (AHP) and RUSLE. Model Earth Syst Environ 1–14. https://doi.org/10.1007/s40808-023-01845-9

Benzougagh B, Boudad Larbi P, Dridri Abdallah P, Driss S (2016). Utilization of GIS in morphometric analysis and prioritization of sub-watersheds of Oued Inaouene (Northeast Morocco). Eur Sci J 12(6). https://doi.org/10.19044/esj.2016.v12n6p266

Benzougagh B, Dridri A, Boudad L, Kodad O, Sdkaoui D, Bouikbane H (2017) Evaluation of natural hazard of Inaouene Watershed River in Northeast of Morocco: application of morphometric and geographic information system approaches. Int J Innov Appl Stud 19(1):85

Benzougagh B, Baamar B, Dridri A, Boudad L, Sadkaoui D, Mimich K (2020a) Relationship between landslide and morpho-structural analysis: a case study in Northeast of Morocco. Appl Water Sci 10(7):1–10. https://doi.org/10.1007/s13201-020-01258-4

Benzougagh B, Sarita GM, Dridri A, Boudad L, Sadkaoui D, Mimich K, Khaled MK (2020b) Mapping of soil sensitivity to water erosion by RUSLE model: case of the Inaouene watershed (Northeast Morocco). Arab J Geosci 13(21):1–15. https://doi.org/10.1007/s12517-020-06079-y

Benzougagh B, Dridri A, Boudad L, Baamar B, Sadkaoui D, Khedher KM (2022) Identification of critical watershed at risk of soil erosion using morphometric and geographic information system analysis. Appl Water Sci 12:1–20. https://doi.org/10.1007/s13201-021-01532-z

Benzougagh B, Fellah BE (2023) Mapping of land degradation using SAM approach: the case of Inaouene watershed (Northeast Morocco). Model Earth Syst Environ. https://doi.org/10.1007/s40808-023-01711-8

Brandolini F, Compostella C, Pelfini M, Turner S (2023) The evolution of historic agroforestry landscape in the Northern Apennines (Italy) and its consequences for slope geomorphic processes. Land 12(5):1054. https://doi.org/10.3390/land12051054

Cavicchioli R, Ripple WJ, Timmis KN, Azam F, Bakken LR, Baylis M, Webster NS et al (2019) Scientists' warning to humanity: microorganisms and climate change. Nat Rev Microbiol 17(9):569–586. https://doi.org/10.1038/s41579-019-0222-5

Chakravarty S, Paikaray BK, Mishra R, Dash S (2021) Hyperspectral image classification using spectral angle mapper. In: 2021 IEEE international women in engineering (WIE) conference on electrical and computer engineering (WIECON-ECE), December. IEEE, pp 87–90. https://doi.org/10.1109/WIECON-ECE54711.2021.9829585

Chamizo S, Rodríguez-Caballero E, Román JR, Cantón Y (2017) Effects of biocrust on soil erosion and organic carbon losses under natural rainfall. CATENA 148:117–125. https://doi.org/10.1016/j.catena.2016.06.017

Christovam LE, Pessoa GG, Shimabukuro MH, Galo MLBT (2019) Land use and land cover classification using hyperspectral imagery: evaluating the performance of spectral angle mapper, support vector machine and random forest. Int Arch Photogramm Remote Sens Spat Inf Sci 42:1841–1847. https://doi.org/10.5194/isprs-archives-XLII-2-W13-1841-2019

Costanza R, De Groot R, Sutton P, Van der Ploeg S, Anderson SJ, Kubiszewski I, Turner RK et al (2014) Changes in the global value of ecosystem services. Glob Environ Chang 26:152–158. https://doi.org/10.1016/j.gloenvcha.2014.04.002u

Crystal-Ornelas R, Thapa R, Tully KL (2021) Soil organic carbon is affected by organic amendments, conservation tillage, and cover cropping in organic farming systems: a meta-analysis. Agr Ecosyst Environ 312:107356. https://doi.org/10.1016/j.agee.2021.107356

Eswaran H, Lal R, Reich PF (2001) Land degradation. In: An overview conference on land degradation and desertification. Oxford, New Delhi, Khon Kaen, Thailand

Fadhil AM (2009) Land degradation detection using geo-information technology for some sites in Iraq. J Al-Nahrain Univ Sci 12:94–108. https://doi.org/10.22401/jnus.12.3.13

FAO (2015) Global soil organic carbon map (GSOCmap). Food and Agriculture Organization, Rome, Italy

Farah A, Algouti A, Algouti A, Ifkirne M, Ezziyani A (2021) Mapping of soil degradation in semi-arid environments in the ouarzazate basin in the south of the central High Atlas, Morocco, using sentinel 2A data. Remote Sens Appl: Soc Environ 23:100548. https://doi.org/10.1016/j.rsase.2021.100548

Ganasri BP, Ramesh H (2016) Assessment of soil erosion by RUSLE model using remote sensing and GIS: a case study of Nethravathi Basin. Geosci Front 7:953–996. https://doi.org/10.1016/j.gsf.2015.10.007

Gong X, Li Y, Wang X, Zhang Z, Lian J, Ma L, Cao W et al (2022) Quantitative assessment of the contributions of climate change and human activities on vegetation degradation and restoration in typical ecologically fragile areas of China. Ecol Indic 144:109536. https://doi.org/10.1016/j.ecolind.2022.109536

Gulzar Y (2023) Fruit image classification model based on MobileNetV2 with deep transfer learning technique. Sustainability 15(3):1906. https://doi.org/10.3390/su15031906

Hossain A, Krupnik TJ, Timsina J, Mahboob MG, Chaki AK, Farooq M, Hasanuzzaman M et al (2020) Agricultural land degradation: processes and problems undermining future food security. In: Environment, climate, plant and vegetation growth. Springer International Publishing, Cham, pp 17–61. https://doi.org/10.1007/978-3-030-49732-3_2

Hossini H, Karimi H, Mustafa Y, Al-Quraishi AMF (2022) Role of effective factors on soil erosion and land degradation: A review. In: Environmental degradation in Asia. Springer, Cham

Isa SFM, Azhar ATS, Aziman M (2018) Design, operation and construction of a large rainfall simulator for the field study on acidic barren slope. Civil Eng J 4(8):1851–1857. https://doi.org/10.28991/cej-03091119

Kader S, Novicevic R, Jaufer L (2022) Soil management in sustainable agriculture: analytical approach for the ammonia removal from the diary manure. Agric For 68(4):69–78. https://doi.org/10.17707/AgricultForest.68.4.06

Kader S, Raimi MO, Spalevic V, Iyingiala AA, Bukola RW, Jaufer L, Butt TE (2023a) A concise study on essential parameters for the sustainability of Lagoon waters in terms of scientific literature. Turk J Agric For 47(3):288–307. https://doi.org/10.55730/1300-011X.3087

Kader S, Jaufer L, Bashir O, Olalekan Raimi M (2023b) A comparative study on the stormwater retention of organic waste substrates biochar, sawdust, and wood bark recovered from Psidium guajava L. species. Agric For 69(1):105–112. https://doi.org/10.17707/AgricultForest.69.1.09

Karimi H, Mustafa YT, Hossini H, Al-Quraishi AMF (2022a) Assessment of land degradation vulnerability using GIS-based multicriteria decision analysis in Zakho District, Kurdistan Region of Iraq. In: Al-Quraishi AMF, Mustafa YT, Negm AM (eds) Environmental degradation in Asia. Earth and environmental sciences library. Springer, Cham. https://doi.org/10.1007/978-3-031-12112-8_3

Karimi Z, Abdi E, Deljouei A, Cislaghi A, Shirvany A, Schwarz M, Hales TC (2022b) Vegetation-induced soil stabilization in coastal area: an example from a natural mangrove forest. CATENA 216:106410. https://doi.org/10.1016/j.catena.2022.106410

Kathwas AK, Saur R (2022) Assessment of land degradation dynamics using spectral angle mapper method and demographic analytics. In: Chakravarthy VVSSS, Flores-Fuentes W, Bhateja V, Biswal B (eds) Advances in micro-electronics, embedded systems and IoT. Lecture notes in electrical engineering, vol 838. Springer, Singapore. https://doi.org/10.1007/978-981-16-8550-7_21

Khalis H, Sadiki A, Jawhari F, Mesrar H, Azab E, Gobouri AA, Bourhia M et al (2021) Effects of climate change on vegetation cover in the Oued Lahdar watershed. Northeastern Morocco. Plants, 10(8):1624. https://doi.org/10.3390/plants10081624

Kruse FA, Boardman JW, Lefkoff AB, Heidebrecht KB, Shapiro AT, Barloon PJ, Goetz AFH (1993) The spectral image processing system (SIPS)—interactive visualization and analysis of imaging spectrometer data. Remote Sens Environ 44:145–163

Leblanc D (1977) Stratigraphie et structure du Rif externe oriental au Nord de Taza (Maroc). Bull Soc Geol Fr 7(2):319–330. https://doi.org/10.2113/gssgfbull.S7-XIX.2.319

Mathur A, Foody GM (2008) Crop classification by a support vector machine with intelligently selected training data for an operational application. Int J Remote Sens 29:2227–2240

Meshram SG, Meshram C, Santos CAG, Benzougagh B, Khedher KM (2022) Streamflow prediction based on artificial intelligence techniques. Iran J Sci Technol, Trans Civ Eng 46(3):2393–2403. https://doi.org/10.1007/s40996-021-00696-7

Montanarella L, Panagos P (2021) The relevance of sustainable soil management within the European Green Deal. Land Use Policy 100:104950. https://doi.org/10.1016/j.landusepol.2020.104950

Nair KP, Nair KP (2019) Soil fertility and nutrient management. In: Intelligent soil management for sustainable agriculture: the nutrient buffer power concept, pp 165–189. https://doi.org/10.1007/978-3-030-15530-8_17

Peigné J, Ball BC, Roger-Estrade J, David CJSU (2007) Is conservation tillage suitable for organic farming? A review. Soil Use Manag 23(2):129–144. https://doi.org/10.1111/j.1475-2743.2006.00082.x

Perez L, Wang J (2017) The effectiveness of data augmentation in image classification using deep learning. arXiv preprint arXiv:1712.04621. https://doi.org/10.48550/arXiv.1712.04621

Rawat KS, Mishra AK, Bhattacharyya R (2016) Soil erosion risk assessment and spatial mapping using LANDSAT-7 ETM+ RUSLE, and GIS—a case study. Arab J Geosci 9(2016):288. https://doi.org/10.1007/s12517-015-2157-0

Searchinger T, Hanson C, Ranganathan J, Lipinski B, Waite R, Winterbottom R, Ari TB et al (2014) Creating a sustainable food future. A menu of solutions to sustainably feed more than 9 billion people by 2050. World resources report 2013–14: interim findings. World Resources Institute (WRI); World Bank Groupe-Banque Mondiale; United Nations Environment Programme (UNEP); United Nations Development Programme (UNDP); Centre de Coopération Internationale en Recherche Agronomique pour le Développement (CIRAD); Institut National de la Recherche Agronomique (INRA), p 154

Seifollahi-Aghmiuni S, Kalantari Z, Egidi G, Gaburova L, Salvati L (2022) Urbanisation-driven land degradation and socioeconomic challenges in peri-urban areas: insights from Southern Europe. Ambio 51(6):1446–1458. https://doi.org/10.1007/s13280-022-01701-7

Sestras P, Mircea S, Roşca S, Bilaşco Ş, Sălăgean T, Dragomir LO, Kader S (2023) GIS based soil erosion assessment using the USLE model for efficient land management: a case study in an area with diverse pedo-geomorphological and bioclimatic characteristics. Not Bot Horti Agrobot Cluj-Napoca 51(3):13263–13263. https://doi.org/10.15835/nbha51313263

Shahfahad, Talukdar S, Naikoo MW, Rahman A, Gagnon AS, Islam ARMT, Mosavi A (2023) Comparative evaluation of operational land imager sensor on board Landsat 8 and Landsat 9 for land use land cover mapping over a heterogeneous landscape. Geocarto Int 38(1):2152496. https://doi.org/10.1080/10106049.2022.2152496

Shi D, Yang X (2017) A relative evaluation of random forests for land cover mapping in an urban area. Photogramm Eng Remote Sens 83(8):541–552. https://doi.org/10.14358/PERS.83.8.541

Singh RB (2000) Environmental consequences of agricultural development: a case study from the Green Revolution state of Haryana, India. Agr Ecosyst Environ 82(1–3):97–103. https://doi.org/10.1016/S0167-8809(00)00219-X

Soil management in sustainable agriculture: analytical approach for the ammonia removal from the diary manure. https://www.researchgate.net/publication/366290282_Soil_Management_in_Sustainable_Agriculture_Analytical_approach_for_the_Ammonia_removal_from_the_Diary_manure#fullTextFileContent. Accessed 28 Nov 2023

Tribak A, Arari K, Abahrour M, El Garouani A, Amhani Z (2017) Evaluation quantitative de l'érosion hydrique dans un environnement méditerranéen: cas de quelques basins marneux du Prérif oriental Maroc, dans Eau et climat en Afrique du Nord et au Moyen-Orient. Târgoviste: Editions Transversal (Roumanie), pp 101–112

UNCCD (2017) The global land outlook. United Nations Convention to Combat Desertification, Bonn, Germany

Wassie SB (2020) Natural resource degradation tendencies in Ethiopia: a review. Environ Syst Res 9(1):1–29. https://doi.org/10.1186/s40068-020-00194-1

Weeraratna S (2022) Factors causing land degradation. In: Understanding land degradation: an overview. Springer International Publishing, Cham, pp 5–22. https://doi.org/10.1007/978-3-031-12138-8_2

Wong JL, Lee ML, Teo FY, Liew KW (2021) A review of impacts of climate change on slope stability. In: Climate change and water security: select proceedings of VCDRR 2021, pp 157–178. https://doi.org/10.1007/978-981-16-5501-2_13

Zhang K, Feng R, Zhang Z, Deng C, Zhang H, Liu K (2022) Exploring the driving factors of remote sensing ecological index changes from the perspective of geospatial differentiation: a case study of the Weihe River Basin, China. Int J Environ Res Public Health 19(17):10930. https://doi.org/10.3390/ijerph191710930

Zheng K, Tan L, Sun Y, Wu Y, Duan Z, Xu Y, Gao C (2021) Impacts of climate change and anthropogenic activities on vegetation change: evidence from typical areas in China. Ecol Ind 126:107648. https://doi.org/10.1016/j.ecolind.2021.107648

Zuazo VHD, Pleguezuelo CRR (2009) Soil-erosion and runoff prevention by plant covers: a review. Sustain Agric 785–811. https://doi.org/10.1007/978-90-481-2666-8_48

Remotely Sensed Data and GIS for Long-Term Monitoring of the Ghout Oases Degradation in the Region of Oued Souf (Northeastern Algerian Sahara)

Nouar Boulghobra

Abstract For 12 centuries, Ghout oases in the region of Oued Souf constituted an efficient and sustainable agro-system, based on optimal adaptation to the specific local environment and limited socioeconomic needs. This study aims to spatiotemporally assess and monitor the Ghout oases dynamics in 18 municipalities in the El Oued governorate, using medium-resolution imagery of Landsat and GIS from 1987 to 2018, based on spectral indices and change detection method. The obtained results confirm the official statistics and highlight the considerable decrease in the Ghout-occupied areas from 4,462 ha (1987) to 1,033 ha (2018), especially in the northern areas, as well as increasing of the built-up superficies and extension of new irrigated perimeters throughout the entire region. The recent mutation from a traditional Oasian system based on durability and self-sufficiency into an evolved system controlled by a free economy mechanism engendered the imbalance of Oasian ecosystem functioning, which manifests in groundwater level fluctuation leading, principally, to the disappearance of the Ghout oases. For sustainable and effective Saharan agro-development, it is necessary to adopt appropriate economic strategies based on the rational and controlled exploitation of the soil and water resources.

Keywords Ghout Oasian system · Degradation · Oued Souf · Landsat · GIS · Agricultural mutation

1 Introduction

Historically, Saharan agriculture was principally based on date palm (*Phoenix dactylifera L.*) cultivation, which constitutes the principal component in arid ecosystem biodiversity and the principal human and socioeconomic activity (Al-Khayri et al. 2015). In the Algerian Sahara, due to the absence of precipitation,

N. Boulghobra (✉)
Scientific and Technical Research Center on the Arid Regions CRSTRA, University Campus Mohamed Kheider, BP 1682 RP, 07000 Biskra, Algeria
e-mail: boulghobra.n@gmail.com

© The Author(s), under exclusive license to Springer Nature Switzerland AG 2024
A. M. F. Al-Quraishi and Y. T. Mustafa (eds.), *Natural Resources Deterioration in MENA Region*, Earth and Environmental Sciences Library,
https://doi.org/10.1007/978-3-031-58315-5_3

groundwater resources constitute the main source for agricultural land irrigation (Senoussi 2000). The existence of a shallow phreatic aquifer is the main factor determining the spatial distribution of date palm oases in Algeria. They occur in lowland territories such as the Tidikelt, Saoura, Touat, Ziban, and the low Algerian Sahara, including Oued M'ya, Oued Righ, and Oued Souf (Bouguedoura et al. 2015). Date palm oases in Oued Souf are an ingenious ancestral Oasian system. It is classified by the Food and Agriculture Organization (FAO) as a Globally Important Agricultural Heritage System (GIAHS). The Ghout's originality lies in its sustainable character and adaptation to the local bioclimatic and hydrogeological conditions (Remini and Miloudi 2021). Date palm trees are planted in groups of 20–100 inside an artificial basin of 10 m depth and 80–200 m large, 1 m at least above the phreatic aquifer, to ensure their constant self-irrigation, which preserves the water level balance (Cote 1998).

In order to promote Saharan agriculture, Algeria adopted new agro-development programs in the 1980s as an alternative politic to the rent-based economy (Dubost 1986; Daoudi and Colin 2017; Otmane 2019). Similarly, the traditional Oasian system in the region of El Oued has undergone a profound change; a modern drilling system (deep wells) is adopted to extract groundwater from the continental intercalary and complex terminal aquifers and this to meet the increasing water demand for irrigating the new land reclamations. Beside the risk of irrigated soils salinization (Tigrine and Boutiba 2023), and the decline of groundwater quality (Bouselsal and Saibi 2022), the over-extraction of these hydro-resources engendered high fluctuation of the piezometric level, which manifests on Ghout oases degradation and disappearance, by asphyxiation as a result of water table rise or, by drought following the water table decline (Cote 1998). State of the art shows that several previous studies have addressed rising water concerns in the region of Oued Souf (Nesson 1975; Cote 1998; Guendouz et al. 2006; Remini and Kechad 2011; Saibi et al. 2009; Khechana and Derradji 2012; Khezzani and Bouchemal 2018). These research studies addressed the causes, mechanisms, and environmental impacts of the groundwater depth fluctuation in Oued Souf using purely hydraulic, hydrogeological, or hydrochemical approaches. Besides, satellite imagery and GIS were used to assess and monitor date palm plantations in Saharan ecosystems (Otmane and Kouzmine 2010; Fadhil 2013; Mihi et al. 2019; Ait Lamqadem et al. 2019; Mihi et al. 2022; Belhadj et al. 2023).

The objective of this chapter is to dress spatiotemporal assessment and mapping of the Ghout oases dynamics using remotely sensed data and GIS, spectral indices, and change detection methods for diachronic monitoring of the Ghout-occupied areas, built-up sprawling and new irrigated lands extension during the period 1987–2018 over 18 municipalities of El Oued governorate. Spatial distribution of the Ghout degradation rates depending on eventual influent factors (population, built-up areas, increased irrigated cropland, and groundwater depth change) was assessed, mapped, and discussed.

2 Data and Methods

2.1 Study Area

El Oued, as a governorate, is located 600 km southeastern Algiers. It extends over 44,586 km² and includes 30 municipalities (18 included in this study), between the longitudes 6° and 8° east and latitudes 33° and 34° north. From the north, it is limited by the immense Chotts of El-Djérid, Merouane, Melrhir, and El-Rharsa, from the west by the Oued Righ valley and Oued M'ya in the south (Fig. 1). Oued Souf, as a natural entity, corresponds to one of four territories constituting the Low Algerian Sahara (700,000 km²): Ziban, Oued Righ, Oued Souf and Oued M'ya. Low topography, quaternary geology, and paleo-hydrology are the main physiographic characteristics of these regions (Ballais 2010).

Local topography decreases from 140 m height in the south, which is occupied by recent eolian sand dunes (above 8 m height), to less than 15 m in the north, where to occur the Chott Merouane depression. The two regions are connected with aligned quaternary interdune areas oriented south-north (Voisin 2004; Beauzée 1952), corresponding to the Ghout oases-occupied areas. In addition to the Quaternary shallow water table (1–60 m depth), the hydrogeology of Oued Souf is characterized by the existence of two Tertiary and Cretaceous important aquifers (Bel and Demargne 1996; Bel and Cuche 1970; OSS 2003): (1) the complex terminal

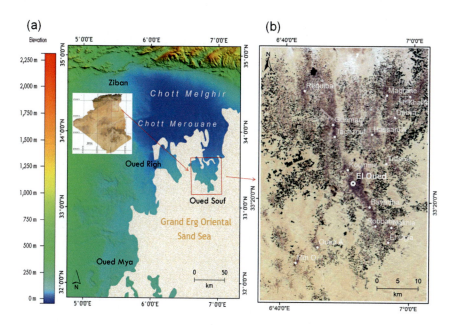

Fig. 1 a Situation of the low Algerian Sahara in Algeria. b Study area as a color composite of the bands RGB/432 of the Landsat image dated 2018, with the municipality's limits

aquifer CT (350,000 km², 100–500 m depth), is principally composed of Pliocene sand, Miocene Gypseous clays and Paleocene limestone, and (2) the intercalary continental CI aquifer (700,000 km²; 1,000 m depth), which is mainly composed of Albian sand and sandstone. These aquifers belong to the northern Sahara aquifer system and contain about 60,000 billion cubic meters (Khadraoui 2006).

Oued Souf is a hyper-arid climate region. Meteorological records from the station of Guemar airport (Lat. 33.5; Long. 6.11; Alt. 63 m) during the period 2000–2015 show that mean temperature is about 22.6 °C, rainfalls is scarce and irregular (61 mm), humidity is relatively low (46%), annual sunshine duration is long (3,373 h) and, the mean wind speed is about 2.6 m/s with eastern prevailing direction.

2.2 Remotely Sensed Data

For global analysis, medium resolution remotely sensed images (30 m) were used for extracting the Ghout date palm oases in the region of El Oued, and monitoring their temporal dynamics for 32 years; the diachronic analysis was conducted using Landsat Thematic Mapper images dated January 1, 1987, and Landsat Operational Land Imager dated June 15, 2018. Each dated scene is spatially located on the tile-center path/row 193/037. The bi-dated optical images used in this study are open-access and available on the United States Global Survey website (www.usgs.gov); all the scenes present minimum atmospheric distortion and are clear of clouds (0%). They are already orthorectified and projected according to the Universal Transverse Mercator zone 32.

2.3 Extraction of the Ghout Date Palm Oases

The vegetation indices express the density and activity of green vegetation as a response to chlorophyll existence. Derived from the image multispectral bands, multiple vegetation indexes exist; the Normalized Difference Vegetation Index (NDVI) developed by Rouse et al. (1974) is known to be sensitive to biomass content. It refers to the ratio of the near-infrared band NIR, with a wavelength ranging from 0.76 to 0.90 μm (Landsat TM) and a wavelength ranging from 0.851 to 0.879 μm (Landsat OLI) as follows:

$$NDVI = \frac{NIR - R}{NIR + R} \quad (1)$$

Under Geographic Information System (GIS) software, the NDVI was calculated for both 1987 and 2018 images; date palm oases corresponding to the Ghout system were identified and extracted by thresholding the calculated NDVI values. To monitor the Ghout Oases dynamics from 1987 to 2018, a geometric mask was applied on the

Landsat image dated 2018 to allow a reliable survey and to avoid any confusion between the pre-existing Ghout date palm and the newly established land reclamation; the relative statistics of the derived classes were then summarized, analyzed, discussed and mapped by considering all municipalities belonging to the study area.

2.4 Assessment of Ghout Oases and Agricultural Land Dynamics

Image difference, change vector analysis, principal component analysis–based, and even image regression; refer to various methods often used in detecting land cover change. Besides, models as single dynamic degree SDD and bidirectional dynamic degree BDD could effectively be used to detect, assess and monitor the spatiotemporal land cover changes during a defined period (Xie et al. 2015; Lv et al. 2017; Ait Lamqadem et al. 2019; Belhadj et al. 2023). In this study, the dynamics assessment of Ghout oases in the region of El Oued was performed by computing the difference between the two date palm classes, with 1987 as the initial date and 2018 as the final date. After applying the municipality's geometric mask, produced results correspond to the simple difference of Ghout oases superficies, which allows the determination of the Ghout increasing or decreasing during the considered period. Besides, the maps' positive differences between 1987 and 2018 were also considered. They correspond to newly planted areas mainly occupied by irrigated agricultural reclamations.

2.5 Detecting and Monitoring the Urban Sprawl

Detection and extraction of the urban built-up areas in the region of El Oued were conducted using the Normalized Difference Built-up Index NDBI (ranging from −1 to 1), which is based on the Near Infrared (NIR) and the Shortwave Infrared (SWIR1) bands of the Landsat images (Jensen 2009; Khudair et al. 2019) as follows:

$$NDBI = \frac{NIR - SWIR1}{NIR + SWIR1} \qquad (2)$$

Superficies of areas occupied by urban built-up were extracted by thresholding the resulting rasters referring to both Landsat images (1987 and 2018); geometric masks were also applied to obtain results on a municipality scale.

2.6 Additional Data

In addition to bi-dated satellite imagery covering the Oued Souf region, complementary datasets were acquired from different national administrations in El Oued Governorate; these data could be delivered or extracted from the Landsat imagery and correspond to each administrative entity (municipality):

Population census data relative to 18 municipalities during the years 1998 and 2018 from the National Office of Statistics, these data were then used to calculate the population-increasing rate PIR using the formulae:

$$PIR_{1998-2018} = \frac{Pop_{2018} - Pop_{1998}}{Pop_{1998}} * 100 \quad (3)$$

- Superficies of agricultural irrigated lands by the municipality for the years 2009 and 2019, acquired from the agricultural services directorate, then calculated as well the increasing rate according to Eq. 2;
- The contour map shows changes in the groundwater depth in the region of El Oued from 2010 to 2014, according to Khezzani and Bouchemal (2018).

The data mentioned above was used to discuss the results and confirm/affirm the research hypothesis presuming that the degradation or disappearance of Ghout oases could be due to multiple hydrogeological and socioeconomic factors.

3 Results and Discussion

3.1 The Ghout Oases Degradation

The satellite images processing allowed extracting the Ghout date palm superficies in the region of El Oued in 1987 and 2018 (Table 1). In 1987, Ghout oases constituted the main agricultural component of the traditional Oasian system (Fig. 2a). They occur throughout the quaternary inter-dunes lowlands and occupy about 4,462 ha. In the north, municipalities such as Debila and Reguiba contain respectively 322 ha and 280 ha of date palm oases, but the largest Ghout-occupied areas occur in the southern municipalities, especially Bayadha (458 ha), El Oued (398 ha) and Oued El-Alenda (371 ha).

From 1987 to 2018, Ghout oases superficies have seriously decreased: about 3,429 ha representing 77% of the pre-existing Ghout date palm, have been subject to severe degradation (Fig. 2b); the superficies of disappeared Ghout oases range from 26 ha in Magrane to 341 ha in Bayadha. Ranging from 62 to 93%, rates of Ghout oases disappearance throughout the region are notably high; northern municipalities such as Magrane (93%), Hassani Abdelkrim (91%), Trifaoui (87%), Debila, Sidi Aoun (85%), and Ourmes (82%) have lost very significant superficies of their

Table 1 Summary of Ghout oases superficies and degradation percentage in the Oued Souf municipalities during the period 1987–2018

Municipality	1987	2018	
	Existing Ghout (ha)	Disappeared Ghout (ha)	Rate (%)
Mih Ouensa	324	250	77
Oued El-Alenda	371	275	74
Robbah	315	223	71
El-Ogla	204	130	64
Nakhla	157	98	62
Bayadha	458	341	74
El-Oued	398	300	75
Trifaoui	285	249	87
Ourmes	161	132	82
Kouinine	270	200	74
Taghzout	201	159	79
Hassani Abdelkrim	210	192	91
Debila	322	275	85
Hassi Khalifa	167	132	79
Magrane	28	26	93
Sidi Aoun	110	94	85
Guemar	201	147	73
Reguiba	280	206	74
Total	4462	3429	77

phoenicultural resources. Superficies of Ghout-occupied areas derived from satellite images confirm the real statistics provided by the Agricultural Services Directorate that are obtained by field counting; among 9,500 Ghout oases previously existing in the region of El Oued, 500 were already flooded in 1994, and 1,000 are completely disappeared in 2000, corresponding to 230,000 date palm trees perished from water excess (suffocation due to asphyxiation) or deficit (hydrogeological insufficiency), highest rates of degradation were recorded in El Oued, Robbah, Debila, and Bayadha.

3.2 Rise of the Phreatic Aquifer: Principal Cause of the Ghout Degradation

Degradation of the Ghout oases in the region of El Oued dates back over 50 years, where the interaction of multiple conditions (natural/man-made, direct/indirect) led to the progressive and continuous decline of all the traditional Oasian system (Fig. 3).

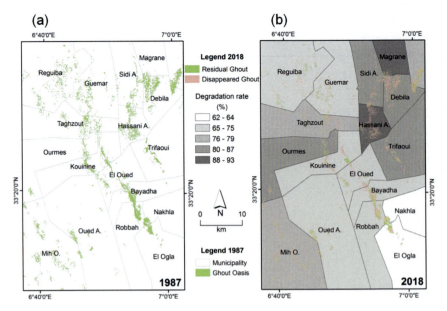

Fig. 2 a Spatial distribution of Ghout oases in 1987 over the study area. b Spatial pattern of disappeared and residual Ghout oases and rates of Ghout degradation in 2018

Before the 1960s, Ghout system, as ancestral know-how, presented an optimal adaptation to the region's topographic and hydrogeological context. Since the date palm trees are planted 1–1.5 m above the shallow phreatic aquifer, they are auto-irrigated directly from the shallow phreatic aquifer, ensuring the population and also the population's drinking water needs. In this period, the water table level was relatively stationary because water withdrawals were compensated by discharges (Dubost 2002). From the 1960s to the late 1980s, the first deeper well was drilled down to 250 m depth in the terminal complex (CT) aquifer to meet the growing need for drinking water. Later, from 1970 to 1980, almost two deep wells per year were drilled in the same aquifer (Cote 2006).

During the 1990s decade and later, hundreds of supplementary wells were drilled in the complex terminal and two deeper ones in the intercalary continental aquifer CI. In parallel, groundwater resource exploitation was confronted with the severe deficiency of natural drainage due to the local topography and the lack of a sanitation network. These factors generated a progressive rising of the water table by 1–2 m leading consequently to the immersion of hundreds Ghout oases in Mih Ouensa (southwest), Hassani Abdelkrim (northeast), also in the middle Oued Souf valley over 25 km length from Kouinine to El Ogla (Cote 1998; Voisin 2004). The extensive extraction of groundwater resources in the region of Oued Souf has been adopted to meet the agricultural and socioeconomic needs of the growing population. According to Voisin (2004) in Khezzani and Bouchemal (2018), the Souf territory population increased from 21,000 in 1887 to 87,400 in 1954, and the water table

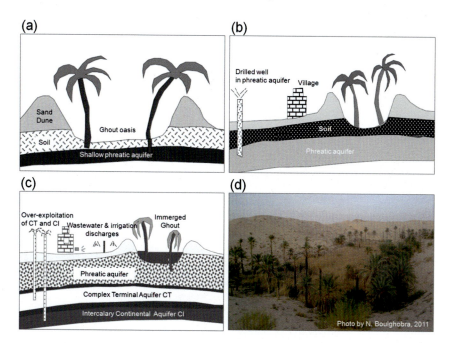

Fig. 3 **a** Schematic cross-section of Ghout oases before the 1960s. **b** Schematic cross-section of Ghout since the 1960s. **c** Process of the water table rising after the 1980s (Remini 2006). **d** Photo of flooded Ghout oasis in Kouinine (El Oued)

level consequently decreased during the periods 1940–1942; after that, 50–150 cm of level decline was recorded from 1953 to 1956 (Bataillon 1955; Dervieux 1956).

More recently, the population of the 18 municipalities included in the study area (part of the larger Governorate of El Oued) increased considerably during the last four decades: from 363,107 inhabitants in 1998 to 464,978 in 2008, and from 538,220 in 2013 to 623,530 in 2018 (national office of statistics). During 1998–2018, the population increase rates range from 43 to 97%, with the highest values percentages in Mih Ouensa, Guemar, and Nakhla municipalities. Since the population growth and Ghout oase degradation have considerably increased during the period, no clear statistical correlation between the two growing rates was detected. However, demographic pressure is the main reason driving the authorities to expand development programs, including over-drilling for domestic, industrial and agricultural needs. According to Khechana et al. (2010), and based on the annual population growth rates from 1999 to 2006 in El Oued, demand for water resources will increase from 901 hm^3 in 2006 to 1,093 hm^3 in 2025 and 1,845 hm^3 horizons 2040, and that could amplify the actual concerns and requires the adoption of more effective development strategies. As a result of the demographic increase and cultural and socioeconomic mutations, the principal city of Oued Souf and its peripheries have experienced significant urban and architectural changes (Fig. 4). The extraction, monitoring, and mapping of the

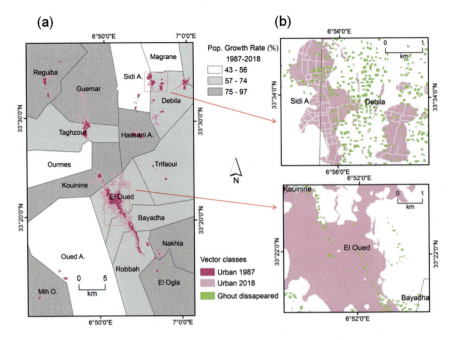

Fig. 4 a Spatial pattern of the urban built (1987 and 2018) and the population increasing rate in percent (1998 to 2018). **b** Zoom-in of two excerpt areas showing the built-up sprawling on areas initially occupied by Ghout oases

urban built areas through the included municipalities, have been conducted using the Normalized Difference Built-up Index NDBI.

Results highlighted the increase in the built-occupied areas throughout the study region. The urban fabric of El Oued city as the main urban agglomeration and its attached municipalities increased from 1,586 ha in 1987 to 5,173 ha in 2018, i.e. an extension rate exceeding 200%. Most spectacular extension rates correspond to Kouinine, Guemar, Nakhla, Ourmes, Bayadha, Mih Ouensa, and El-Oued, cities start sprawling out from the traditional medina towards external areas, to the detriment of lands initially occupied by date palm trees, which also contributed to the disappearance of multiple Ghout oases.

3.3 Agricultural System Mutation

In addition to the Ghout date palm oases, post-1987 planted vegetation was derived and assessed (Table 2). This class corresponds to the positive change that occurred at the end of the defined period that happened in the specified period. From 1987 to 2018, a total area of 10,232 ha was newly planted everywhere in the study area, especially in the northern parts, municipalities such as Reguiba (1,134 ha), Oued El-Alenda

(1,091 ha), Kouinine (993 ha), Guemar (923 ha), Ourmes (900 ha) and Taghzout (836 ha) have been marked by the most significant increase in new agricultural lands reclamation (Fig. 5), these increasing superficies corresponds to recent irrigate agricultural lands occupied by cereals, livestock feed, arboriculture, vegetable and industrial crops in association with new irrigated date palm oases.

The traditional Oasian system based on self-irrigated date palm was the unique agrarian system in the region of El Oued for 12 centuries. The Ghout oasis, as ancestral know-how, was the best adapted and sustainable operating mode to the local bioclimatic conditions, and it was founded on self-subsistence economy to meet limited socioeconomic demand. In the last decades, a new economic mechanism was established in the region, leading to the adoption of an evolved Oasian system based on a different irrigation system. A spectacular increase in new irrigated land reclamations results from the adoption of two agricultural programs:

In the early 1980s, the Agricultural Land Ownership Access Program (APFA) was adopted, allowing agricultural land ownership following their exploitation; thus, farmers introduced new crops such as vegetables, olives and new date palm using their overpumping equipment to extract the necessary groundwater. The National Program for the Development of Agriculture (PNDA) was applied the early 2000s. The authorities provide considerable financial support to the farmers to develop further agricultural land reclamation, especially date palm and potato (Solanum tuberosum), which requires drilling more deep wells to meet the increasing demand for irrigation.

Table 2 Summary of bi-date socio-agricultural factors amplifying the Ghout oases degradation and disappearance in the region of Oued Souf

Population (inhab.)*	1998	2008	2013	2018
	363,107	464,978	538,220	623,530
Urban built-up (ha)**	1987	–	–	2018
	1,586	–	–	5,173
Increase. veg. post 1987 (ha)**	1987	–	–	2018
	0	–	–	10,233
Irrig. land areas (ha)*	–	2009	–	2019
	–	25,900	–	61,825
Date palm plantation (ha)*	1999	–	–	2019
	25,720	–	–	38,147
Dates prod. (quintal)*	1999	–	–	2019
	1,441,231	–	–	2,752,100
Potato prod. (quintal)*	–	2009	–	2019
	–	3,588,962	–	12,140,000

*Data provided by official administrations
**Data derived from satellite images

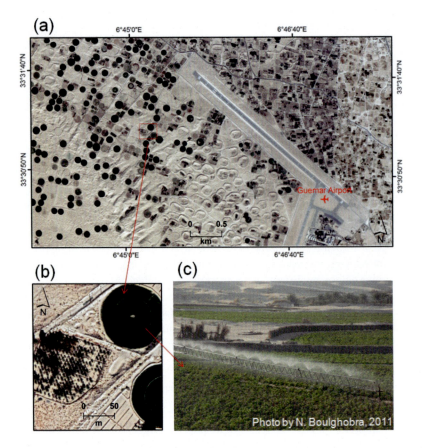

Fig. 5 a Aerial view by Google Earth showing the extension of new land reclamations in Guemar (north El Oued). b Zoom-in showing the association of irrigated croplands with irrigated date palm cultivation. c Photo showing intensive overhead irrigation system in potato parcel

As a result of the state support, new irrigated superficies have highly extended throughout the entire region of El Oued. According to the Directorate of Agricultural Statistics (2019), total irrigated lands increased from 25,900 ha in 2009 to 61,825 ha in 2019, especially in Magrane (1,255–5,639 ha), Bayadha (251–1,024 ha), Trifaoui (1,734–6,951 ha), Hassi Khalifa (3,027–10,878 ha), Nakhla (518–1,754 ha) and Ourmes (1,874–5,629 ha). From 1999 to 2019, date palm plantation superficies continuously increased from 25,720 ha in 1999 to 38,147 ha in 2019, which resulted in a large increase in agricultural production: date fruits production increased from 1,441,231 quintals (1999) to 2,752,100 (2019), and potatoes from 3,588,962 quintals (2009) to 12,140,000 quintals (2019).

In 2010, El Oued involves more than 27,471 agricultural exploitations; most of them are located north of the region in Hassi Khalifa (4,958), Reguiba (2,974),

Trifaoui (2,870), Debila (2,825) and Guemar (2,121), this exploitation insured important financial returns exceeding 20 M$/year especially Hassi Khalifa (2.8 M$) and Reguiba (2.5 M$). This makes intensive agriculture the main economic activity consuming 69.5% (592 hm^3) of mobilized water resources, in comparison with drinking water (30%—255 hm^3) and the industrial sector (0.5%—4 hm^3). Figure 6a highlights the spatial distribution of disappeared Ghout oases from 1987 to 2018 superimposed on the map of groundwater depth changing between 2010 and 2014, according to Khezzani and Bouchemal (2018). Obviously, areas with the highest rates of Ghout disappearance correspond to regions suffering from a substantial decline of the water table level exceeding 2.4 m, especially in the northern part of the region (Reguiba, Guemar, and Debila) and the west southern region as Mih Ouensa.

The extracted geo-information from satellite images, as the data provided by the agriculture administration, confirm that these regions are the most marked by the extension of post-1987 land reclamations and the increase of new irrigated agricultural perimeters (Fig. 6b). The multiplication of newly drilled wells and the continuous extraction of groundwater from deeper aquifers (CT and CI) since the 2000s, led to a significant decrease of the water table which avoid the self-irrigation of date palm trees (requiring at least 0.5–1 m of the aquifer depth) and this manifests on the Ghout drought and progressive disappearance especially in the north (Fig. 7). In the opposite, Ghout oases disappearance in areas located in the center of the region (as El Oued and Kouinine) is principally due to the water table rising because of the

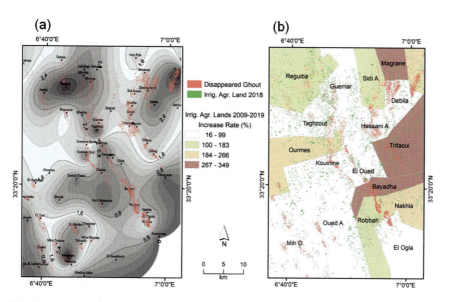

Fig. 6 a Disappeared Ghout as vector class superimposed on the map of groundwater depth changing (2010–2014) by Khezzani and Bouchemal (2018). **b** Change detection map showing the post-1987 increased vegetation, disappeared Ghout, and the increasing rates in percent of irrigated agricultural lands (2009–2019)

Fig. 7 **a** Extraction of groundwater using overpumping equipment. **b** The over-exploitation of groundwater for irrigating new cropland (**b1**) led to the decline of piezometric levels and the drought of Ghout Oasis (**b2**). **c** The Ghout oasis disappeared due to drought in Reguiba (north of El Oued). **d** Photo of healthy Ghout oasis located near El Ogla southeast of El Oued, within a balanced water table

return of the irrigation water, leading to the immersion of date palm trees (Saibi et al. 2009).

4 Conclusions

Ghout Oasis is an ancestral agricultural technique of date palm tree plantation in Oued Souf (northeast of the Algerian Sahara). Classified by the Food and Agricultural Organization FAO as a Globally Important Agricultural Heritage System, it is continuously endangered due to multiple natural and man-made factors. This research is conducted on El Oued's mid-region and includes 18 municipalities. The adopted approach consisted of applying spectral indices tool and change detection method for assessing, monitoring, and mapping the Ghout date palm oases dynamics, the new irrigated land reclamation, and the built-up areas throughout the study area from 1987 to 2018. The results show that the Ghout-occupied areas have considerably decreased from 4,462 ha in 1987 to 1,033 ha in 2018. Ghout degradation occurred in the entire region, especially in the northern areas with the highest disappearance rates.

On the other hand, the urban superficies have significantly sprawled from 1,586 ha (1987) to (5,173 ha) in 2018, similar to the post-1987 agricultural land reclamations (up to 10,233 ha). The study concludes that Ghout date palm degradation in El Oued had previously resulted from the increase in socioeconomic pressure and posteriorly from the over-exploitation of groundwater resources to meet the increasing agricultural needs on irrigation water demand. The traditional agricultural system, the oasian Oasian agro-ecosystem in El Oued, has experienced a deep mutation during the last five decades. New free economic mechanisms are established, leading to the degradation of the Ghout date palm as an important and sustainable ancestral system.

5 Recommendations

In Algeria, Saharan agriculture is a real alternative to fossil energy sources, and food security is a strategic national objective. In Saharan regions, ecosystems are particularly fragile, and the abundance and quality of natural resources are both sensitive to climatic extremes and global changes, where it is necessary to assess the Saharan region's potentials and limits. Scientific research should provide reliable and integrated decision-making–support in order to control the development process better and rationalize the exploitation of natural resources for sustainable and efficient development.

References

Ait Lamqadem A, Afrasinei GM, Saber H (2019) Analysis of Landsat-derived multitemporal vegetation cover to understand drivers of oasis agro-ecosystems change. J Appl Remote Sens 13(1)

Al-Khayri JM, Jain SM, Johnson DV (2015) Date palm genetic resources and utilization. Africa and the Americas. Springer, Netherlands

Ballais JL (2010) From mythical wadis to artificial rivers: the hydrography of the Algerian Lower Sahara. Physio-Géo IV:107–127

Bataillon C (1955) The Souf, study of human geography. Works of the Saharan Researchs Institute, Algiers

Beauzée G (1952) The Oases of El-Oued. Study of phreatic aquifer renewal conditions. Mission report carried out from March 19 to April 3, 1952. Building and Public Works Laboratory, Algiers, 31 p

Bel F, Cuche D (1970) Study of the aquifers of the Terminal Complex of the lower Sahara. Geological and hydrogeological data for the reconstruction of the mathematical model. Technical Report, Direction of Hydraulics of the Wilaya of Ouargla, Algeria

Bel F, Demargne F (1996) Geological study of the Continental Terminal, ANRH, Algiers, Algeria, 22 p

Belhadj A, Boulghobra N, Demnati Allache F (2023) Multi-temporal Landsat imagery and MSAVI index for monitoring rangeland degradation in arid ecosystem, case study of Biskra (southeast Algeria). Environ Monit Assess 195:656

Bouguedoura N, Bennaceur M, Babahani S, Benziouche SE (2015) Date palm status and perspective in Algeria. In: Al-Khayri JM, Jain SM, Johnson D (eds) Date palm genetic resources and utilization. Springer, Netherlands, pp 125–168

Bouselsal B, Saibi H (2022) Evaluation of groundwater quality and hydrochemical characteristics in the shallow aquifer of El-Oued region (Algerian Sahara). Groundw Sustain Dev 17:100747

Cote M (1998) Oases suffering from too much water? Sécheresse 9:123–130

Cote M (2006) If the Souf was told to me: how a landscape is made and unmade? Saïd Hannachi, Média-Plus, Constantine

Daoudi A, Colin JP (2017) Construction and transfer of land ownership in the new steppe and Saharan agriculture in Algeria. In: Guignard D (ed) Property and society in contemporary Algeria. Iremam, Open Edition Books

Dervieux F (1956) The Souf phreatic aquifer (Algeria) study of the aquifer renewal: contribution to the study of capillary phenomena in a powdery environment. J Terre Eaux 29:11–39

Dubost D (1986) New agricultural perspectives of the Algerian Sahara. Rev Mondes Musulmans Mediterr 41–42:339–356

Dubost D (2002) Ecology, planning and agricultural development of the Algerian Oases. Scientific and Technical Research Center on Arid Regions CRSTRA, 423 p

Fadhil AM (2013) Sand dunes monitoring using remote sensing and GIS techniques for some sites in Iraq. In: Proc. SPIE 8762, PIAGENG 2013: intelligent information, control, and communication technology for agricultural engineering

Guendouz A, Moulla AS, Remini B, Michelot JL (2006) Hydrochemical and isotopic behaviour of a Saharan phreatic aquifer suffering severe natural and anthropic constraints (Case of Oued-Souf region, Algeria). Hydrogeol J 14:955–968

Jensen JR (2009) Remote sensing of the environment: an earth resource perspective 2/e. Pearson Education India

Khadraoui A (2006) Water and soil in Algeria, management and impact on the environment. EMPAC, Constantine, Algeria, 392 p

Khechana S, Derradji E (2012) Management of water resources in a hyper-arid area: strategy and issues (Case of Oued-Souf Valley-South Eastern of Algeria). J Water Resour Prot 4:922–928

Khechana S, Derradji F, Derouiche A (2010) Integrated management of water resources in the valley of Oued-Souf (Algeria): issues fitness for a new strategy. J Fundam Appl Sci 2(2):22–36

Khezzani B, Bouchemal S (2018) Variations in groundwater levels and quality due to agricultural over-exploitation in an arid environment: the phreatic aquifer of the Souf oasis (Algerian Sahara). Environ Earth Sci 77(4):142

Khudair D, Fadhil A, Alauldeen H (2019) Spatiotemporal monitoring and modeling of urban sprawl using remote sensing and GIS: a case study Al-Karkh, Baghdad, Iraq. J Adv Res Dyn Control Syst 11(6 Special Issue)

Lv Z, Shi W, Zhou X, Benediktsson JA (2017) Semi-automatic system for land cover change detection using bi-temporal remote sensing images. Remote Sens 9:1112

Mihi A, Tarai N, Chenchouni H (2019) Monitoring dynamics of date palm plantations from 1984 to 2013 using Landsat time-series in Sahara Desert Oases of Algeria. In: El-Askary HM et al (eds) Advances in remote sensing and geo informatics applications. Advances in science, technology & innovation

Mihi A, Tarai N, Benaradj A, Chenchouni H (2022) Spatiotemporal changes in date palm oases of Algeria over the last century. In: El-Askary H, Erguler ZA, Karakus M, Chaminé HI (eds) Research developments in geotechnics, geo-informatics and remote sensing. CAJG 2019. Advances in science, technology & innovation. Springer, Cham

Nesson C (1975) The evolution of resources in the oases of the Low Sahara. In: Research on Algeria. Mémoires et documents, vol 17. C.N.R.S., Paris, 291 p

OSS Sahel and Sahara Observatory (2003) Northern Sahara Aquifer System. Internal report, Tunis, Tunisia, 229 p

Otmane T (2019) From water ownership to land ownership: shifting logics in access to agricultural land in the southwest of the Algerian Sahara. Dev Durable Territ 10(3):1–20

Otmane T, Kouzmine Y (2010) Spatial review of agricultural development in the Algerian Sahara. Cybergeo: European Journal of Geography, Space, Society, Territory, document 632

Remini B (2006) The disappearance of the Ghouts in the region of El Oued (Algeria). Larhyss/J 5:49–62

Remini B, Kechad R (2011) Impact of the water table razing on the degradation of El Oued palm plantation (Algeria) mechanisms and solutions. Geogr Tech 1:48–56

Remini B, Miloudi A (2021) Souf (Algeria), the revolution of crater palm groves (Ghouts). Larhyss J 47:161–188

Rouse JW, Haas RW, Schell JA, Deering DW, Harlan JC (1974) Monitoring the vernal advancement and retrogradation (Greenwave effect) of natural vegetation. NASA/GSFCT Type III Final report, Greenbelt, MD, USA

Saibi H, Semmar A, Mesbah M, Ehara S (2009) Variographic analysis of water table data from the Oued-Souf phreatic aquifer, northeastern part of the Algerian Sahara. Arab J Geosci 2:83–93

Senoussi A (2000) The date palm in the region of Ouargla: eternal culture and incredible development prospects. In: Study day on date palm cultivation. University of Laghouat, Algeria

Tigrine CDA, Boutiba M (2023) Salinization risk assessment of irrigated soils in the Souf Valley (South-East Algeria) using chemical analysis, multivariate statistics, and GIS. Arab J Geosci 16:438

Voisin AR (2004) The Souf, monograph. El-Walid, El-Oued

Xie Y, Zhao H, Wang G (2015) Spatio-temporal changes in oases in the Heihe River Basin of China: 1963–2013. Écoscience 22(1):33–46

Tectonic and Erosion in the Zagros Fold-and-Thrust Belt (ZFTB)

Ahmed K. Obaid, Arsalan A. Othman, Sarkawt G. Salar, Varoujan K. Sissakian, and Salahalddin S. Ali

Abstract Denudation involves weathering mountains and soil formation over long periods and is classified as soil degradation. This study focuses on the relationship between uplift and denudation using the hypsometric index (HI) and river profile steepness as key points to describe erosional processes and their relationship with environmental land degradation in the Zagros Fold and Thrust Belt (ZFTB). The interplay between tectonics, climate, and geomorphological features has been investigated to shed light on the erosional processes in the ZFTB. There is uniformity between regions of higher precipitation, steep slopes, higher river steepness, and higher HI, particularly, in areas to the northeast of the Kirkuk Embayment (KE) and areas of Bakhtyari Culmination (BCu). Higher-strength lithologies show consistency with the areas of higher HI, steepness, and higher precipitation, implying that the erodibility of exposed lithology does not control higher geomorphic indices values.

A. K. Obaid (✉)
Department of Geology, University of Baghdad, Baghdad, Iraq
e-mail: ahmedobaid@uobaghdad.edu.iq; ahmed.k.obaid@durham.ac.uk

Department of Earth Sciences, University of Durham, Durham DH1 3LE, UK

A. A. Othman
Iraq Geological Survey, Sulaymaniyah Office, Sulaymaniyah, Iraq
e-mail: arsalan.aljaf@geosurviraq.iq; arsalan.aljaf@komar.edu.iq

Department of Petroleum Engineering, Komar University of Science and Technology, Sulaimaniyah 46001, Iraq

S. G. Salar
Department of Geography, College of Education, University of Garmian, Sulaymaniyah, Iraq
e-mail: sarkawt.ghazi@garmian.edu.krd

V. K. Sissakian
Komar University of Science and Technology, Sulaymaniyah, Iraq
e-mail: varoujan.sissakian@komar.edu.iq

S. S. Ali
Civil Engineering Department, College of Engineering, Komar University of Science and Technology, Sulaimaniyah 46001, Iraq
e-mail: salah.saeed@komar.edu.iq

© The Author(s), under exclusive license to Springer Nature Switzerland AG 2024
A. M. F. Al-Quraishi and Y. T. Mustafa (eds.), *Natural Resources Deterioration in MENA Region*, Earth and Environmental Sciences Library,
https://doi.org/10.1007/978-3-031-58315-5_4

The results showed a coupling between climate (e.g., orographic rain) and tectonic uplift controlling higher erosion areas (land degradation), where we found steep slopes, high HI, and steepness values.

Keywords Tectonic · Erosion · Zagros region · Geomorphic indices · Kirkuk embayment · Bakhtyari culmination

1 Introduction

In the last decades, enormous efforts have been made to quantify and understand the relationship between the two competing forces (1) tectonic, which leads to topography building, and (2) erosion, which leads to lowering the topography. The process that controls the evolution of the earth's crust structure and its characteristics during mountain building is known as tectonics (Jon et al. 2007). Erosion is the reaction to the tectonic through the process of mountains building, creating instability within the earth's surface. Thus, erosion redistributes the materials of the earth's surface from the higher to the lower altitudes to achieve the equilibrium state between uplift and erosion. Without chronological tests, it is very hard to understand the competition between tectonics and erosion. Thus, most landscape evolution models from geomorphic markers are still speculative (Burbank and Anderson 2011).

Erosion is a primary cause of the degradation of land, which is responsible for about 84% of land degradation worldwide (Toy et al. 2002; Blanco-Canqui and Lal 2008). Therefore, investigating the identity of the tectonic situation of a region will increase our understanding of erosion processes, which in turn contributes to the recognition of the land degradation of a region. The correlation between focused orographic precipitation areas and high topography with the Apatite Fishing Tracking cooling age distribution across the Sutlej region in the Himalayas suggested a strong interplay between erosion and uplift. In the internal part of the Himalayas, climatically controlled surface processes have been proven to determine tectonic deformation (Thiede et al. 2004). However, there is no similar study in the Zagros to generalize this hypothesis for all fold-and-thrust belts.

The relation between uplift and denudation has been studied in terms of using the hypsometric index (HI) (Strahler 1952). This index works well with the roughness of the surface to describe the erosional cycle and, consequently, landscape evolution through the youthful, mature, and old stages (Keller and Pinter 2002), thereby allowing the determination of the tectonic activity of regions. The HI with other geomorphic indices (e.g., mountain front sinuosity, stream length gradient index, and drainage basin asymmetry) (Strahler 1952; Hack 1973; Bull 1977; Cox 1994); are affected by the river action, which climatic and /or tectonic drivers may govern. To maintain a tectonic history of land structures, the amount of surface uplift and terrain elimination must be dynamically equilibrium (Dietrich et al. 2003).

Landscape relief increases with the increase of erosion due to river action. The stream channel network controls the landscape relief (Whipple and Tucker 1999).

Thus, an area's drainage network contains important information that can help interpret past and present tectonics. The susceptibility of the profile of a river to uplifting can explore features of active deformation (Seeber and Gornitz 1983). It is crucial to analyze a river longitudinal profile to be in steady-state conditions (Whipple and Tucker 1999; Snyder et al. 2000; Wohl and Merritt 2001; Kirby et al. 2003; Wobus et al. 2006). So, the occurrence of knickpoints along river profiles can be recognized as variations in the slope of river profiles (Wobus et al. 2006; Kirby and Whipple 2012).

Fluvial erosion has been described in three conditions: (1) detachment-limited, (2) Transport-limited, and (3) hybrid river models. In the first model, erosion = uplift, where the gradient of a river is governed by uplifting or the lowering of the base level of a river and substrate erodibility. In the second model, the channel slope is determined by the channel's strength to transport eroded materials. In the third model, the gradient of a channel is controlled by substrate erodibility and sediment (Whipple 2002; Tucker and Whipple 2002).

The power law (Hack's law) describes the connection between the local gradient (S) and the area upstream (A) of a river (Flint 1974).

$$S = k_s A^{-\theta} \qquad (1)$$

k_s is the steepness of the river segment, and θ is the concavity index, which shows the change in slope along the profile of a river. Slope (S) and upstream area (A) can be calculated from the Digital Elevation Model (DEM) when applying linear regression on the S-A plot.

Different types of knickpoints can be developed as a result of tectonism, which is either uplifting that results from folding and/or faulting or erosion to reach the base level (Wobus et al. 2006; Goldrick and Bishop 2007; Kirby and Whipple 2012). Therefore, the interpreter must be careful about which type of knickpoint will depend upon for tectonic interpretation. Tectonic history in active orogens has been identified using the distribution of knickpoint (Miller et al. 2012; Schildgen et al. 2012; Morell et al. 2012).

The slope of a river basin and its forms are defined as hypsometry (Langbein 1947). An individual basin can be represented by multiple levels of altitude, confining multiple areas. These levels of altitude and areas form the curve of hypsometry, which can be changed into a HI (Strahler 1952). The HI has been extensively used to represent erosional phase and landform evolution (Strahler 1952; Schumm Stanley 1956). When the HI value is near 1, this refers to the lowest terrain elimination, which refers to the youth time of land surface (uplifting > erosion). While the HI value near 0, refers to higher terrain elimination and maturity of land surface (erosion > uplifting). The dimensionless form of the HI permits the differentiation between different basins of different areas.

The dissection of basin topography can be characterized, supported by the use of HI as a robust tool to determine the relative differences of tectonic action of a region (Keller and Pinter 2002). The rapid evolution of DEMs has simplified the method of HI calculation automatically using Eq. (2) (Pike and Wilson 1971; Keller and Pinter 2002).

$$HI = \frac{H_{\text{mean}} - H_{\text{min}}}{H_{\text{max}} - H_{\text{min}}} \qquad (2)$$

H_{max} = the ultimate elevation, H_{min} = least elevation, and H_{mean} = average elevation.

Recently, the map forms of geomorphic indices have been used to assess tectonic activity and landscape maturity in many regions of the globe (Domínguez-González et al. 2015; Obaid and Allen 2017; Zebari et al. 2019). River profile has been used to document surface uplift (Al-Attar et al. 2022; Seeber and Gornitz 1983; Clark et al. 2004; Whipple et al. 2016; Cannon et al. 2018; Whipple et al. 2016). Gao et al. (2016) and Obaid and Allen (2019) have shown the efficiency of basin-scale HI in understanding the surface uplift of the Longmenshan in SE Tibet and the ZFTB, respectively. Therefore, we think that the key point in setting the relation between tectonics and erosion is to analyze river basin hypsometry and the steepness of the river profile in addition to other geomorphic indices.

Several studies investigate the relation between sediment fluxes (Zhang et al. 2021) and denudation (Wang et al. 2021) with tectonic. The results show consistency between the higher rate of erosion and tectonic uplift across active faults in Longmenshan. Also, sediment fluxes during the Neogene are compatible with rapid tectonic uplift in Myanmar. In addition, tectonic erosion and crustal removal show similar patterns in the evolution of Japanese forearc sediment provenance (Pastor-Galán et al. 2021) as well as with tectonic accretion processes (Mandal et al. 2021).

During its geologic evolution history, the ZFTB has been subjected to a variety of tectonic events, which might be preserved in landforms. Therefore, this chapter focuses on the geomorphic indices from fluvial origin to investigate the interplay between tectonics, climate, and the earth's surface to shed light on the erosional processes in the ZFTB.

1.1 Regional Tectonics of the ZFTB

The Zagros range (Fig. 1) comprises the main part of the collisional area of the Arabian Plate (Arabia) and the Eurasian Plate (Eurasia). It accommodates the northern direction motion of Arabia at a speed of nearly 16–26 mm/year (Vernant et al. 2004). Spatially, the orogen covers ~1800 km in length, starting from the southeastern part of Turkey, the north part and the north-east borders of Iraq, through south-east south of Iran to approach the Oman Line (Alavi 2007; Vera and Gines 2009; Othman and Gloaguen 2013; Othman et al. 2018). Several collisional and rifting events occurred to initiate the Zagros collision zone. Several basement faults occurred and the successive tectonic events caused the reactivation of some of these faults until ended with the Arabia and Eurasia plates collision (Ameen 1992; Jassim and Goff 2006; Stern and Johnson 2010; Burberry 2015). The ongoing convergence of the Arabian and Eurasian plates has initiated groups of folds in different orientations (Fig. 1).

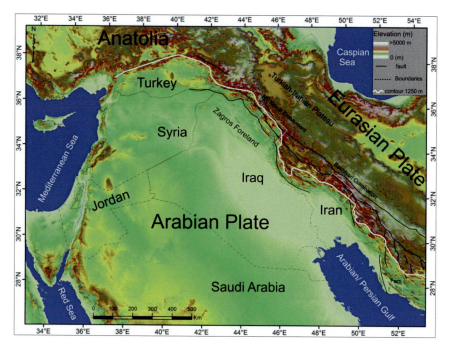

Fig. 1 Map shows the areal extent of the ZFTB

1.2 The Topography of the ZFTB

Different topographic regions comprise the ZFTB: the higher east and northeast elevation topography and the low river dissection across the Turkish-Iranian Plateau (TIP), the west and low southwest relief and low topography of the Zagros Foreland, and the higher river dissection (relief) areas of the ZFTB between the TIP and the Zagros foreland (ZF) (Allen et al. 2013). The growth of the TIP is debated; it is unclear why there are differences in the rate and the process of the TIP growth from one area to another within the Zagros. There is a southwest growth of the TIP in the arc of Fars, and presumably in the arc of Lurestan, but maybe have a northeastward retarding in the Dezful/Bakhtyari (D/B) region. The TIP growth is plausible in relation to the thickening of the crust and the upper level of seismicity cut-off at an altitude of 12.5 km (Nissen et al. 2011; Allen et al. 2013).

Few studies have dealt with river and fold interplay in the ZFTB to understand the effect of uplift on earth material removal by rivers (Burberry et al. 2008, 2010; Ramsey et al. 2008; Bretis et al. 2011; Walker et al. 2011; Bahrami 2013; Zebari and Burberry 2015). Different regions across the Zagros show different styles of interaction between folds and rivers, which might cause differences in climate and topographic uplift. Figures 2 and 3 show a specific style of folds and river interaction, where rivers cut the anticlines in a rugged and high topographic uplift in the north

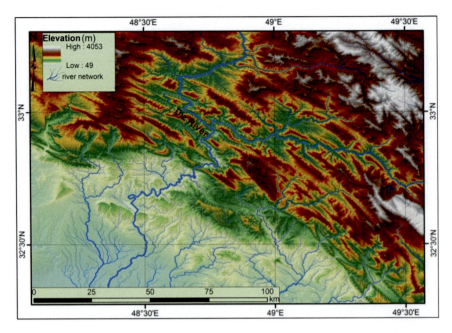

Fig. 2 Rivers and folds the interactions in the Dezful region, where there is a youthful landscape (after Obaid (2018))

of the KE. The Fars region shows an opposite scenario, as rivers generally deflect around anticlines (Fig. 4).

1.3 The Climate of the ZFTB

The climatic conditions of the ZFTB are characterized by aridity and relatively high-temperature conditions in summer. The aridity continued in winter despite cold weather (Kottek et al. 2006). Rainfall irregularity in place and time is attributed to the interplay between two synoptic systems with the elevation variation in the ZFTB. These synoptic systems are the Mediterranean and Sudan synoptic systems (Katiraie-Boroujerdy et al. 2013). The outcome of analyzing the data of the Tropical Rainfall Measurement Mission (TRMM) satellite 3B43 from 1998 to 2016 shows the high fluctuation in rainfall in the ZFTB (Fig. 5). The highest rainfall (0.16 m·yr^{-1}) was in the BCu, Lurestan, and areas located to the northeast of the KE. The lowest rainfall (0.035 m·yr^{-1}) occurred in the Fars region and adjacent areas, the TIP, and the ZF. Rainfall produces significant sediment transports from higher to lower slopes areas in the ZFTB (Othman et al. 2021, 2023).

Tectonic and Erosion in the Zagros Fold-and-Thrust Belt (ZFTB)

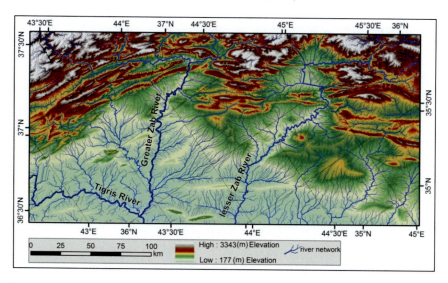

Fig. 3 Rivers and folds the interactions in the KE (after Obaid (2018))

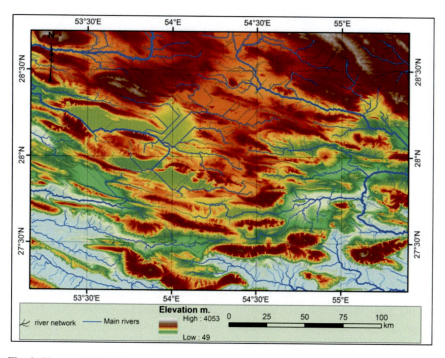

Fig. 4 Rivers and folds the interactions in the Fars region (after Obaid (2018))

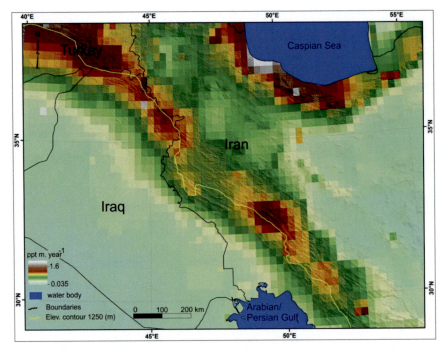

Fig. 5 Average yearly rainfall from the (TRMM) satellite of the 3B43 version from 1998 to 2016 with a resolution of 0.25°. Note the areas of higher precipitation coincide with the areas of higher steepness (Fig. 6) and higher HI (see Fig. 5 in (Obaid and Allen 2019))

2 Geomorphic Indices

2.1 Normalized River-Channel Steepness (k_{sn}) Distribution Across the ZFTB

The values of the k_{sn} have been used relatively to describe the changes in the topography across the Zagros range. All rivers segments with a length of >1 km have been used to generate river longitudinal profiles (Fig. 6). k_{sn} value <50, 50–100, 100–150, and >150 are relatively low, intermediate, high, and very high respectively. The TIP and the ZF rivers exhibit $k_{sn} < 50$ m$^{0.9}$ (Fig. 6a). The k_{sn} values from 50 to 100 m$^{0.9}$ located in the rugged zones of the BCu, and in areas near the altitude of 1.25 km in the northwest part of the ZFTB (Fig. 6b). A likewise pattern occurs over the rugged zones for the k_{sn} values range between 100 and 150 m$^{0.9}$ (Fig. 6). Few rivers segments hold values of $k_{sn} \geq 150$ m$^{0.9}$ are distributed in the higher relief zones of the BCu and the northwest part of the ZFTB (Fig. 6a, 6b). The relatively lowest k_{sn} values occur across the ZF and TIP (Fig. 6a, b).

Over the entire Zagros, it seems that the uplift is not uniform, with the implication of the applicability of the steady-state approach to analyze k_{sn} (Snyder et al. 2000).

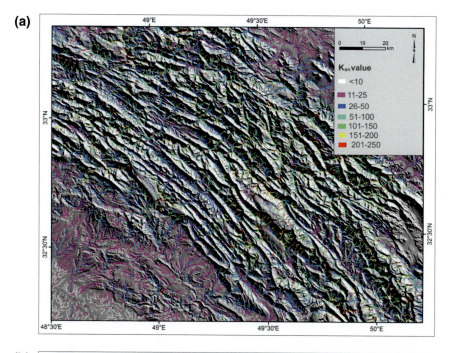

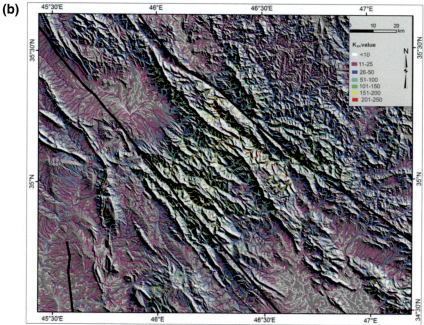

Fig. 6 a Distribution of k_{sn} values across the BCu. b Distribution of k_{sn} values across the northeast of the KE

The general pattern of relatively higher k_{sn} values are distributed around the 12.5 km contour of elevation; except in the Fars region and also from the inner aseismic zone towards the zone of seismogenic thrusting of the ZFTB (Fig. 6). There is no clear sharp boundary of higher k_{sn} zones across the main faults of the Zagros, in contrast with the Longmenshan in SE Tibet (Gao et al. 2016) and the Himalaya (Kirby and Whipple 2012). This difference might be attributed to the tectonic variations between many non-continuous, non-emergent thrusts of the ZFTB, and the non-segmented and long-extended thrusts of the Himalayas such as Main Central Thrust and SE Tibet.

2.2 Hypsometric Index (HI) Distribution in the ZFTB

The relatively low HI values (<0.3) from the 3rd-order watershed in the ZFTB revealed two main zones (Obaid and Allen 2019). The first zone is located in the TIP where topographic slopes are very gentle (Allen et al. 2013). The second zone represents the ZF and Mesopotamian Plain. The mountainous areas of the Zagros represent moderate and somewhat high HI values (>0.3), which are characterized by steep slopes and rugged areas. The KE, the BCu, and the areas near the coast in the Fars region exhibit the highest values of the HI (Obaid and Allen 2019).

Over most of the ZFTB, there is a consistency between the lower to higher HI transition boundary (HI = 0.3) and the upper limit of thrust seismicity, confined below the elevation contour of 12.5 km. This style occurs northeast of the KE and the Lurestan Arc. Relatively higher values of HI in the BCu continue to the northeast passing the 12.5 km altitude contour. In the Fars region, the limit of thrust seismicity continues to the north passing the high-to-low HI boundary (HI = 0.3) (Obaid and Allen 2019). The region of Fars has the highest values of HI located southwestwards of the upper limit of seismogenic thrusting, in a reverse style to the Dezful /Bakhtyari region. Based on the HI value distribution, there is a big similarity between the Kirkuk and the Dezful /Bakhtyari regions, which might be attributed to the similarity in their structural configuration (Obaid and Allen 2019).

3 Tectonic and the (HI) and (k_{sn}) Distribution Across the ZFTB

Keller and Pinter (2002) mentioned that the lithological factor is one of the three important parameters used in tectonic interpretation (tectonic, climate, and lithology). The main exposed lithologies in BCu "for instant" are limestones, marls, conglomerates, patches of ophiolites, and the Bakhtyari Formation (sandstones and conglomerates). A series of igneous and metamorphic rocks occur to the northeast of the BCu. There are considerable variances in the HI value for the same lithology. Contrarily, variance lithologies including ophiolites and limestones, exhibit similarity in the HI values, which range from 0.3 to 0.4 (Obaid and Allen 2019). The distribution of HI values is independent of the exposed lithologies, which means that the effects of the variations in rock types on the HI values are limited (Gao et al. 2016; Obaid and Allen 2019).

The Dezful/Bakhtyari regions are characterized by a relatively wet climate (Fig. 5). This climate condition accelerates landscape erosion by the action of water in an area where a narrow zone of high strain-controlled deformation (Allen et al. 2013). The high HI zone extends to the area located northeast of the upper limit of thrust seismicity. Tectonically, this zone is part of the TIP, as there is no active (seismogenic) shortening. Geomorphologically, it is not developed to be a plateau of low relief as a result of the river network, which preserved the current relief (Obaid and Allen 2019). It seems that the Dezful/Bakhtyari and KE are experiencing climatic positive feedback producing a somewhat moist climate in an area of a higher elevation ridge (Figs. 1 and 5). In the Dezful/Bakhtyari we can see transverse rivers so as in the KE regions as a consequence of somewhat higher rainfall and hard thrust fault action, which enables streams to carve when intersecting many anticlines in these regions.

Based on the regional analysis, no sharp changes of basin scale-based HI values for individual structures as it (could be) if there is a domination of active deformation by a few main thrust faults in the ZFTB. This style contrasts with the Longmenshan style (east part of the Tibetan Plateau), where sudden changes have been noticed in the HI value (Gao et al. 2016). We compared the HI result with the lithologies exposed in the BCu. There are no noticeable changes in k_{sn} value with the lithological changes, specifically at the BCu. Exposed limestone (Fig. 7) shows a clear difference in the values of k_{sn}, whilst the river profile shows similar values of k_{sn} in various lithologies. Figure 7 shows k_{sn} values of 100–150 m$^{0.9}$ for a specific river segment regardless of changes in lithology (limestone, dolomite, and serpentinite).

4 Interaction Between Tectonic and Erosion Across the ZFTB

The Zagros shows a considerable variance in the HI values for the same lithology and nearly similar values of the HI for various rock types (e.g., the ophiolitic assemblages and carbonate rocks) (Obaid and Allen 2019). In addition, the steepness index showed no correlation; neither with the change in lithologies nor with major tectonic features, such as Main Recent Fault or Main Frontal Fault. The relatively higher HI and steepness values have coincided with an elevation contour of 1250 m (Fig. 1 and 6), which is the upper limit of the thrust seismicity. The difference in the HI and k_{sn} distribution, representing higher uplift and focused erosion, might be attributed to tectonic differences between regions across the Zagros. In the Dezful/Bakhtyari regions, a difference has been assigned by Allen and Talebian (2011) to the various tectonic events in the adjacent areas prior to the collision of the Arabian Plate and

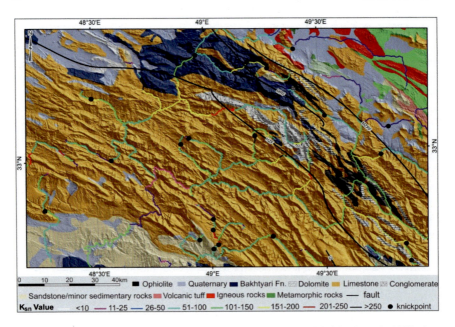

Fig. 7 Exposed rock types of the ZFTB (redrawn after (Afaghi and Salek 1975a, b; 1977a, b, c; Afaghi et al. 1978)). Note: the changes in k_{sn} values of river segments with the changes in lithology (after Obaid (2018))

Eurasian Plate, which did not occur in the Fars region (Allen et al. 2013; Talebian and Jackson 2004).

The tectonic differences have positive climatic feedback as a moist climate in the Dezful/Bakhtyari consistently has higher topographic barriers. Contrary, the Fars region is represented somewhat by a dry climate and low topography (Figs. 5). There is a possible similarity between the origin of the Kirkuk Embayment and the Dezful Embayment, from the point that high strain might be focused at the Imbricated Zone to the northeast of the KE where higher k_{sn}, HI and higher relief occur. Depending on the geomorphic indices and the Erosion Potential Model (EPM) (Salar 2013), the western part of Zagros probably exhibits a northeastward increase in tectonic activity and deformation.

There is a consistency between regions with high levels of precipitation (Fig. 5), higher river steepness (Fig. 6), and Higher HI; specifically, areas to the northeast of the KE and areas of the BCu. The rock strength map (Fig. 8) reveals that areas of higher HI, higher steepness, and higher precipitation are correlated with high-strength lithology. This means that the dissection, which leads to higher steepness and higher HI (higher erosion rate), is not controlled by the erodibility degree of exposed lithology. Tectonic uplift and steep slope areas trigger higher HI and higher steepness values. These areas, in turn, are correlated with high land degradation, where a high incision rate (erosion) occurs. There seems to be a coupling between

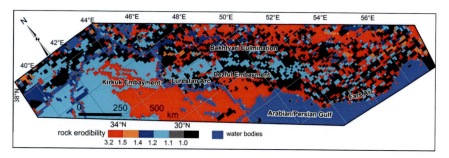

Fig. 8 Distribution of the rock's erodibility classes in the ZFTB (after (Moosdorf et al. 2018))

climate (e.g., orographic rain) and tectonic uplift with higher erosion areas where we found steep slopes, high HI, and steepness values.

5 Conclusions

Geomorphic indices from fluvial origin such as k_{sn} and HI have been used to investigate the interaction between uplift and environmental land degradation in ZFTB. Results reveal uniformity between areas of higher precipitation, steep slopes, higher k_{sn}, and higher HI. There is no correlation between ksn and HI with the lithological changes or the numerous segmented active thrust faults. There is a geomorphological similarity between the Kirkuk Embayment and the Dezful Embayment, where the intense thrusting focused to the northeast (at BCu); this probably can be interpreted as the similarity in the style of deformation between the two embayments.

Areas of higher HI, steepness, and precipitation show consistency with higher-strength lithologies, which can be interpreted as the erodibility of exposed lithology does not control the higher geomorphic indices values. Results show a conjunction between climate and tectonic uplift controlling higher erosion areas (land degradation) where we found steep slopes, high HI, and higher steepness values.

6 Recommendations

Based on the result of this chapter, we highly recommend a detailed study of the change in the Hypsometric Curve (HC) of basins. The changes in the form of the HC might reflect several physical and fluvial processes, and this will help in understanding the tectonic and fluvial competition, which shapes the landscape and plays an important role in land degradation. Also, it is important to study the geomorphology and erosion models of the KE and BCu, which might shed light on the similarities of the subsurface forces.

Acknowledgements We extend our gratitude to Prof. Dr. Salam I. Marhoon, the esteemed head of the Geology department at the University of Baghdad, for his invaluable encouragement and support in creating this chapter. We express our appreciation to the USGS for generously providing the data used in this study. Additionally, we would like to acknowledge the editors of this book and chapter reviewers for their valuable feedback, which has greatly enhanced the quality of this chapter.

References

Afaghi A, Salek MM (1975a) Geological map of Iran, sheet No., 5 North-west Iran, scale 1:1,000,000. Tehran, Iran
Afaghi A, Salek MM (1975b) Geological map of Iran, sheet No., 4 South-West Iran, scale 1:1,000,000. Tehran, Iran
Afaghi A, Salek MM (1977a) Geological map of Iran, sheet No., 6 South-East Iran, scale 1:1,000,000. Tehran, Iran
Afaghi A, Salek MM (1977b) Geological map of Iran, sheet No., 3 North-East Iran, scale 1:1,000,000. Tehran, Iran
Afaghi A, Salek MM (1977c) Geological map of Iran, sheet No., 2 North-Central Iran, scale 1:1,000,000. Tehran, Iran
Afaghi A, Salek MM, Moazami J (1978) Geological map of Iran, sheet No., 1 North-west Iran, scale 1:1,000,000. Tehran, Iran
Al-Attar ZT, Othman AA, Al-Hakari SHS, et al (2022) A Neotectonic statistical assessment through watershed geomorphic analysis: a case study in the greater Zab River Basin, Turkey–Iraq. Environ Earth Sci 81. https://doi.org/10.1007/s12665-022-10478-7
Alavi M (2007) Structures of the Zagros fold-thrust belt in Iran. Am J Sci 307:1064–1095. https://doi.org/10.2475/09.2007.02
Allen MB, Saville C, Blanc E-P et al (2013) Orogenic plateau growth: expansion of the Turkish-Iranian plateau across the zagros fold-and-thrust telt. Tectonics 32:171–190
Allen MB, Talebian M (2011) Structural variation along the zagros and the nature of the dezful embayment. Geol Mag 148:911–924. https://doi.org/10.1017/S0016756811000318
Ameen MS (1992) Effect of basement tectonics on hydrocarbon generation, migration, and accumulation in Northern Iraq. Am Assoc Pet Geol Bull 76:356–370
Bahrami S (2013) Analyzing the drainage system anomaly of zagros basins: implications for active tectonics. Tectonophysics 608:914–928
Blanco-Canqui H, Lal R (2008) Principles of soil conservation and management. Springer, Netherlands
Bretis B, Bartl N, Grasemann B (2011) Lateral fold growth and linkage in the zagros fold and thrust belt (Kurdistan, NE Iraq). 615–630. https://doi.org/10.1111/j.1365-2117.2011.00506.x
Bull WB (1977) Tectonic Geomorphology of the Mojave Desert: US, Geological Survey Contract Report 14-08-001-G-394, Office of Earthquakes, Volcanoes, and Engineering. Calif Menlo Park, p 188
Burbank DW, Anderson RS (2011) Tectonic geomorphology. Wiley
Burberry CM (2015) The effect of basement fault reactivation on the triassic-recent geology of kurdistan, North Iraq. J Pet Geol 38:37–58
Burberry CM, Cosgrove JW, Liu J-G (2010) A study of fold characteristics and deformation style using the evolution of the land surface: Zagros simply folded belt, Iran. Geol Soc London, Spec Publ 330:139–154
Burberry CM, Cosgrove JW, Liu JG (2008) Spatial arrangement of fold types in the zagros simply folded belt, Iran, indicated by landform morphology and drainage pattern characteristics. J Maps 4:417–430

Cannon JM, Murphy MA, Taylor M (2018) Segmented strain accumulation in the high himalaya expressed in river channel steepness. Geosphere 14:1131–1149

Clark MK, Schoenbohm LM, Royden LH, et al (2004) Surface uplift, tectonics, and erosion of eastern tibet from large-scale drainage patterns. Tectonics 23

Cox RT (1994) Analysis of drainage-basin symmetry as a rapid technique to identify areas of possible quaternary tilt-block tectonics: an example from the mississippi embayment. Geol Soc Am Bull 106:571–581

Dietrich WE, Bellugi DG, Heimsath AM et al (2003) Geomorphic transport laws for predicting landscape form and dynamics. Geophys Monogr 135:1–30. https://doi.org/10.1029/135GM09

Domínguez-González L, Andreani L, Stanek KP, Gloaguen R (2015) Geomorpho-tectonic evolution of the jamaican restraining bend. Geomorphology 228:320–334. https://doi.org/10.1016/j.geomorph.2014.09.019

Flint JJ (1974) Stream gradient as a function of order, magnitude, and discharge. Water Resour Res 10:969–973. https://doi.org/10.1029/WR010i005p00969

Gao M, Zeilinger G, Xu X et al (2016) Active tectonics evaluation from geomorphic indices for the central and the southern longmenshan range on the Eastern Tibetan Plateau, China. Tectonics 35:1812–1826

Goldrick G, Bishop P (2007) Regional analysis of bedrock stream long profiles: evaluation of Hack's SL Form, and formulation and assessment of an alternative (the DS form). Earth Surf Process Landfs 32:649–6714. https://doi.org/10.1002/esp

Hack JT (1973) Stream-profile analysis and stream-gradient index. J Res Us Geol Surv 1:421–429

Jassim SZ, Goff JC (2006) Geology of Iraq. Dolin, Prague, and Moravian Museum. Brno p 341

Jon G, Jordan, H T, Siever R (2007) Understanding Earth. W. H. Freeman, New York

Katiraie-Boroujerdy P-S, Nasrollahi N, Hsu K, Sorooshian S (2013) Evaluation of satellite-based precipitation estimation over Iran. J Arid Environ 97:205–219

Whipple KX, Tucker GE (1999) Dynamics of the stream-power river incision model: implications for height limits of mountain ranges, landscape response timescales, and research needs. 104:661–674

Keller EA, Pinter N (2002) Active tectonics earthquakes, uplift, and landscape, Second. Printice Hall, New Jersy

Kirby E, Whipple KX (2012) Expression of active tectonics in erosional landscapes. J Struct Geol 44:54–75

Kirby E, Whipple KX, Tang W, Chen Z (2003) Distribution of active rock uplift along the eastern margin of the tibetan plateau: inferences from bedrock channel longitudinal profiles. J Geophys Res Solid Earth 108

Kottek M, Grieser J, Beck C et al (2006) World Map of the Köppen-Geiger climate classification updated. Meteorol Zeitschrift 15:259–263

Langbein WB (1947) Topographic characteristics of drainage basins

Mandal SK, Scherler D, Wittmann H (2021) Tectonic accretion controls erosional cyclicity in the Himalayas. AGU Adv 2:1–27. https://doi.org/10.1029/2021av000487

Miller SR, Baldwin SL, Fitzgerald PG (2012) Transient fluvial incision and active surface uplift in the woodlark rift of eastern papua New Guinea. Lithosphere 4:131–149

Moosdorf N, Cohen S, von Hagke C (2018) A Global Erodibility Index to Represent Sediment Production Potential of Different Rock Types. Appl Geogr 101:36–44. https://doi.org/10.1016/j.apgeog.2018.10.010

Morell KD, Kirby E, Fisher DM, Van Soest M (2012) Geomorphic and exhumational response of the central american volcanic arc to cocos ridge subduction. J Geophys Res Solid Earth 117:. https://doi.org/10.1029/2011JB008969

Nissen E, Tatar M, Jackson JA, Allen MB (2011) New views on earthquake faulting in the zagros fold-and-thrust belt of Iran. Geophys J Int 186:928–944. https://doi.org/10.1111/j.1365-246X.2011.05119.x

Obaid A, Allen M (2019) Landscape expressions of tectonics in the zagros fold-and-thrust belt. Tectonophysics 766. https://doi.org/10.1016/j.tecto.2019.05.024

Obaid AK (2018) Tectonic and fluvial geomorphy in the zagros fold-and-thrust belt. Durham

Obaid AK, Allen MB (2017) Landscape maturity, fold growth sequence and structural style in the kirkuk embayment of the Zagros, Northern Iraq. Tectonophysics 717. https://doi.org/10.1016/j.tecto.2017.07.006

Othman AA, Ali SS, Salar SG, et al (2023) Insights for estimating and predicting reservoir sedimentation using the RUSLE-SDR approach: a case of darbandikhan lake basin, Iraq–Iran. Remote Sens 15. https://doi.org/10.3390/rs15030697

Othman AA, Gloaguen R (2013) Automatic extraction and size distribution of landslides in Kurdistan region, NE Iraq. Remote Sens 5:2389–2410. https://doi.org/10.3390/rs5052389

Othman AA, Gloaguen R, Andreani L, Rahnama M (2018) Improving landslide susceptibility mapping using morphometric features in the mawat area, kurdistan region, NE Iraq: comparison of different statistical models. Geomorphology 319:147–160. https://doi.org/10.1016/j.geomorph.2018.07.018

Othman AA, Obaid AK, Amin Al-Manmi DAM et al (2021) New insight on soil loss estimation in the northwestern region of the zagros fold and thrust belt. ISPRS Int J Geo-Information 10:1–24. https://doi.org/10.3390/ijgi10020059

Pastor-Galán D, Spencer CJ, Furukawa T, Tsujimori T (2021) Evidence for crustal removal, tectonic erosion and flare-ups from the japanese evolving forearc sediment provenance. Earth Planet Sci Lett 564:116893. https://doi.org/10.1016/j.epsl.2021.116893

Pike RJ, Wilson SE (1971) Elevation-relief ratio, hypsometric integral, and geomorphic area-altitude analysis. Bull Geol Soc Am 82:1079–1084. https://doi.org/10.1130/0016-7606(1971)82[1079:ERHIAG]2.0.CO;2

Ramsey LA, Walker RT, Jackson J (2008) Fold evolution and drainage development in the zagros mountains of fars province, SE Iran. Basin Res 20:23–48

Salar SG (2013) Geomorphic analysis for water harvesting using GIS Technique in selected basins/Garmian Iraqi Kurdistan Region. University of Sulaimani

Schildgen TF, Cosentino D, Bookhagen B et al (2012) Multi-phased uplift of the southern margin of the central anatolian Plateau, Turkey: a record of tectonic and upper mantle processes. Earth Planet Sci Lett 317–318:85–95. https://doi.org/10.1016/j.epsl.2011.12.003

Schumm Stanley A (1956). Evolution of drainage systems and slopes in badlands. https://doi.org/10.1130/0016-7606(1956)67

Seeber L, Gornitz V (1983) River profiles along the himalayan arc as indicators of active tectonics. Tectonophysics 92:335–367

Snyder NP, Whipple KX, Tucker GE, Merritts DJ (2000) Stream profiles in the mendocino triple junction region, Northern California. GSA Bull 112:1250–1263. https://doi.org/10.1130/0016-7606(2000)112%3c1250:lrttfd%3e2.3.co;2

Stern RJ, Johnson P (2010) Continental lithosphere of the arabian plate: a geologic, petrologic, and geophysical synthesis. Earth-Sci Rev 101:29–67

Strahler AN (1952) Hypsometric (area–altitude) analysis of erosional topography. Geol Soc Am Bull 63:1117–1142

Talebian M, Jackson J (2004) A reappraisal of earthquake focal mechanisms and active shortening in the zagros mountains of Iran. Geophys J Int 156:506–526. https://doi.org/10.1111/j.1365-246X.2004.02092.x

Thiede RC, Bookhagen B, Arrowsmith JR et al (2004) Climatic control on rapid exhumation along the southern himalayan front. Earth Planet Sci Lett 222:791–806

Toy TJ, Foster GR, Renard KG (2002) Soil erosion: processes, prediction, measurement, and control. Wiley

Tucker GE, Whipple KX (2002) Topographic outcomes predicted by stream erosion models: sensitivity analysis and intermodel comparison. J Geophys Res Solid Earth 107:ETG 1-1-ETG 1-16. https://doi.org/10.1029/2001JB000162

De Vera J, Gines J (2009) Structure of the Zagros fold and thrust belt in the Kurdistan Region, Northern Iraq. Trab. 217:213–217

Vernant P, Nilforoushan F, Hatzfeld D et al (2004) Present-day crustal deformation and plate kinematics in the middle east constrained by GPS measurements in Iran and Northern Oman. Geophys J Int 157:381–398. https://doi.org/10.1111/j.1365-246X.2004.02222.x

Walker RT, Ramsey LA, Jackson J (2011) Geomorphic evidence for ancestral drainage patterns in the zagros simple folded zone and growth of the Iranian Plateau. Geol Mag 148:901–910

Wang W, Godard V, Liu-Zeng J et al (2021) Tectonic Controls on Surface Erosion Rates in the Longmen Shan, Eastern Tibet. Tectonics 40:1–21. https://doi.org/10.1029/2020TC006445

Whipple KX (2002) Implications of sediment-Flux-dependent river incision models for landscape evolution. J Geophys Res 107. https://doi.org/10.1029/2000jb000044

Whipple KX, Shirzaei M, Hodges KV, Arrowsmith JR (2016) Active sortening within the himalayan orogenic wedge implied by the 2015 gorkha earthquake. Nat Geosci 9:711

Wobus C, Whipple KX, Kirby E et al (2006) Tectonics from topography: procedures, promise, and pitfalls. Spec Pap Soc Am 398:55

Wohl EE, Merritt DM (2001) Bedrock channel morphology. Geol Soc Am Bull 113:1205–1212

Zebari M, others (2019) Supplement of relative timing of uplift along the zagros mountain front flexure (Kurdistan Region of Iraq): constrained by geomorphic indices and landscape evolution modeling

Zebari MM, Burberry CM (2015) 4-D evolution of anticlines and implications for hydrocarbon exploration within the Zagros fold-thrust belt, Kurdistan region, Iraq. GeoArabia 20:161–188

Zhang P, Mei L, Jiang SY et al (2021) Erosion and sedimentation in SE tibet and myanmar during the evolution of the burmese continental margin from the late cretaceous to early neogene. Gondwana Res 95:149–175. https://doi.org/10.1016/j.gr.2021.04.005

Characterization of Post-uprising Impacts on Landcover and Land Use: Al Wasita-Satish—Area Northeast Libya Case Study

Salah Hamad and Attia Alsanousi

Abstract Spatial and temporal processes primarily drive land use and land cover (LULC) changes through human activities, such as agricultural expansion and demographic change. This study aims to quantify the LULC change resulting from the civil war in Al Wasita-Satish Plateau, NE Libya. Using two satellite images from the Landsat 5 Thematic Mapper (TM) in 2010 and Landsat 8 OLI-TIRs in 2022, LULC changes were evaluated. Agriculture and arid, urban, dense, and sparse vegetation were the predominant LULC types in the study area. In 2010, sparse vegetation comprised the most significant percentage of the land area (37.15%), followed by dense vegetation (26.45%), agriculture (18.69%), and barren land. The constructed category had a relatively small area (1, 31%). According to the LULC maps for 2022, the proportion of barren land increased by 30.50% and sparse vegetation decreased by 29.09% and dense vegetation decreased by 15.81%. In addition, there was a slight increase in the agricultural (18.69%) and urban classes (4.25%). The findings demonstrate that the Libyan Civil War of 2011 affected the LULC, where land and environmental protection laws and law enforcement tools were rendered ineffective. Despite the absence of armed conflict in the study area, the displacement of people from conflict zones and their settlements contributed to the change in LULC. Local communities that practice agriculture and grazing have created alternative economic activities with a quick and comfortable financial return, such as the construction of tourist settlements, due to the area's popularity as a tourist destination and lack of financial resources.

Keywords QGIS · Remote sensing · Vegetation cover · GIS · Libya

---

S. Hamad (✉) · A. Alsanousi
Faculty of Natural Resources and Environmental Science, Omar Al-Mukhtar University, Al Baydah, Libya
e-mail: salah.hamad@omu.edu.ly

A. Alsanousi
e-mail: attiaalsanousi@yahoo.com

© The Author(s), under exclusive license to Springer Nature Switzerland AG 2024
A. M. F. Al-Quraishi and Y. T. Mustafa (eds.), *Natural Resources Deterioration in MENA Region*, Earth and Environmental Sciences Library,
https://doi.org/10.1007/978-3-031-58315-5_5

1 Introduction

Land cover (LC) is an observable biophysical cover on Earth's surface. The structures and processes describe land use (LU) and the inputs that people use to create, alter, or sustain a particular land cover. Thus, the definition of LU provides a clear relationship between land cover and people's actions in their climate (Gregorio and Jansesn 2000). Changes in land use/land cover (LULC) play a crucial role in environmental change studies at local and regional levels. These changes have become an essential and fundamental component of current strategies for the environmental monitoring and management of natural resources. It also models the process of land change and seeks to provide a better forecast of the dynamics of land transitions and scenarios for future change (Vivekananda et al. 2022; Hussein et al. 2022; Twisa and Buchroithner 2019; de Sousa-Neto et al. 2018; Robinson et al. 2013; Gupta and Munshi 1985). Traditionally, LULC transformation has been driven by two types of forces: anthropogenic factors alternatively defined as land-use change drivers and natural processes, and their impacts are nevertheless seen as a direct driver of LULC change and modification, but also as a direct driver of environmental changes (Wilson and Wilson 2013; Meyer and Turner 1992). In addition, LULC changes significantly impact climate through various mechanisms that alter surface temperature and rainfall (Alqasemi et al. 2020; Gaznayee et al. 2022).

The region's climate also plays a crucial role in LULC change (Gogoi et al. 2019; Hubbard et al. 2014; Kim et al. 2013). Past and ongoing anthropogenic activities worldwide are responsible for large-scale land surface changes and several ecological, physical, and socioeconomic impacts. Urbanization, agricultural expansion, and other socio-economic activities, such as industry and tourism, are the standard local and regional drivers of LULC changes (Munthali et al. 2019; Twisa and Buchroithner 2019; Muttitanon and Tripathi 2005; Pitman and Noblet-ducoudre 2012). Civil conflicts and wars also impact LULC, as the displacement of people from one area to another leads to the creation of new human settlements and, thus, anthropogenic activities. Evacuation of land also leads to the recovery of the ecosystem. In a study of temporal changes in LULC for 29 years (1986–2015) post-civil war and Ebola epidemics in Liberia's coastal areas, the results showed that bare land and sediment classes have decreased. In contrast, the water, vegetation, and residential classes increased during the 29-year evaluation period. However, vegetation coverage decreased during the post-civil war period (2002–2015) (Awange et al. 2018).

In addition, assessing the impacts of the civil war (1991–2002) on LULC change in the Kono and Sierra Leone districts revealed that grassland and agriculture decreased during the war, while the rest of the LULC showed expansion. Grassland cover gradually decreased during the war and post-war periods, while bare land increased (Wilson and Wilson 2013). Another example is Halgurd Sakran National Park (HSNP), Kurdistan Region of Iraq, where a significant change in the LULC was noticed in the park's lower part, particularly in pasture land, cultivated, and cultivated forested areas. Changes in socio-economic factors occurred after the post-war economic sanctions by the United Nations during 1991–2003. Thus, changes increased during the sanctions and then decreased (Hamad et al. 2018).

Many scholars, organizations, and institutions worldwide have conducted LULC change analyses, mainly focusing on the application of remote sensing (RS) data and Geographical Information Systems (GIS) techniques, which are practical tools to precisely assess temporal RS data on the spatial distribution of LULC changes (Hussein et al. 2022; Khwarahm et al. 2021; Afrin et al. 2019; Rawat and Kumar 2015; Reis 2008). In contrast to time-consuming and costly traditional methods based on field surveys and ancillary data, multispectral, temporal, and medium–high spatial resolution remote sensing data have emerged as an essential source of information processed in the GIS environment, which analyzes and classifies LULC features in a short time and in a cost-effective manner (Vivekananda et al. 2022; Estoque and Murayama 2015). High-resolution satellite imagery or aerial photos are essential to analyze LULC changes in large cities. However, these datasets are limitedly accessible owing to financial considerations.

Medium-resolution data, such as Landsat datasets of the Multi-Spectral Scanner (MSS), TM, and Landsat 8 Operational Land Imager (OLI), have been commonly used for LULC change detection research (Divya et al. 2021; Gadrani et al. 2018; Sundarakumar et al. 2012). Several LULC change detection methods have been developed, the most common of which is the post-classification image comparison of different dates. The satellite image classification process involves grouping image pixel values into different categories, for which several methods and approaches are available to classify satellite images. The most widely used methods are unsupervised (K-means and ISODATA), supervised, Maximum Likelihood (ML), and object-based classification (Mishra et al. 2022; Abburu and Babu Golla 2015). Furthermore, remote sensing indices such as the vegetation groups detected by the Normalized Difference Vegetation Index (NDVI) can also be applied. In addition, the evaluation of built-up areas and normalized difference built-up index (NDBI) can be used (Hussain et al. 2019; Khudair et al. 2019; Rahman et al. 2017; Kaptué et al. 2015).

Since 2011, civil conflict in Libya has disrupted land and environmental protection laws and enforcement mechanisms, leading to the imposition of customary tribal rules governing land tenure and use. The aftermath of the conflict spurred local communities to seek alternative economic activities for swift financial gains amid financial constraints and economic disruptions. Consequently, the region witnessed widespread disturbances to natural vegetation, driven by recurring short-interval fires aimed at clearing forested areas for recreational settlements and road networks, extensive woodcutting for charcoal production, unsustainable agricultural practices, and the establishment of new settlements within forested zones (Abdalrahman et al. 2010; Mnsur and Abdalrahman 2015). This study aims to quantitatively assess land use and land cover changes, particularly in natural vegetation, following the Libyan civil conflicts in the Al Wasitaa-Satiah plateau of northeast Libya from 2010 to 2022. Utilizing open-source GIS software Quantum GIS (QGIS) and multitemporal Landsat imagery, the research seeks to elucidate the extent of LULC transformations in the region. The aftermath of the 2011 civil conflict profoundly disrupted land management and environmental conservation efforts, resulting in significant environmental degradation and resource exploitation. Customary tribal rules exacerbated these pressures, leading to increased exploitation of natural resources through activities such as short-interval fires, woodcutting, and unsustainable agricultural practices. Moreover, the expansion of settlements further encroached upon forested areas, exacerbating habitat loss and fragmentation. These post-conflict dynamics highlight the urgent need for targeted interventions to mitigate environmental degradation and promote sustainable land management practices in the region, as emphasized by reports from the International Organization for Migration (IOM 2023).

2 Materials and Methods

2.1 The Study Area

The study area in the northern middle part of Al Jabal Al Akhdar (Green Mountain), as shown in (Fig. 1), is an upland area along the northeastern Libyan coast, as shown in Fig. 1. The Sirte Gulf bounds to the west, the Sirte Basin to the southwest, the Bomba Gulf to the east, and Marmarica to the southeast. Al Jabal Al Akhdar is a crescent-shaped ridge in its central part, which reaches more than 870 m above mean sea level. The northern flank comprises step-like plateaus bordered by two main escarpments, further apart in the west but gradually drawing eastward closer together and roughly parallel to the coast. A large portion of the two plateau benches, particularly the second, is dissected by the wadis (Hamad 2019; Suleiman et al. 2016). The study area lies on the first plateau, bordered by the second escarpment in the south, where Al Baydah and Shahat cities are located, and in the north by the Jar Jaroumah-Sousa coastal plain. The topographic elevation ranged from 50 to 450 m. The surface geology of the exposed rocks consists mainly of sedimentary marine carbonate formations composed of limestone and marl, ranging from the Eocene to the Quaternary (Hydrogeo 1985; IRC 1974). The predominant soil types are the red Mediterranean soils of Ferrisiallitic Red (Rhodoxeralfs) and shallow calcareous Rendzinas (Rendolls Lithic) soils (Mahmoud 1995; Selkhozpromexport 1980).

The climate in the study area is classified as semi-arid, where precipitation mainly occurs from October to March and April, and the maximum rainfall values are observed in December and January. In addition, extended dry periods frequently occur during wet seasons, where the mean annual precipitation ranges from 300 to 450 mm. In addition, rain is characterized by spatial and temporal variability. The mean annual temperature ranges from to 10–20 °C. The coldest temperatures occur in January and February is 0–11 °C, and the highest temperatures occur from June to August, with maximum temperatures exceeding 35 °C (Hamad 2022; Peng et al. 2022; Fick and Hijmans 2017; OMU 2005a, b; Hydrogeo 1985). The local communities practice grazing and rain-fed agriculture. Some people use irrigation for various crops because they can drill water wells, which are costly because of the

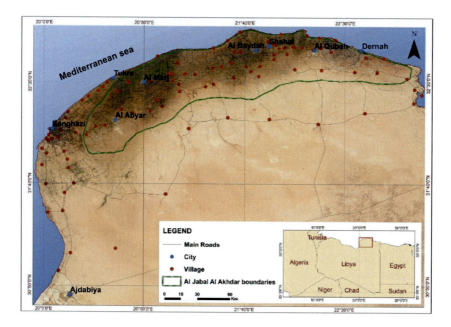

Fig. 1 Location map of Al Wasitaa-Satiah Plateau

deep groundwater levels, ranging from 150 to 300 m below ground level (Hamad 2012). The land cover type of the study area was as follows:

1. The dense vegetation of forest and dry Mediterranean woodlands, which is limited only to wadis of the study area and consists mainly of Phoenician Juniper (*Juniperus phoenicea*), Mediterranean Cypress (*Cupressus sempervirens*), Aleppo Pine (*Pinus halepensis*), and other deciduous tree species like *Olea europaea, Ceratonia siliqua, Quercus coccifera* and *Arbutus pavarii*.
2. Grasslands and steppes consist of plant communities of dwarf shrubs, grasses, and annual herbs sparsely distributed in the study area.
3. Barren Land consists of mixed soil and consolidated and nonconsolidated marine carbonate rocks.
4. Cultivated and uncultivated lands, such as farmlands and crop fields represent agriculture.
5. Built-up residential, commercial and service, industrial, infrastructure, and mixed urban and other urban roads

According to (OMU 2005a, b; Zunni 2013, El-Baras and Saaed 2013), the natural vegetation of the study area, the Al Wasitaa-Satiah plateau, is typical of the Al-Jabal Al-Akhdar Region, a diverse combination of conifer and broad-leaved endemic trees and shrubs of the Mediterranean region, alternatively known as "Maquis vegetation" in some scientific textbooks. The most common tree species based on crown cover percentage are Phoenician Juniper *Juniperus phoenicea* (the most widespread tree species in the region with around 70% of the stocking area), and Mastic Tree *Pistacia lentiscus* (Mostly growing as large shrubs, or small trees in rare cases, occupying significant portions of the Maqui Formation of the study area) in addition to species like Carob Tree *Ceratonia siliqua,* Olive tree *Olea europae,* and Libyan Strawberry-tree *Arbutus pavarii.* Small shrubs and woody perennials such as Calicotome villosa, Rhamnus lyciodes, Sarcopoterium spinosum, and Phlomis floccosa with low cover percentages occupy the understory and open gaps between trees.

The study area is known for its suitable land for growing crops and fruit trees, where hundreds of hectares of natural land have been shifted to crop fields and flowering tree farms. These fields and farms have been sheltered by single to multiple rows of shelter belts consisting mainly of large evergreen trees, such as Aleppo pine *Pinus halepensis,* Mediterranean cypress *Cupressus sempervirens,* and in other cases, some species of *Eucalyptus spp.* and *Acacia spp.* These species are not endemic to the Al-Wasitaa-Satiah Plateau. However, they represent a large portion of the number of individuals per unit area, and therefore, the total biomass of the study area. Agricultural and human expansions, which have occurred on a large scale in recent years, have led to the introduction and establishment of new exotic species, especially near settlement areas. The natural vegetation of the Al Wasitaa-Satiah Plateau is vulnerable to low-frequency fires (5–10-year fire intervals), which often occur during hot summer seasons where the average temperature is high with plenty of dry and dead fuel. The increased number of potential fires affects the growth and distribution of Juniperus phoenicea, which is considered a sensitive species to fire because of its low reproduction profile and the lack of ability to recover post-fire, in contrast to Pistacia lentiscus, in which the occurrence of fire helps the species to occupy more spaces due to its successful reproduction methods and the ability to reproduce new suckers' days after fire occurrence.

2.2 Data

To assess LULC changes, satellite images from Landsat 5 Thematic Mapper (TM) for 2010 and two Landsat 8 OLI-TIRs acquired in 2013 and 2022 were used in this study, as shown in Table 1.

Table 1 Data used in the study

Year	Satellite	Sensor	Path/row	Resolution (m)	Acquisition date	Land cloud cover
2010	LANDSAT 5	TM	138/037	30	2010/08/12	0
2013	LANDSAT 8	OLI-TIRS	138/037	30	2013/06/01	3.19
2022	LANDSAT 8	OLI-TIRS	138/037	30	2022/11/17	0.09

2.3 Methodology

2.3.1 LULC Classification

The Quantum GIS (QGIS) software was used to derive classifications of the satellite images for two periods (2010–2022) with a Semi-automatic Classification Plug-in (SCP), which is an open-source plug-in for QGIS developed by (Congedo 2016). The SCP Plug-in enabled downloading freely available satellite images (Landsat, Sentinel-2, Sentinel-3, ASTER, and MODIS). In addition, it provides several functions for pre-processing, classification, and pixel quantification of LULC categories, and post-processing of the classifications. Pre-processing was carried out using DOS1 for atmospheric correction to eliminate imperfections that may damage the satellite image by enabling surface reflectance values obtained without the effects of atmospheric interference (da Silva et al. 2022; Chapa et al. 2019; Hamad 2019; Correia et al. 2018).

The Plug-in allows the creation of regions of interest (ROI) as training input to define LULC classes and conduct semi-automatic supervised and unsupervised classification, which is implemented by several algorithms: Parallelepiped, Minimum Distance, Maximum Likelihood, Spectral Angle Mapping raster, and Land Cover Signature (Congedo 2016). In this study, supervised classification was performed using Maximum Likelihood (ML) classification algorithms, in which the classifier assumes that the statistics for each class for each band are typically distributed and computes the likelihood that a given pixel belongs to a particular class (Zulhaidi et al. 2007). Accuracy assessment for the created LULC classification involves a comparative analysis with the independent validation dataset, which results in a contingency matrix (error matrix) for computation of descriptive statistics regarding the accuracy expressed by parameters such as the overall accuracy and individual accuracy, either as producer or user. Accuracy was assessed using the SCP accuracy tool, which uses Regions of Interest (ROIs) as stratified random points. In addition, the accuracy tool provides an error matrix to compute the overall, producer, and user accuracy.

2.3.2 NDVI

NDVI is one of the most widely used spectral indices for satellite imagery analysis, vegetation studies, fires, desertification, landslides, and other natural phenomena. NDVI is an indicator that describes the greenness, relative density, and health of vegetation, where the green leaf is measured as a significant indicator of photosynthetically active biomass (Huete 2012). NDVI is determined by the difference in the ratio between the measured canopy reflectance in the red and near-infrared bands (Gandhi et al. 2015; Liu et al. 2015). NDVI was computed for each pixel of the satellite image according to the following equation:

Table 2 Accuracy assessment of the LULC classification

	Producer's accuracy		User's accuracy		Overall accuracy (%)		Kappa coefficient	
LULC	2010	2022	2010	2022	2010	2022	2010	2022
Agriculture	86.4	85.7	88.2	89.0	0.86	0.87	0.81	0.83
Barren	83.7	85.2	92.5	88.9				
Built up	82.6	85.4	89.3	81.4				
Dense vegetation	80.0	88.8	88.2	92.3				
Sparse vegetation	93.5	93.4	81.2	83.6				

$$NDVI = (NIR - RED)/(NIR + RED)$$

where

NIR—reflection in the near-infrared spectrum
RED—reflection in the red range of the spectrum.

3 Results and Discussion

3.1 Accuracy Assessment

The accuracy assessments for the imageries (TM) and (OLI/TIRS), the maximum likelihood classifier, showed an overall accuracy of 0.86% to 0.87%, respectively. The Kappa coefficient, user accuracy (UA), and producer accuracy (PA) are listed in Table 2.

3.2 LULC Change

The results of mapping LULC changes are illustrated in Figs. 2, 3, and 4 and Table 3. LULC represents the study area categorized into five classes: agriculture, barren, built-up, dense, and sparse. In 2010, sparse vegetation (37.15%) accounted for the majority of the total land area, followed by dense vegetation (26.45%), agriculture (18.69%), and barren vegetation. The built-up class had a comparatively smaller size (1.31%). According to the LULC maps for 2022, the proportion of the barren class increased significantly (30.50%), whereas sparse vegetation decreased (29.09%), as did the dense vegetation class (15.81%). Furthermore, there was a slight increase in agriculture (18.69%) and built-up classes (4.25%) (Fig. 5).

Based on these results, the most significant shift in LULC occurred in the barren class, with an increase from 84.39 km^2 in 2010 to 157.03 km^2 in 2022, with more than

Characterization of Post-uprising Impacts on Landcover and Land Use …

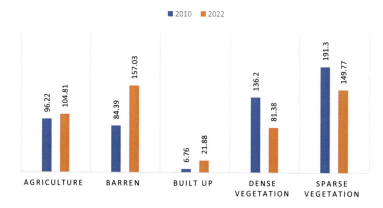

Fig. 2 Area graph for 2010, 2022 LULC

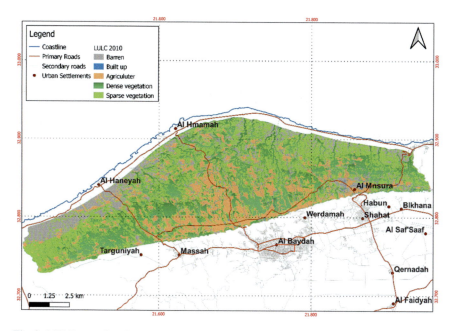

Fig. 3 LULC maps for 2010

72 km² in only ten years. Many ecological and anthropogenic factors contributed to this change, but the mid-May 2013 fire whipped out the area from SW to NW and represented the most significant contribution. The fire spread northward through the study area and stopped only a few meters from the sea, burning many hectares of dense and sparse vegetation. The cause of this fire has not been documented. Nevertheless, natural causes of fires are rare in this world, and human factors are more likely to cause wildfires in most regions. The average area burned annually by fire in Al Jabal

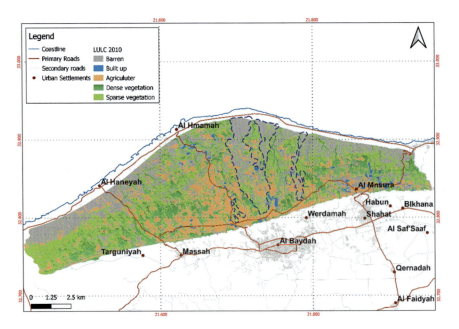

Fig. 4 LULC maps for 2022

Table 3 The LULC classification Results for 2010, and 2022 with the area and percentages of each LULC category

LULC	2010		2022	
	Area (km^2)	%	Area (km^2)	%
Agriculture	96.22	18.69	104.81	20.36
Barren	84.39	16.39	157.03	30.50
Built up	6.76	1.31	21.88	4.25
Dense vegetation	136.20	26.45	81.38	15.81
Sparse vegetation	191.30	37.15	149.77	29.09
Total	514.87	100.00	514.87	100.00

Al Akhdar is approximately 30 ha/year (El-Baras and Saaed 2013), and *J. phoenicea* is considered the species most affected by wildfires. Humans cause more than 90% of these fires, and occur during hot, dry seasons, especially with the help of a southern windy dust storm known locally as "Qiblie". Overgrazing is another factor that should be considered since livestock and domesticated animals are an essential part of the local ecosystem of the region. In several cases, the pastoral capacity exceeded the allowable limit. Cattle, sheep, and goats were the most common livestock raised by locals in the study area. Other possible drivers for this change are woodcutting for charcoal production and the growing expansion of secondary roads through inner-forested areas away from main roads. This may increase the possibility of having

Characterization of Post-uprising Impacts on Landcover and Land Use ... 83

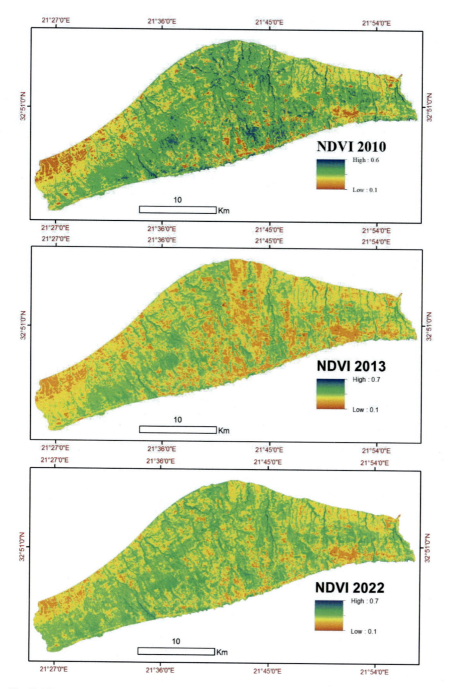

Fig. 5 NDVI maps for the years 2010, 2013 and 2022

barren, built-up, and agricultural regions within these long-lasting, self-maintained forested areas.

3.2.1 Vegetation Cover Change

The results revealed that dense and sparse vegetation declined significantly during the study period (54.82 km², and 41.53 km² total decline for dense and sparse vegetation, respectively) (Tables 3 and 4). Excessive human activities, urbanization, and population increase, especially in recent years, were mainly responsible for the continuous vegetation depletion and increase in the other LULC classes. In a study conducted in some parts of Al Jabal Al Akhdar, the total reduction in vegetation cover was approximately 27 ha between 1984 and 2008, with the most significant reduction occurring between 2000 and 2008. Moreover, these results were parallel with those (Masoud 2015), who stated that forested areas decreased from 2080.5 Km² in 1984 to 1571.11 Km2 in 2003, and rangelands decreased similarly from 6779.38 Km² in 1984 to 6066.49 Km² in 2003 in the total area of Al Jabal Al Akhdar.

The natural vegetation cover of the study area is a part of the Mediterranean forest ecosystem, a diverse combination of conifer and broad-leaved endemic trees and shrubs, with species such as *Juniperus phoenicea* and *Pistacia lentiscus* representing approximately 75% of the stocking area, and this is one of the most negatively affected by excessive human activities (Pérez-Trejo 1992). While *P. lentiscus* shows a resilient ability to reestablish itself post such disturbances due to its effective reproductive ability, *J. phoenicea*, on the other hand, suffers from redeeming its natural habitat in the long term. When a major disturbance occurs, *J. phoenicea*, which is the most abundant and frequent species in the area, mostly fails to regenerate and loses its habitat to other shrubs, such as *Rhamnus oleoides* and *Sarcopoterium spinosum*, as a part of a common retrogressive succession in the Al Jabal Al Akhdar region. Other tree species that take advantage of fire occurrence to occupy more space are *Quercus coccifera* and *Arbutus pavarii* (El-Baras and Saaed 2013). Moreover, urbanization

Table 4 Transition matrix showing LULC change

LULC 2010 area Km²	LULC 2022 area Km²					
	Agriculture	Barren	Built up	Dense vegetation	Sparse vegetation	Total
Agriculture	**93.14**	0.09	2.87	0.02	0.10	96.22
Barren	4.01	**69.93**	5.66	0.44	4.35	84.39
Built up	0.84	0.98	**4.63**	0.06	0.25	6.76
Dense vegetation	1.76	15.00	3.35	**71.91**	44.17	136.20
Sparse vegetation	5.06	71.03	5.36	8.95	**100.90**	191.30
Total	104.81	157.03	21.88	81.38	149.77	514.87

has contributed significantly to the pressure and impact on wildlife in general, and on wild animals in particular, by increasing the number of roads and paths in the study area, whether primary or subsidiary, and isolating forests and wilderness areas into small sections. The deterioration of vegetation cover has significant consequences on ecosystems. It serves as a crucial habitat and food source for specific herbivores while limiting hunting grounds for carnivores. Furthermore, the disruption of animal movement paths becomes evident, as many wild creatures must cover considerable distances to find water and sustain themselves due to scarce resources. This scarcity compels animals to scavenge human waste, thereby increasing their exposure to potential diseases and epidemics.

3.2.2 Agricultural Activities

There has been a tenuous increase in agricultural areas on the Al Wasita-Satiah Plateau since 2010 (less than 10 km^2). This partially contributed to the increased number of new settlements and the area's growing population following the 2011 event. Rain-fed crops, especially wheat and barley fields, and orchards of cultivated fruit trees are standard agricultural practices for local inhabitants. The increase in agricultural practices due to LULC in Al Jabal Al Akhdar has been documented nearby, and this increase has escalated in the entire region of Al Jabal Al Akhdar because of the growing demand for food since 1972 (Ahwaidi 2017). A relatively small increase in agricultural activities as demand for the increasing population has been reported in other African countries (Wilson and Wilson 2013), while a reduction was noted for HSNP in the Kurdistan Region of Iraq (Hamad et al. 2018). There has been a significant shift from local farming to irrigated crops in recent decades. There has been a substantial shift from local farming to irrigated crops in recent decades. The number of newly drilled water wells has increased significantly over the last 20 years, encouraging the establishment of irrigated farms in the study area (El-Baras and Saaed 2013). The number of these newly drilled wells continues to grow, especially since the 2011 conflict, and has become a source of profitable income for local inhabitants. However, this may lead to excessive exhaustion of the groundwater level in the area and may lead to hydrological and ecological problems as a long-term effect.

3.2.3 Population and Built-Up Expansion

One of the most noticeable LULC changes in the Al Wasita-Satiah Plateau since 2010 has been the growing expansion of built-up areas. Housing, accommodations, tourist and recreational units, and settlements are common in this LULC class. In the last decade, built-up areas have increased from 6.76 to 21.88 km^2 (1.31—4.25%) in the last decade. This increase was more evident on the western side of the study area near the suburbs of Satiah, Al Mansura, Kharashif, and Al Wasita, as shown in Fig. 6. This may be attributed to the approximation of these suburbs to the main

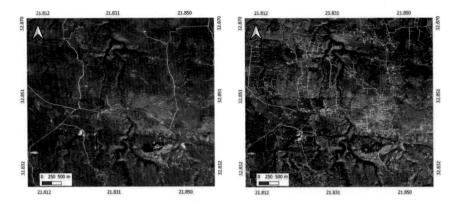

Fig. 6 Aerial view in AlMnsura area (left) 2010 (right) 2022. *Source* Google Earth

regional roads, major cities of Al Jabal Al Akhdar (Albaydah and Shahat), and main touristic orientations of the area (the heritage sites of Cyrene and Susah). However, the increase in built-up areas is relatively small compared to other LULC classes, the socio-economic, demographic, and ecological impacts have not been evaluated, and the long-term consequences are not fully understood.

The increase in built-up areas due to population growth was noted in several countries (Wilson and Wilson 2013; Hussein et al. 2022; Hamad et al. 2018). The 2011 conflict and its outcomes significantly contributed to the study's population growth. Since that year, the region has been stable and is considered a favorite destination for many Libyan citizens, but no major conflict or military actions have occurred nearby. Furthermore, the temporary government, which took over in 2014, had its main headquarters and ministries in major Al Jabal Al Akhdar cities (Albaydah and Shahat). Therefore, these circumstances increased the area's local population and displacement movement, especially in conflicted areas nationwide. As a result, if the current land-use trend continues, more built-up areas and settlements are expected. It has become a profitable, growing business among locals and the primary income source for many of them.

4 Conclusions

In the Al Jabal Al Akhdar region's northern Al Wasita-Satish Plateau, LULC has changed significantly over the past decade. LLC's assessment revealed temporal and spatial changes in land cover during and after the Libyan Civil War. According to the findings of this study, the percentage of barren land has increased significantly since 2011, followed by a slight increase in agricultural activities and urbanized areas. In contrast, dense and sparse vegetation exhibited a more significant decrease. Human influence is responsible for most of these changes and natural land-cover degradation. As a land-cover-altering agent, fires contribute to the rising proportion of barren land. Unless effective regulations and management measures are applied to regulate the excessive deterioration of natural land cover, the current trends in LULC changes are anticipated to expand. In addition, the absence of law enforcement tools, the deactivation of land and environmental protection laws, and the introduction of tribal customary laws since the start of the Libyan civil war in 2011 have facilitated the current patterns of LULC. The findings of this study can be utilized in future research applications to examine and assess the extent to which this LULC change can influence and alter the natural land cover of the study area and the entire Al Jabal Al Akhdar region.

5 Recommendations

The study's findings suggest several recommendations to tackle the ongoing challenges associated with land use and land cover (LULC) changes and to mitigate their adverse effects. These recommendations cover various areas, including policy interventions, community engagement initiatives, fire management strategies, promotion of sustainable agriculture practices, urban planning measures, and continued monitoring and research efforts. Implementing these recommendations can help stakeholders in preserving the natural habitats and biodiversity of the Al Wasita-Satish Plateau while fostering sustainable development and enhancing resilience to environmental challenges. Here are the proposed recommendations:

1. Implementing and enforcing robust land and environmental protection laws are imperative to mitigate further degradation of natural land cover. Efforts should be made to reactivate or strengthen existing regulations that may have become inactive during or after the civil war.
2. Engaging local communities in conservation efforts and raising awareness about the importance of preserving natural habitats can foster a sense of responsibility among residents. Educational campaigns can highlight the ecological significance of maintaining diverse vegetation cover and the consequences of unsustainable land use practices.
3. Given the significant contribution of fires to the increase in barren land, implementing effective fire management strategies is crucial. This includes early detection systems, fire suppression measures, and public awareness campaigns on fire prevention practices. Additionally, investigating the causes of fires and addressing human factors contributing to their occurrence can help mitigate their impact on vegetation cover.
4. Encouraging sustainable agricultural practices that minimize land degradation and water resource depletion is essential. This may involve promoting rain-fed agriculture over irrigated crops to reduce pressure on groundwater resources. Additionally, supporting initiatives for agroforestry and soil conservation can help maintain soil fertility and prevent erosion.

5. Urban expansion should be carefully planned to minimize its ecological footprint and preserve natural habitats. Incorporating green infrastructure and sustainable building practices into urban development projects can help mitigate the loss of vegetation cover and enhance biodiversity within built-up areas.
6. Continued monitoring of LULC changes using remote sensing techniques can provide valuable insights into ongoing trends and inform future conservation efforts. Additionally, investing in research to understand the socio-economic drivers of land use change and their environmental impacts can support evidence-based decision-making.

6 Conflicts of Interest

The authors declare no conflicts of interest.

References

Abburu S, Babu Golla S (2015) Satellite image classification methods and techniques: a review. Int J Comput Appl 119(8):20–25. https://doi.org/10.5120/21088-3779

Abdalrahman Y, Spence K, Rotherham ID (2010) The main causes of direct human-induced land degradation in the Libyan Al-jabal Alakhdar region. Landscape Archaeol Ecol 8(2):7–21

Afrin S, Gupta A, Farjad B, Razu Ahmed M, Achari G, Hassan Q (2019) Development of land-use/land-cover maps using landsat-8 and MODIS data, and their integration for hydro-ecological applications. Sensors (Switzerland) 19(22). https://doi.org/10.3390/s19224891

Ahwaidi GMA (2017) Factors affecting recent vegetation change in north-east Libya. University of Salford, Salford, M5 4WT, UK

Alqasemi AS, Hereher ME, Al-Quraishi AMF (2020) Retrieval of monthly maximum and minimum air temperature using MODIS aqua land surface temperature data over the United Arab Emirates. Geocarto Int 37:2996–3013. https://doi.org/10.1080/10106049.2020.1837261

Di Gregorio Jansen, Jansen LJM (2000) Land cover classification system (LCCS): classification concepts and user manual (SOFTWARE V). Food and Agriculuter organization

Awange JL, Saleem A, Konneh SS, Goncalves RM, Kiema JBK, Hu KX (2018) Liberia's coastal erosion vulnerability and LULC change analysis: post-civil war and Ebola epidemic. Appl Geogr 101:56–67. https://doi.org/10.1016/j.apgeog.2018.10.007

Chapa F, Hariharan S, Hack J (2019) A new approach to high-resolution urban land use classification using open access software and true color satellite images. Sustainability (Switzerland) 11(19). https://doi.org/10.3390/su11195266

Congedo L (2016) Semi-automatic classification plug-in semi-automatic classification plug-in documentation. Technical report. https://doi.org/10.13140/RG.2.2.29474.02242/1

Correia R, Duarte L, Teodoro AC, Monteiro A (2018) Processing image to geographical information systems (PI2GIS)—a learning tool for QGIS. Educ Sci 8(83). https://doi.org/10.3390/educsci8020083

da Silva VS, Salami G, da Silva MIO, Silva EA, Monteiro Junior JJ, Alba E (2022) Methodological evaluation of vegetation indexes in land use and land cover (LULC) classification. Geol Ecol Landscs 4(2):159–169. https://doi.org/10.1080/24749508.2019.1608409

de Sousa-Neto ER, Gomes L, Nascimento N, Pacheco F, Ometto JP (2018) Land use and land cover transition in brazil and their effects on greenhouse gas emissions. In: Soil management and climate change: effects on organic carbon, nitrogen dynamics, and greenhouse gas emissions. https://doi.org/10.1016/B978-0-12-812128-3.00020-3

Divya Y, Gopinathan P, Jayachandran K, Al-Quraishi AMF (2021) Color slices analysis of land use changes due to urbanization in a city environment of Miami Area, South Florida, USA. Model Earth Syst Environ 7:537–546 (2021). https://doi.org/10.1007/s40808-020-00883-x

El-Baras YM, Saaed MW (2013) Threats to plant diversity in the northeastern part of Libya (El-Jabal El-Akahdar And Marmarica plateau). a 2 41–58

Estoque RC, Murayama Y (2015) Intensity and spatial pattern of urban land changes in the megacities of southeast Asia. Land Use Policy 48:213–222. https://doi.org/10.1016/j.landusepol.2015.05.017

Fick SE, Hijmans RJ (2017) WorldClim 2: new 1-km spatial resolution climate surfaces for global land areas. Int J Climatol. https://doi.org/10.1002/joc.5086

Gadrani L, Lominadze G, Tsitsagi M (2018) Assessment of land use/landcover (LULC) change of Tbilisi and surrounding area using remote sensing (RS) and GIS. Ann Agrar Sci 16(2):163–169. https://doi.org/10.1016/j.aasci.2018.02.005

Gandhi GM, Parthiban S, Thummalu N, Christy A (2015) Ndvi: vegetation change detection using remote sensing and Gis – a case study of Vellore District. Procedia Comput Sci 571199–1210. https://doi.org/10.1016/j.procs.2015.07.415

Gaznayee HAA, Al-Quraishi AMF, Mahdi K, Ritsema C (2022) A geospatial approach for analysis of drought impacts on vegetation cover and land surface temperature in the Kurdistan region of Iraq. Water 14:927. https://doi.org/10.3390/w14060927

Gogoi PP, Vinoj V, Swain D, Roberts G, Dash J, Tripathy S (2019) Land use and land cover change effect on surface temperature over Eastern India. Sci Rep 2018. 1–10. https://doi.org/10.1038/s41598-019-45213-z

Gupta DM, Munshi MK (1985) Urban change detection and land-use mapping of Delhi. J Remote Sens 6(3–4):529–534. https://doi.org/10.1080/01431168508948474

Hamad R, Kolo K, Balzter H (2018) Post-war land cover changes and fragmentation in Halgurd Sakran national park (HSNP). Kurd RegN Iraq Land 7(1):38. https://doi.org/10.3390/land70 10038

Hamad S (2012) Status of groundwater resource of Al Jabal Al Akhdar region, North East Libya. Int J Environ Water 2(1):68–78

Hamad S (2019) Spatial characteristics of the southern Al Jabal Al Akhdar watersheds: remote sensing approach. Hydrospatial Anal. https://doi.org/10.21523/gcj3.19030104

Hamad S (2022) Surface runoff estimation of Wadi Ba Al-Arid watershed NE Libya using SCS-CN, GIS and RS data. Iran J Earth Sci 12(3):168–175

Hubbard KG, Lawrence PJ, Mcalpine C (2014) Impacts of land use/land cover change on climate and future research priorities. Am Meteorol Soc. https://doi.org/10.1175/2009BAMS2769.1

Huete AR (2012) Vegetation indices remote sensing and forest monitoring. Abs Geography Compass 6(9):513–532. https://doi.org/10.1111/geco.2012.6.issue-9, https://doi.org/10.1111/j.1749-8198.2012.00507.x

Hussain S, Mubeen M, Ahmad A, Akram W, Hammad HM, Ali M, Masood N, Amin A, Farid HU, Sultana SR, Fahad S, Wang D, Nasim W (2019) Using GIS tools to detect the land use/land cover changes during forty years in Lodhran District of Pakistan. Environ Sci Pollut Res. https://doi.org/10.1007/s11356-019-06072-3

Hussein K, Alkaabi K, Ghebreyesus D, Liaqat MU, Sharif HO (2022) Land use/land cover change along the Eastern Coast of the UAE and its impact on flooding risk. Geomat Nat Haz Risk 11(1):112–130. https://doi.org/10.1080/19475705.2019.1707718

Hydrogeo (1985) Groundwater resource evaluation of Al Baydah—Al Bayadah Area

IRC (1974) Geological maps of Libya

International Organization for Migration (IOM) (2023) Aftershock: an assessment of how climate change is influencing migration and vulnerability in Libya. https://reliefweb.int/report/libya/aft ershock-assessment-how-climate-change-influencing-migration-and-vulnerability-libya-nov ember-2023

Kaptué AT, Prihodko L, Hanan NP, Turner BL (2015) On regreening and degradation in Sahelian watersheds. Proc Natl Acad Sci USA 112(39):12133–12138. https://doi.org/10.1073/pnas.1509645112

Kim J, Choi J, Choi C, Park S (2013) Impacts of changes in climate and land use/land cover under IPCC RCP scenarios on streamflow in the Hoeya River Basin, Korea. Sci Total Environ 452–453(March):181–195. https://doi.org/10.1016/j.scitotenv.2013.02.005

Khudair D, Al-Quraishi AMF, Hasan AA (2019) Spatiotemporal monitoring and modeling of urban sprawl using remote sensing and GIS: a case study Al-Karkh, Baghdad, Iraq. J Adv Res Dyn Cont Sys 11(6):1691–1698

Khwarahm NR, Qader S, Ararat K, Al-Quraishi AMF (2021) Predicting and mapping land cover/land use changes in Erbil /Iraq using CA-Markov synergy model. Earth Sci Inform 14:393–406. https://doi.org/10.1007/s12145-020-00541-x

Liu H, Gong P, Wang J, Clinton N, Bai Y, Liang S (2020) Annual dynamics of global land cover and its long-term changes from 1982 to 2015. Earth Syst Sci Data 12(2): 1217–1243. https://doi.org/10.5194/essd-12-1217-2020

Mahmoud K (1995) Libyan soils (1st ed)

Masoud M (2015) Monitoring land use/land cover using multi-temporal Landsat images in Al-Jabal Al-Akhdar area in Libya between 1984 and 2003. Al Mukhtar J Sci 31(1):12–22

Mishra PK, Rai A, Rai SC (2022) Land use and land cover change detection using geospatial techniques in the Sikkim Himalaya, India. Egypt J Remote Sens Space Sci 23(2):133–143. https://doi.org/10.1016/j.ejrs.2019.02.001

Munthali MG, Botai JO, Davis N, Adeola AM (2019) Multi-temporal analysis of land use and land cover change detection for Dedza district of Malawi using geospatial techniques. 14(5):1151–1162

Mnsur S, Abdalrahman YF (2015) Monitoring land cover changes using remote sensing in Al-Jabal Alakhdar, North-East Libya. Libyan J Basic Sci (LJBS) 3(1):13–21

Muttitanon W, Tripathi NK (2005) Land use/land cover changes in the coastal zone of Ban Don Bay, Thailand using Landsat 5 TM data. Int J Remote Sens 26(11):2311–2323. https://doi.org/10.1080/0143116051233132666

OMU (2005a) Studying and evaluating the natural vegetation in Al-Jabal Al-Akhdar area (In Arabic)

OMU (2005b) Studying and evaluating the natural vegetation in Al Jabal Al Akhdar area. Al Bieda, Libya

Peng J, Dadson S, Hirpa F, Dyer E, Lees T, Miralles DG, Vicente-Serrano SM, Funk C (2022) A pan-African high-resolution drought index dataset. Earth Syst Sci Data. https://doi.org/10.5194/essd-12-753-2022

Pérez-Trejo F (1992) Desertification and land degradation in the European Mediterranean. (Report EUR 14850 EN)

Pitman A, De Noblet-ducoudre N (2012) Human effects on climate through land-use-induced land-cover change. In: The future of the world's climate. 77–95. https://doi.org/10.1016/B978-0-12-386917-3.00004-X

Rahman MTU, Tabassum F, Rasheduzzaman M, Saba H, Sarkar L, Ferdous J, Uddin SZ, Zahedul I (2017) Temporal dynamics of land use/land cover change and its prediction using CA-ANN model for southwestern coastal Bangladesh. Environ Monit Assess 189(565). https://doi.org/10.1007/s10661-017-6272-0

Rawat JS, Kumar M (2015) Monitoring land use/cover change using remote sensing and GIS techniques: a case study of Hawalbagh block, district Almora, Uttarakhand, India. Egypt J Remote Sens Space Sci 18(1):77–84. https://doi.org/10.1016/j.ejrs.2015.02.002

Reis S (2008) Analyzing land use/land cover changes using remote sensing and GIS in Rize, Nort-East Turkey. Sensors 8(10):6188–6202. https://doi.org/10.3390/s8106188

Robinson DT, Brown DG, French NHF, Reed BC (2013) Land use and the carbon cycle: advances in integrated science, management, and policy, 1st edn. Cambridge University Press

SELKHOZPROMEXPORT (1980) Soil studies in the eastern zone of the socialist people. In: Agricultural reclamation and land development, SPLAJ. VNESHTORGREKLAMA. Publishing House, USSR

Suleiman BM, Mohamed M, Hamad S, Elmehdy S (2016) Assessment of forest and Juniperus Phoenicea decline in Al Jabal Al Akhdar using NDVI-remote sensing and GIS data (2006–2013). Int J Remote Sens Appl 6(0):159. https://doi.org/10.14355/ijrsa.2016.06.016

Sundarakumar K, Harika M, Aspiya Begum SK, Yamini S, Balakrishna K (2012) Land use and land cover change detection and urban sprawl analysis of vijayawada city using multitemporal landsat data. Int J Eng Sci Technol 4(1):166–174

Twisa S, Buchroithner MF (2019) Land-use and land-cover (LULC) change detection in Wami river basin, Tanzania. Land 8(9). https://doi.org/10.3390/land8090136

Vivekananda GN, Swathi R, Sujith A (2022) Multi-temporal image analysis for LULC classification and change detection. Eur J Remote Sens 00(00):1–11. https://doi.org/10.1080/22797254.2022.1771215

Meyer WB, Turner BL II (1992) Human population growth and global land-use/cover change. Annu Rev Ecol Syst 23:39–61. https://doi.org/10.2307/2097281

Wilson SA, Wilson CO (2013) Modelling the impacts of civil war on land use and land cover change within Kono District, Sierra Leone: a socio-geospatial approach. Geocarto Int 28(6):476–501. https://doi.org/10.1080/10106049.2012.724456

Zulhaidi H, Shafri M, Suhaili A, Mansor S, Sensing R, Branch FO, Alam WS, Jaya P (2007) The performance of maximum likelihood, spectral angle mapper, neural network and decision tree classifiers in hyperspectral image analysis. J Comput Sci 3(6):419–423

Zunni SA (2013) Forest of Al Jabal Al Akhdar, current and prospective challenges. First Workshop on the Al Jabal Al Akhdar Vegetation Cover. 200

Construction Industry Role in Natural Resources Depletion and How to Reduce It

Bayan Salim Obaid Al-Numan

Abstract Due to the limited natural resources on our planet, we must take into account three essential factors, including environment, society, and economy, in all design and construction aspects. The environment has suffered because of the devastating human impact on our planet caused by industries that pollute, destroy, and produce waste. Rising levels of carbon dioxide and other greenhouse gases (GHG) have contributed to global warming, which has caused climate change that affects human life and threatens the coming generations. Among industries, construction has an aggressive impact on the environment. This is responsible for the detrimental consumption of raw materials from Earth. It is responsible for extracting a large amount of natural resources worldwide. The construction sector contributes to air pollution, climatic change, water pollution, and landfill waste. It is responsible for high energy use and carbon dioxide emissions. This study views the construction industry's effects on the environment, energy spending, and natural resource depletion to lighten the impact the construction industry has on the environment and how to reduce it. This chapter considers the construction industry's impact on natural resources and ways to control and mitigate this impact. The chapter tracks several trends in the construction industry: sustainable design is becoming increasingly mainstream and growing to seek appropriate ways to conserve natural resources and protect the environment. Construction materials continue to evolve with higher-strength steel and concrete, high-performance glazing, waste reduction, and material recycling.

Keywords Construction industry · Natural resources · Sustainability · Environment · Material consumption

---

B. S. O. Al-Numan (✉)
Civil Engineering Department, Faculty of Engineering, Tishk International University, Kurdistan Region, Erbil 44001, Iraq
e-mail: bayan.salim@tiu.edu.iq

© The Author(s), under exclusive license to Springer Nature Switzerland AG 2024
A. M. F. Al-Quraishi and Y. T. Mustafa (eds.), *Natural Resources Deterioration in MENA Region*, Earth and Environmental Sciences Library,
https://doi.org/10.1007/978-3-031-58315-5_6

1 Construction as a Natural Resource-Depleting Industry

Huge quantities of Earth's resources are expended because of building construction. This industry is a major cause of environmental pollution. A report by The U.S. Green Building Council indicates that 35% of the energy spent worldwide is associated with building construction. The same percentage was true for GHG emissions. In the United States, the industry is responsible for more than 33% of the US total energy use, 66% of its electricity spending, 30% of its raw materials use, 25% of its harvested wood, and 12% of its fresh water. Building construction and operation accounts for 50% of US GHG emissions and approximately 33% of its solid waste stream (World Business Council for Sustainable Development 2009). Cities and buildings are expanding, and the impact on the construction industry is great and will extend to the next generation. It has been estimated (Hawken et al. 1999) that each of energy and water as global resources used in building construction has spent 50% of the total spent in all human activities. For global pollution that can be attributed to building construction, it has been estimated (Brown and Bardi 2001) that building construction is responsible globally on 50% of climate change gases.

More than 30–50% (different sources give different numbers) of the total material use in Europe goes to housing, mainly consisting of iron, aluminum, copper, clay, sand, gravel, limestone, wood, and building stone. Minerals have the highest share of all the materials in buildings. Approximately 65% of the total aggregates (sand, gravel, and crushed rock) and approximately 20% of the total metals are used by the construction sector (Christensen and Huebsch 2012). Buildings are also a major source of air pollutant emissions. In short, building construction and operation cause environmental deterioration, greatly reducing the Earth's resources and jeopardizing the building industry's future and people's health and welfare.

1.1 Sustainability

Sustainability may be defined as meeting the needs of the present generation without compromising the ability of future generations to meet their needs (American Society of Civil Engineers 2017). Civil engineers should ensure that the buildings they design and construct have sufficient durability to maintain the use of the building's valid maximum lifespan. At the same time, they should use as few natural resources as possible with the least amount of embodied energy when producing the elements of the project while still meeting client, economic, and social demands and code requirements. The American Society of Civil Engineers ASCE adopted a Code of Ethics, the "model for professional conduct" for ASCE members (American Society of Civil Engineers 2017). The ASCE code includes four fundamental principles and seven fundamental canons. The first principle indicates the importance of engineers' knowledge and skills for promoting the environment and people's welfare. The sustainable principles listed in the Code of Ethics encourage civil engineers to understand and incorporate sustainable measures into the field (Kibert 2016; Abraham 2017).

By consuming irreplaceable fossil fuels and other nonrenewable resources and building in sprawling urban patterns that cover extensive areas of prime agricultural land, we have been building in a manner that will make it increasingly difficult for our children and grandchildren to meet their needs for communities, buildings, and healthy lives (Allen and Iano 2019). There are many difficulties in applying sustainability practices in many steps in construction projects, as indicated by some local research works (Al-Numan and Thamir 2016a, b). This may be attributed to the lack of related laws and comprehensive application of sustainable designs.

1.2 The Embodied Energy

The EE of a building is the calculation of all the energy used to produce the materials that make up the building. This includes the energy used in mining, manufacturing, and transporting materials and services in the economy that support these processes. The total embodied energy of a building is the total energy required for

- production of all the materials used in the initial construction (initial embodied energy)
- production of all the materials used in repairs or renovations over the life of the building (recurrent embodied energy)
- transport of materials to the site and
- Energy used onsite during construction, repairs, or renovations.

Estimating the EE of a material, component, or building is difficult. The subject includes the combined effects of various materials, their history of production, and their contributions to the EE of a building. EE can also differ based on the efficiency of the process, energy sources, and materials transportation (Craig et al. 2008). Life cycle assessment (LCA) is the basis for calculating the EE (ISO 14040 2006; Saling et al. 2002, 2005). An alternative environmental assessment can be performed using the "Leadership in Energy and Environmental Design" or "LEED" method. LEED (World Business Council for Sustainable Development 2009) does not rely on EE and embodied CO_2 but evaluates buildings by a credit mark between 0 and 70, ranging from a minimum evaluation of poor environmental design to a high evaluation of excellent environmental design.

1.3 Embodied Greenhouse Gases Emissions

The embodied GHG emissions of greenhouse gases are the cumulative quantity of GHG (CO_2, methane, nitric oxide, and other global warming gases), which are produced during the direct and indirect processes related to the creation of the building, its maintenance, and end-of-life. The CO_2 equivalent indicates that all GHG emission effects are considered. ACI 130-19 (ACI 130-19 2019) provides the embodied energies of selected construction materials versus embodied CO_2. The embodied energy of various construction materials ranges approximately from 0.6 to 100 MJ/kg versus embodied CO_2 range approximately from 0.05 to 10 kg CO_2/kg. (ACI 130-19 2019).

Knowledge of the environmental problems created by the construction industry, how to solve them, and skills in sustainable design and construction of buildings based on this knowledge are crucial in controlling the depletion of natural resources.

2 Stages of Construction Industry Damage to the Natural Resources and Environment

2.1 Obtaining and Producing Materials for a Building

There are concerns about whether construction materials remain plentiful or rare. Most of these are nonrenewable. Parts of these materials are recycled for other uses. Most of these materials have a high EE, expended in obtaining and producing them, and plenty of water is expended. Pollutants from various construction processes contaminate air, water, and soil. CO_2 and other GHG are emitted. Large quantities of waste were generated. Currently, the material requirements of buildings represent one of the greatest resource use challenges in terms of the mass of resources used. Extensive material manufacturing is associated with climate change, desertification, and soil erosion.

2.2 Construction of the Building

High energy and water are expended on building sites and in transporting materials to the building site from distant places, thereby generating pollutants. A large amount of waste is generated.

2.3 Operation and Maintenance of the Building

High energy and water are used in building maintenance during their lifetime owing to the materials used in their maintenance operations. Energy and time are spent on building maintenance.

2.4 Demolition of the Building

When demolition is inevitable, it will result in a large amount of waste. Materials from demolished buildings can be used in recycled production, new construction, or other uses, rather than being treated as waste.

3 Building Sites

Building sites can destroy natural resources and increase pollution and energy consumption. The construction of prime agricultural land causes the loss of land and its production. The same is true for forests and lands close to rivers, lakes, and seashores, which support recreational use. Buildings should be constructed to be well-connected to transportation networks. This would save fuel and mitigate air pollution. New construction projects may destroy land features and recreational facilities, which can be lost or never replaced. The building site layout can be designed for optimum exposure to sunlight and wind to obtain the highest solar heat in winter and the lowest in summer to save energy spent on heating, air conditioning, and electrical lighting (Allen and Iano 2019). In addition, the building design should consider that the site controls and improves the environment through the best sun and wind orientation, and minimizes fuel and electricity spending as much as possible. Soils left after excavation should be reused at the same or neighboring site. Air pollution and soil erosion caused by water and wind should be prevented during construction. Recycling of construction wastes should be performed. Pollution due to the use and maintenance of construction machinery should be minimized.

4 Brick, Stone, and Concrete Masonry

4.1 Brick Masonry

The raw materials used for the bricks were clay and shale. They are plentiful and usually cause problems with drainage, vegetation, and wildlife habitat disruption. Brick dust can be included in the manufacture of clay bricks, in addition to a variety of wastes such as fly ash. Manufacturing plants for bricks are usually constructed near their raw material sources and produce few waste materials. Unusable bricks are ground and recycled during manufacturing. However, this method requires a relatively large amount of water. Much energy is spent on brick production because it depends on firing. Its EE may range from 2.3 to 9.3 MJ/kg (Crawford 2019). The calculation of the EE includes material production, mining, manufacturing, and transportation. Brick kilns use natural gas, oil, and coal as their energy sources. As a source of pollution to the environment, the emission of fluorine, chlorine, and other types of air pollution results from clay masonry firing.

4.2 Sustainability Features

Waste from brick masonry work, such as broken bricks, unsatisfactory bricks, and excess mortar, is relatively scarce at construction sites. They are usually used in landscaping, landfilling, or burial at the same site. Brick construction is known as a heat insulator and is useful in saving fuel spent on cooling and heating. Brick buildings have very long lifespans because of their high durability. This requires minimal maintenance. When a brick building is demolished, sound bricks can be reused, and crushed bricks can be used for landscaping and onsite filling. It is recommended that the brick manufacturing and construction industries consider the application of sustainability measures for their products during manufacturing and construction and encourage and apply reuse and recycling when demolition takes place at the end of the design life of the brick building. Industries should investigate the possibility of recycling and reusing masonry materials (Straka 2015).

4.3 Stone and Concrete Masonry Materials

Stones are abundant and finite natural resources. Obtaining stones from open-pit quarries disrupts the habitat of humans, wildlife, and vegetation. Concrete used in the manufacture of blocks used in masonry construction may involve recycled waste materials from various industries, such as fly ash, crushed glass, slag, and other wastes. Energy is also spent through the mortar used for construction, concrete, brick, and stone masonry. Mortar which is made from portland cement and lime is of high embodied energy during manufacture and construction. Stones are heavy, expensive, and require high energy for transportation. Stones may be brought from local quarries or imported from many places worldwide. This may require long-distance shipping, particularly when certain unique types of stones are desired or when special stone-building skills are required. Large quantities of water are used in the cutting, shaping, and polishing operations during stone fabrication. Large quantities of stones may be wasted during fabrication. However, these wastes may be used as fill materials on construction sites or as recycled aggregate in asphalt or concrete construction. However, stone waste is often used in landfills. The EE of a building stone can vary significantly with the source of the stone, fabrication processes, distances, and shipping methods. Stones that are easily quarried, fabricated, and used locally may have an embodied energy of as little as 0.7–0.9 MJ/kg.

Most concrete masonry units are produced in local plants relatively near construction project sites. The embodied energy of concrete masonry units (approximately 0.6 MJ/kg) is slightly higher than that of the concrete from which they are made owing to the additional energy consumed in curing the units (Allen and Iano 2019). Construction waste from concrete and stone masonry work, such as broken blocks, unsatisfactory blocks, and excess mortar, is relatively scarce at construction sites. They are usually used as recycled aggregates, landscaping, landfills, or buried at the same site. Concrete and stone buildings have very long lifespans because of their high durability. This requires minimal maintenance. When demolishing concrete blocks and stone buildings, crushed blocks and stones can be reused and recycled for landscaping, paving, and on-site filling. It is recommended that the concrete block and stone manufacturing and construction industries consider the application of sustainability measures for their products during manufacturing and construction, and encourage and apply reuse and recycling when demolition takes place at the end of the design life of concrete and stone masonry buildings. Industries should therefore investigate the possibility of recycling and reusing masonry materials.

5 Steel Construction

Iron ore, coal, and limestone are raw steel materials, in addition to air and water. Mining and quarrying of these minerals cause loss of wildlife habitat, land disruption and pollution of rivers and waterways. These raw materials are plentiful worldwide, but there is concern regarding the depletion of high-grade iron ore in many places. In steel construction, alloys include various proportions of alloying metals such as manganese, chromium, and nickel. Thus, the sources of these materials have become depleted. The EE of steel produced from ore was high (approximately 33 MJ/kg). Recently, scrap steel has been a typical ingredient added during this process, resulting in steel with a recycled material content of approximately 30%. Currently, the electric arc furnace process produces most structural steel made from recycled scrap; its EE is about (9.3 MJ/kg), approximately one-third of that of steel made from ore. Almost 90% or more of the steel was recycled when using the electric arc furnace method. The electric arc furnace method produced hot-rolled structural steel shapes used in bridges and multistory structures. Steel plates and sheets can be manufactured using either an electric arc furnace or basic oxygen processes. These are used to construct hollow structural sections, light-gauge steel members, and decking. Eventually, almost all structural steel from demolished buildings is recycled or reused, which is a significant sustainability feature. In recent years, (430 million metric tons) of scrap steel have been consumed worldwide. (Allen and Iano 2019).

Mini mills usually receive scrap used to manufacture structural steel from locations far from the mini mill. It is considered a process that deserves credit when the steel construction site that uses steel from mills is not too far from the mill. This is because the steel is locally extracted, processed, and produced. However, this may not be true for some produced steel alloys that are not available nearby but from a limited number of places or may be produced abroad. Steel manufacturing and construction are relatively clean. However, construction requires paints and oils used on steel members, which can cause air pollution. Concrete frames are heavier than comparable steel frames. This means that concrete buildings generally have larger foundations and require more excavation. Some fireproofing materials spread on steel members are sources of air pollution. Steel buildings require little or no maintenance and have a very long lifetime, especially if well protected from corrosion and fire. Periodic paintings are required for steel members exposed to the weather. This is not required if the steel is galvanized, stainless steel, or polymer-coated. Insulation processes are often used for steel construction on building walls and roofs. The insulation should be such that it does not allow heat to be conducted indoors and outdoors. When demolition occurs at the end of a steel building's design life, the material is typically recycled. Steel construction is desirable because of its low environmental impact. This is desirable, to a greater extent, than before. Because steel is almost always recyclable or reused, with minimal use of natural resources and a limited amount of waste, it is an environmentally friendly material. Moreover, steel structures can adapt to the changing requirements of users. The energy consumption for the production and assembly of steel structures represents only 3% of the total energy demand of a building throughout its lifespan (Broniewicz and Broniewicz 2020).

6 Concrete Construction

Concrete production spends approximately 1.5 billion metric tons of Portland cement annually, 9 billion metric tons of aggregates, and 0.9 billion metric tons of water. In terms of natural resource use, this industry is the largest among other industries (Di Stefano 2022). Open-pit quarrying to obtain concrete raw materials causes habitat loss, soil erosion, pollutant runoff, and landscape damage. Concrete construction also uses large quantities of other materials, such as wood, wood panel products, steel, aluminum, and plastics for formwork and reinforcement. The embodied energy of the concrete varies depending on its design strength. On average, EE may be taken 0.6 MJ/kg. Several measures can be taken to mitigate the impact of concrete construction, one of which is the use of supplementary cementitious materials (SCMs) (ACI 130-19 2019), which are waste materials from several industries, such as fly ash from power plants and slag from iron furnaces. These materials can replace up to 50% of the cement content, reduce transportation costs by creating local concrete production plants, reduce the use of materials for reinforcement and formwork, reduce energy spending, and reduce waste and emissions that harm the environment from all steps of concrete construction, starting from obtaining raw materials for building demolition. Concrete buildings have a higher lifespan when the quality of concrete and other construction materials is higher. This reduces the demolition sequence, waste, and concrete demand (Allen and Iano 2019).

6.1 Portland Cement

Due to its quarrying, production, and transportation, Portland cement is the largest energy consumer in the concrete construction process. More than 80% of the total energy required for concrete production is spent on Portland cement (PC). Portland cement production contributes to approximately 5% of all carbon dioxide gas generated by human activities worldwide. Dust and air pollutants are produced in large amounts during cement production, and it is estimated that approximately one ton of carbon dioxide (a major GHG) is released into the atmosphere to produce one ton of cement clinker. Certain industrial waste materials with cementing properties can be used to replace part or all the Portland cement used as a concrete ingredient, and therefore, it will significantly reduce the energy spent on concrete production. Providing such supplementary cementitious materials to substitute up to 50% of Portland cement in concrete may reduce the embodied energy by up to 33%, and certain SCMs may save up to 1300 Mt of CO_2-equivalent emissions (Shah et al. 2022).

Wood ash and rice husk ash are two examples of waste materials from other industries that can also be used as cementing materials. Motor oil and rubber vehicle tires can be employed as fuels in cement kilns. While consuming waste products from other industries, a cement manufacturing plant can, if operated efficiently, generate virtually no solid waste (Allen and Iano 2019). The cement industry in Japan (Kawai and Osako 2012) has worked hard to maximize waste utilization. With no waste to be recycled in the cement industry, limestone consumption would have been 18% higher, clay consumption would have been 2,899% higher, and energy consumption would have been 22% higher. The waste recycled in cement production's mixing and burning process directly reduced GHG emissions by 12%. Waste utilization for cement production contributed to an increase in resource productivity of approximately 8,000 Japanese Yen/ton.

6.2 Aggregates

Aggregates are available worldwide, but some places suffer from shortages in high-quality aggregates. Aggregates normally used in concrete can be partially substituted by recycled waste materials such as crushed parts of concrete from demolished buildings, crushed glass, and waste plastics, some of which may produce heavier concrete.

6.3 Water

Water of a quality suitable for concrete is scarce in many developing countries. Concretes that use less water with superplasticizers, air entrainment, and fly ash could be helpful.

6.4 Demolition and Recycling

Reinforcing steel from demolished concrete buildings is typically recycled. In many cases, new concrete results from crushed parts of demolished concrete that are sorted and used as aggregates. Demolished concrete is usually buried at a site or dumped in landfills. It was confirmed that the quality of the produced eco-friendly recycled concrete is less than that of ordinary concrete. However, there are continuous efforts to improve it by mechanical or other treatments of the recycled concrete used as aggregates for the new eco-friendly concrete (Kumar and Singh 2023; Kumar et al. 2024). Using demolition concrete or bricks as aggregates in eco-friendly concrete may raise concerns about durability issues, such as a higher risk of reinforcing steel corrosion. More precautionary measures should be applied to protect the eco-friendly concrete (Mohammedameen and Al-Numan 2013).

6.5 Precast Concrete Construction

The embodied energy of precast concrete is higher than that of cast-in situ concrete because it is usually of higher strength; therefore, its concrete mixes and production require higher energy. The EE of precast concrete ranges from 1.1 to 1.4 MJ/kg. However, precast concrete production reuses formwork, which can be reused several times. Precast concrete elements are often smaller than cast-in-situ concrete elements because of the optimized design and high-quality production control systems in factories. The reduced cooling and heating costs characterize the precast concrete wall panel systems. This is attributed to the proper sealing of joints, which prevents air leakage and contributes to better indoor air quality compared to cast-in-situ concrete construction. Precast concrete wall panels can be reused when buildings are altered (Allen and Iano 2019).

7 Roofing

The roofs of buildings can contribute to controlling natural resource depletion in many ways. Solar radiation is reflected when using a light-colored roof covering. It may reflect more than 50% of the radiation when it strikes its surface, substantially reducing the heat applied to the building. These roofs reduce the energy costs for the cooling of buildings and, in addition will extend the life of roofing materials. A roof surface can support flat-plate solar heat collectors used to reduce building heating costs or arrays of photovoltaic cells to provide electrical power.

7.1 Roof Insulation

The roof and wall thermal insulation materials are the best energy-saving and cost-effective building components. Significant cooling and heating energy reduction is obtained and great comfort is provided to building users when insulators are properly installed. The energy savings gained from these insulation components would pay for their construction costs quickly.

7.2 Roof Membranes

The impacts of roof membranes on the environment are varied. Bituminous roofing consisting of asphaltic compounds is considered to be highly impacted because it is mainly derived from petroleum and coal. Asphalt shingles on steep roofs mostly consist of petroleum. Cellulose and glass fibers are generally used as roofing felts. Single-ply membranes, in which the general formulations of rubber and plastic are mainly derived from petroleum, also significantly impact the environment. Petroleum, which is utilized as the major raw material in producing these membranes, harms the environment, causes gas emissions, and results in a high embodied energy of the material. Green roofs can be developed by covering membranes with soil and plants. The plantation at the roof generates oxygen and consumes carbon dioxide, protects the roof membrane, reduces the temperature inside the space below the roof, cools the roof surface by evaporation and transpiration, reduces the volume of stormwater, delays stormwater passage into sewers, and creates a pleasant roofscape (Allen and Iano 2019). The impacts of various roofing options were compared through a life-cycle assessment (LCA) (Duong Lea et al. 2019). The roofing options were clay tile, concrete tile, and sheet metal. The lowest carbon footprint for clay tiles was 4.4 t of CO_2 equivalent (CO_2 e-) with EE demand of 52.7 MJ/100 m^2. The highest carbon footprint for sheet-metal is 9.85 t of CO_2 e-, with concrete roof tiles having the highest EE demand (83 MJ). The findings confirm significant savings in

EE when using a sheet-metal roof. In general, sheet-metal roofs can obtain significant carbon and EE savings of (71–73%) compared to clay tile or concrete roof tiles.

8 Nonstructural Components in Buildings

8.1 Glass Production and Uses

Glass is made from natural and abundant raw materials (sand, sodium carbonate and limestone) that are melted at very high temperatures to form glass. The embodied energy of glass manufactured using efficient methods is approximately 5–10 MJ/kg. Glass is a long-aged material with negligible loss of quality, and is clean. However, most of the old glass goes to landfill. In winter, glass can allow sun heat to enter the building, which can be prevented in summer. This would save the energy spent on heating and cooling. It can also reduce the energy spent on lighting by allowing daylight to enter the building without glare and save the energy spent on cooling because of electric lighting. These energy savings continue over the entire building life, and they can be significant. Accordingly, glass is a key parameter for energy saving in building construction and operation, and can help reduce the energy waste used for building operation (Allen and Iano 2019).

8.2 Windows and Doors

Solar heat gain and wintertime heat losses through windows account for a considerable amount of the building's heating and cooling electrical loads. However, windows and doors allow air leakage and thereby play a considerable role in increasing the energy required for the cooling and heating of buildings. The heat leaked in large amounts through doors by conduction through the material of the door. The best doors are foam-core doors, which provide better thermal insulation. All doors were tightly weather-stripped to limit the loss of conditioned air. When buildings with wood windows are demolished, the windows are generally sent to landfills and not recycled. Aluminum frames must be thermally broken to achieve energy efficiency. They are often recycled during demolition in every case. The PVC window frames are thermally efficient. They can be recycled during demolition, and a significant percentage is currently recycled at the present time. The steel window and door frames are made of recycled steel and can be recycled again when a building is demolished. Their thermal performance is moderate and can be significantly improved by the insertion of thermal breaks (Allen and Iano 2019).

8.3 Exterior Walls

The exterior walls of buildings play a greater role in the energy performance than any other building component. The exterior wall should be well insulated. During winter, a large amount of heat is lost, whereas during summer, a large amount of heat is gained when a poorly designed all-glass box system is used. Energy costs may be reduced when windows are opened and closed. Buildings that make the exterior glass that facing south would provide the best solar heat that can be utilized naturally in winter. Simultaneously, overheating, glare, and deterioration of finishes and other interior surfaces resulting from ultraviolet radiation must be carefully avoided when exposed to sunlight. One study calculated the environmental impacts of using various exterior wall systems in residential buildings in a hot and dry climate. In the construction stage, the best impact on the environment was found for insulated concrete buildings. The buildings that follow are concrete block buildings, cast-in-place concrete buildings, and steel stud buildings. Wood frames had the lowest impact on the environment, but only at the construction stage. In the operating stage, insulated concrete buildings' best performance in reducing the environmental impact is recorded because they require less operational energy (Kahhat et al. 2009).

8.4 Interior Finishes

Finished materials such as tiles, carpets, gypsum products, and others may end up as waste when demolished and transported to landfills. If they have a high recycled content, it will reduce the demand for virgin materials and waste. When manufactured locally, the finishing materials require less energy for transport.

8.5 Gypsum Products

Gypsum is widely used in construction and its consumption is continuously increasing. The increase in gypsum in the construction and demolition waste follows. Most gypsum is extracted and quarried in surface quarries, which causes environmental deterioration through erosion, pollution, and loss of wildlife habitats. Gypsum is also manufactured synthetically, instead of being landfilled, when recovered from power plant flue gases. Gypsum manufacturing requires relatively low temperatures. The EE of gypsum is relatively low, approximately 2.8 MJ/kg for plaster and 6.0 MJ/kg for gypsum board (Allen and Iano 2019). Gypsum boards are essentially recycled materials, including synthetic gypsum and recycled wastepaper. Approximately 10% or more of gypsum boards are wasted at construction sites. Gypsum board waste can be stored in the hollow cavities of finished walls. This would eliminate transportation and disposal costs and reduce the amount sent to landfills. New gypsum boards can be manufactured from recycled gypsum board waste.

A study Weimann et al. (2021) made an assessment using life cycle assessment LCA on the environmental impacts of industrial processes for recycling gypsum waste. Original and complementary industrial databases were used for calculations in the LCAs. Comparison targeted the utilization of various types of gypsum, namely, natural gypsum and gypsum derived from coal-fired power plants. In addition to recycled gypsum. The results revealed that using recycled gypsum can be more environmentally friendly than using natural gypsum or coal-fired power plant gypsum.

8.6 Finish Ceilings and Floors

Acoustical ceiling tiles can be a source of volatile organic compound (VOC) emissions as well as a reservoir for emissions from other sources. Generally, no emissions result from hard concrete flooring materials. The same is true for hard stone, masonry, ceramic tile floors, and cementitious mortars and grouts. Sealers used as hard-flooring materials are potential sources of emissions.

8.7 Resilient Flooring

Emissions can potentially result when using subfloors made of self-leveling cement used for resilient flooring coverings. Vinyl (polyvinyl chloride) is a component of many resilient floor coverings and other interior-finish products. It had an embodied energy of approximately 70 MJ/kg. Vinyl manufacturers release large amounts of toxic air pollutants. Vinyl materials are lightweight, strong, and relatively low in terms of embodied energy compared with many available alternative materials.

9 Construction to Save the Environment

Although the chapter describes the role of construction industries in aggressively impacting the environment and their responsibility for the detrimental consumption of raw materials uprooted from the earth and of a large amount of natural resource extraction worldwide, a very recent case shows that the construction industry can participate and be a part of the solution of an environmental risk issue. The state of Louisiana, USA, suffers more storm surge damage under worst-case predictions of rising sea levels. The state has lost more than 5,000 km^2 of its coast since 1900, owing to a range of human activities. Much of that loss could occur if a worldwide failure to reduce greenhouse gases resulted in a higher sea level rise, approximately

0.7 m., during the coming 50 years. New projects must be undertaken to fight land loss. The dozens of projects now under consideration include restoring wetlands along the coast, and structural projects such as levees and gates, which will reduce the risk of hurricane storm surge flooding (Kumar and Singh 2023).

10 Conclusions

Globally, construction-related activities consume approximately 40% of the world's energy resources and more than 10% of the fresh water, while emitting approximately 30% of the world's greenhouse gases and 10% of fine dust, and contributing to more than 40% of the waste in landfills. Construction also typically displaces some of the most productive agricultural lands and contributes to the biodiversity loss and ecosystems. Therefore, implementing sustainable practices, products, and systems is essential to the long-term viability of modern construction practices. The objective of this review is to highlight the solutions towards sustainable construction.

11 Recommendations

The following ways are recommended to mitigate natural resource depletion in the construction industry:

- Saving building energy spending and utilizing natural sources such as sunlight and wind as energy sources for buildings. This will reduce the depletion of fossil fuels.
- Reuse of Existing Buildings.
- New buildings should not be arranged or constructed on valuable lands.
- Design and construction practices should save and protect water resources.
- Reduce GHG emissions and pollute construction and operation of buildings. This will keep the future environment cleaner.

These sustainable measures are possible with little increase in construction costs; however, they preserve the natural resources. Worldwide, there is still an absence or weakness of binding sustainability laws and a lack of adoption of comprehensive designs based on the pillars of sustainability.

References

Abraham A (2017) Fundamentals of sustainability in civil engineering. Taylor and Francis Group LLC. CRC Press, Boca Raton, FL, pp 33487–2742

ACI 130-19, (2019) Report on the role of materials in sustainable concrete construction (ACI 130R-19). American Concrete Institute, p 37

Allen E, Iano J (2019) Fundamentals of building construction: materials and methods, 7th edn. Wiley, p 944

Al-Numan BS, Thamir M (2016) Sustainability in construction projects: part 1 elements. In: Proceedings of the second international conference on development in civil and computer engineering applications. Ishik University, Erbil, Iraq, pp 61–68

Al-Numan BS, Thamir M (2016) Sustainability in construction projects: part 2 case studies. In: Second international conference on development in civil and computer engineering applications. Ishik University, Erbil, Iraq, pp 69–77

Broniewicz, F. and Broniewicz, M (2020) Sustainability of steel office buildings. Presented at the 9th Innovations-Sustainability-Modernity-Openness Conference (ISMO'20), Bialystok, Poland, 20–21 May 2020. MDPI Proceedings

Brown MT, Bardi E (2001) A compendium of data for energy computation issued in a series of folios. In: The Handbook of energy evaluation, Folio #3: Energy of ecosystems. Center for Environmental Policy, Environmental Engineering Sciences, University of Florida, Gainesville

Christensen BJ, Huebsch C (2012) Solutions to support more sustainable construction practices SP-289.13 American Concrete Institute Special Publication SP 289

American Society of Civil Engineers (2017) Code of ethics. https://www.asce.org/code-of-ethics/

Craig I (2008) Jones and Geoffrey Hammond, embodied energy and carbon in construction materials. Proc ICE Energy 161(2):87–98. https://doi.org/10.1680/ener.2008.161.2.87

Crawford RH (2019) Embodied energy of common construction assemblies (Version 1.0). The University of Melbourne, Melbourne

Di Stefano D (2022) Cement: the most destructive material in the world or a driver of progress?. https://www.greenbiz.com/article/cement-most-destructive-material-world-or-driver-progress

Duong Lea AB., Whyte, A. and Biswas, WS (2019) Carbon footprint and embodied energy assessment of roof covering materials, School of Civil and Mechanical Engineering, Curtin University, Australia. https://espace.curtin.edu.au bitstream handle

Hawken P, Lovins E, Lovins H (1999) Natural, capitalism—creating the next industrial revolution, Little Brown and Co.. p 369

ISO 14040 (2006) Environmental management—life cycle assessment—principles and framework

Kahhat R et al (2009) Environmental impacts over the life cycle of residential buildings using different exterior wall systems. J Infrastruct Syst 15(3). https://doi.org/10.1061/(ASCE)1076-0342(2009)15:3(211)

Kawai K, Osako M (2012) Reduction of natural resource consumption in cement production in Japan by waste utilization. J Mater Cycles Waste Manag 14:94–101. https://doi.org/10.1007/s10163-012-0042-4

Kibert JC (2016) Sustainable construction: green building design and delivery, 4th edn. Wiley. ISBN: 978-1-119-05532-7

Kumar A, Singh GJ, Raj P, Kumar R (2024) Performance of quality-controlled recycled concrete aggregates. ACI Mater J 121(1):17–30

Kumar A, Singh GJ (2023) Improving the physical and mechanical properties of recycled concrete aggregate: a state-of-the-art review. Eng Res Express 5(1) 012007

Mohammedameen BAB, Al-Numan BS (2013) Comparative study on corrosion rates in concrete made of recycled- concrete and recylced-brick aggregate.Euroasian J Sci Eng 9(3):38–46

Saling P, Kicherer A, Dittrich-Kraemer B, Wittlinger R, Zombik W, Schmidt I, Schrott W, Schmidt S (2002) Eco-efficiency analysis by BASF: the method. Int J Life Cycle Assess 7(4):203–218

Saling P, Maisch R, Silvani M, Koenig N (2005) Assessing the environmental hazard potential for life cycle assessment, eco-efficiency and SEE balance. Int J Life Cycle Assess 10(5):364–371

Shah IH, Miller SA, Jiang D et al (2022) Cement substitution with secondary materials can reduce annual global CO_2 emissions by up to 1.3 gigatons. Nat Commun 13:5758. https://doi.org/10.1038/s41467-022-33289-7

Straka1 VA (2015) Environmental impact of masonry industry, 10th Canadian masonry symposium, Banff, Alberta

Weimann K, Adam C, Buchert M, Sutter J (2021) Environmental evaluation of gypsum plasterboard recycling. Minerals 11(2):101. https://doi.org/10.3390/min11020101

World Business Council for Sustainable Development (2009) Transforming the market: energy efficiency in buildings. www.wbcsd.org

Soil Erosion

Assessing Soil Erosion Vulnerability in Semi-Arid Haouz Plain, Marrakech, Morocco: Land Cover, Socio-Spatial Mutations, and Climatic Variations

Youssef Bammou, Brahim Benzougagh, Brahim Igmoullan, Ayad M. Fadhil Al-Quraishi, Fadhil Ali Ghaib, and Shuraik Kader

Abstract The most serious type of land degradation is soil water erosion, which negatively affects agricultural production and has important environmental and socio-economic implications. Human activity promotes soil water erosion. Although much research has been conducted on this subject, there are few studies on soil erosion, especially in the study area. The spatiotemporal variation of soil erosion from 1992, 2002, and 2020 was studied using the spatial model of soil erosion risk and mapping the annual rate of soil loss in the Haouz plain in the lower part of the Tensift Basin in Morocco. The Revised Universal Soil Loss Equation (RUSLE) model, analyzed in a Geographic Information System (GIS), was adopted to estimate soil loss. This

---

Y. Bammou · B. Igmoullan
Department of Geology, Faculty of Science and Technology, Laboratory of Geo-Resources, Geo-Environment and Civil Engineering (L3G), Cadi Ayad University, Marrakesh, Morocco
e-mail: youssef.bammou@ced.uca.ma

B. Igmoullan
e-mail: igmoullan@gmail.com

B. Benzougagh (✉)
Geophysics and Natural Hazards Laboratory, Department of Geomorphology and Geomatics (D2G), Scientific Institute, Mohammed V University in Rabat, Avenue Ibn Batouta, Agdal, PO Box 703, 10106 Rabat-City, Morocco
e-mail: brahim.benzougagh@is.um5.ac.ma

A. M. F. Al-Quraishi · F. A. Ghaib
Petroleum and Mining Engineering Department, Faculty of Engineering, Tishk International University, Erbil 44001, Kurdistan Region, Iraq
e-mail: ayad.alquraishi@gmail.com; ayad.alquraishi@tiu.edu.iq

F. A. Ghaib
e-mail: fadhil.ali@tiu.edu.iq

S. Kader
School of Engineering and Built Environment, Griffith University, Nathan, QLD 4111, Australia

Green Infrastructure Research Labs (GIRLS), Cities Research Institute, Griffith University, Gold Coast, QLD 4215, Australia
e-mail: shuraik.mohamedabdulkader@griffithuni.edu.au

© The Author(s), under exclusive license to Springer Nature Switzerland AG 2024
A. M. F. Al-Quraishi and Y. T. Mustafa (eds.), *Natural Resources Deterioration in MENA Region*, Earth and Environmental Sciences Library,
https://doi.org/10.1007/978-3-031-58315-5_7

study aimed to identify the characteristics and variations in soil erosion under the combined effects of climate factors and land use change. The soil erosion rate of the study area experienced erosion at different levels and is estimated to be 9.84, 13.76, and 14.32% in 1992, 2002, and 2020, respectively. Precipitation, soil, land use, and digital terrain model data were used to construct individual model variables. The GIS platform processes and multiplies the raster layers of topography, cover management, soil erodibility, rainfall erosivity, and conservation techniques. At sites with steep slopes and degraded slopes, the predicted soil loss ranged from 0 to 35 t/ha/yr. Based on the spatial distribution of soil erosion risk, 2.90, 7.74, and 19.01% of the plains experienced soil losses greater than 26t/ha/yr in 1992, 2002, and 2020, respectively. The results indicated strong soil erosion. To effectively plan and execute sustainable soil management methods to prevent soil erosion, especially for the sustainability of the plain, an accurate estimate of soil loss was provided by the RUSLE and GIS methodology.

Keywords GIS · Remote sensing · RUSLE · Water erosion · Climate change · Socio-spatial mutations · Haouz plain

1 Introduction

The most significant type of land degradation is soil erosion, which is affected by natural and anthropogenic factors (Bhattacharya et al. 2020; Ganasri and Ramesh 2016; Rosas and Gutierrez 2019; Teng et al. 2019; Bammou et al. 2023a). According to Rozos et al., (2013), soil erosion is a significant environmental issue that diminishes global ecosystem services and functions (Al-Quraishi 2003; Haregeweyn et al. 2015; Mehwish et al. 2024; Hossin et al. 2022). Owing to its local effects, soil erosion has a detrimental influence on soil quality and agricultural productivity, restricting sustainable agricultural land use (Hurni et al. 2008; Molla and Sisheber 2017; Bag et al. 2022; Kader et al. 2023b). Additionally, it has long-lasting off-site effects such as the destruction of downstream structures, the loss of agricultural land, sedimentation that increases the risk of floods, and diffuse pollution sources that cause eutrophication and turbidity (Lal 2001; Hurni et al. 2008; Bewket and Teferi 2009; Kader et al. 2022; Sestras et al. 2023). The effects of soil erosion caused by human activity, such as widespread deforestation, excessive grazing, agricultural intensification, and population increases, have worsened (Wolka et al. 2015; Haregeweyn et al. 2017; Neupane et al. 2023).

Quantitative and geographical information regarding soil loss is crucial for managing, protecting, and reducing soil erosion (Prasannakumar et al. 2011, 2012; Hossini et al. 2022). Land use, plant cover, terrain, climate, and soil properties can influence soil erosion. Soil properties affect soil erosion. Additionally, human interference alters plant cover and accelerates soil erosion (Jinren and Yingkui 2003; Benzougagh et al. 2016, 2017; Bammou et al. 2023b). Geospatial information systems (GIS) and remote sensing (RS) have provided the best methods for modeling soil erosion (Demirci and Karaburun 2012; Al-Quraishi and Negm 2020). Soil erosion models can consider the intricate connections between various influences on soil erosion. Several scholars have modeled soil erosion using RS and GIS technologies (Pandey et al. 2007; Efe et al. 2008; Kouli et al. 2009; Benzer 2010; Prasannakumar et al. 2012; Alexakis et al. 2013; Farhan and Nawaiseh 2015; Benzougagh et al. 2020a, b, 2023; Kader et al. 2023a). Rapid evaluation of probable soil erosion in a particular area is made possible by the capacity of RS and GIS to handle massive spatial datasets (Meliho et al. 2016, 2020).

Studies on water erosion in Morocco have shown that the annual soil erosion rate typically falls within the range of 23–55%, as documented by various researchers (Merzouki 1992; Bonn 1998; Damnati et al. 2004; Tahiri et al. 2014; Benzougagh et al. 2022). Extreme erosion values have also been recorded, varying from 115 to 524 t/ha/year, depending on several factors such as topography, climate conditions, geological characteristics, and improper intensification of agriculture in protected areas. These factors have exacerbated the issue. A projected 100 million tons of soil are lost annually, with more than 15 million hectares of agricultural land in jeopardy (Merzouk et al. 1996; Benmansour et al. 2013; Markhi et al. 2015). This chapter broadens our understanding of the causes and progression of soil erosion in the Haouz Plain by evaluating the effects of land use and climatic conditions on soil erosion. Policymakers and planners should utilize these findings to develop effective land management strategies.

2 Materials and Methods

2.1 Study Area

The Haouz Plain extends over an area of 7215 km^2 between the Atlas chain in the south with an altitude that peaks at 4165 m (Toubkal) and the Jbilet range in the north (with little relief and a maximum of 1061 m). It clashes east with the first slopes of the Middle Atlas and is limited to the west by the Mzoudia range. The latter individualizes to the west of Haouz, the Mejjate Basin, which is limited by the Essaouira Chichaoua Basin (Piqué et al. 2007). The Haouz of Marrakech is presented as a depression 40 km wide, stretching east to west by more than 150 km. It can be subdivided into three parts: Eastern Haouz, with an area of 1657 km^2, located between Wadi Lakhdar and Wadi R'dat; Central Haouz, which extends between Wadi R'dat and Wadi Nfis on an area of 3180 km^2; and Western Haouz or Mejjat, which extends between Wadi Nfis and the edge of the plateau of Essaouira on an area of 2378 km^2 (Fig. 1).

In the climate context, the weather stations (Marrakech, Tahanaout, Aghbalou, Sidi Rahal, Tafériat, Imine El Hammam, Barrage Lalla Takerkoust, Sidi Hssain, Sidi Bouothmane, Chichaoua, and Abadla) spread over the Haouz reveals elements concerning the climate. In the Haouz Plain, the climate is continental arid to semi-arid,

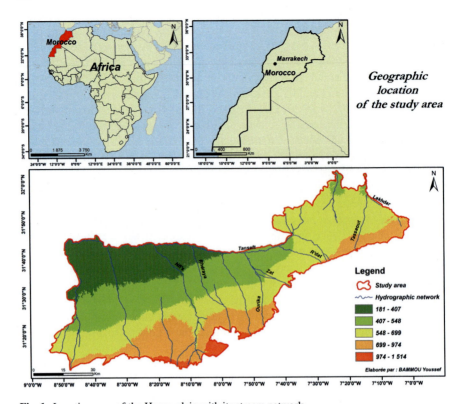

Fig. 1 Location map of the Haouz plain with its stream network

and the rainfall is low and marked by strong spatial and temporal heterogeneity. From 1962 to 2015, the average rainfall recorded at the station of Aghbalou (altitude:985 m) in the High Atlas (Ourika watershed) was approximately 527 mm/year, while it was 184 mm/year at the station of Marrakech located in the plain.

2.2 Methods

The methodology employed in this study for calculating soil erosion is shown in Fig. 2. It encompasses the evaluation of soil erosion, considering rainfall influences, soil quality, topographical features, land-use patterns, and conservation practices. GIS technology has proven highly effective in visually representing these factors in a spatial context. For the processing, storage, analysis, presentation, and visualization of the data and results of this research, ArcGIS 10.5.1 software was utilized. All components were estimated using specific approaches or information from the published literature. An erosion probability map was created by multiplying the model layer data files in the GIS environment. The following sections explain the methods used to evaluate the various factors included in the model.

Assessing Soil Erosion Vulnerability in Semi-Arid Haouz Plain ... 117

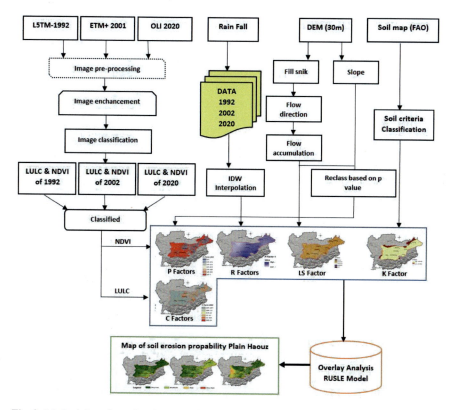

Fig. 2 Methodology for estimating potential erosion using RUSLE approach

2.3 Data Types and Sources

The digital dataset, comprising soil information, land-use land-cover (LULC) data, digital elevation model (DEM), and precipitation data, was compiled from various sources, including field surveys, as detailed in Table 1. Monthly precipitation data from 1992 to 2020 were collected from nine meteorological stations to determine total annual precipitation. Maps depicting land cover and vegetation were generated by analyzing Landsat satellite imagery acquired in 1992, 2002, and 2020. Data on soil types were sourced from FAO's 2007 Digital Database of World Soils (DSMW).

2.4 RUSLE Model

Several models, including the PAP/CAR, MUSLE, and WEPP, can be used to estimate soil erosion. Among these, the RUSLE model, as described by Renard et al. (1997), (Table 1) is one of the most commonly used empirical models in this context.

Table 1 The sources and description of the data types used in this study

Type of data	Associated factor	Source	Description
ASTER DEM	LS	USGS (http://gdex.cr.usgs.gov/gdex/)	30 m resolution
LULC	C	Landsat, image from (1992, 2002 and 2020) (https://earthexplorer.usgs.gov)	30 m resolution, WRS_PATH = 202 WRS_ROW = 038 DATE_ACQUIRED = OLI 2019–09–11, ETM + 2002–09–01, and TM-1992–09–16
Soil data	K	FAO (https://www.fao.org/soils-portal/)	Soil types (FAO-2007)
Rainfall data	R	(ABHT) Tensift hydraulic basin agency	Rainfall data for the period of 1992–2020 from nine stations
Land conservation practices	P	Field observation, literature review, Google earth image	Existing conservation practices in the Haouz

This popularity arises from its relatively low data requirements, easy accessibility to necessary datasets, and compatibility with GIS databases, as noted by Jiang et al. (2015). In Morocco, the RUSLE model has been extensively employed for mapping and assessing the risk of water erosion, as documented in studies by Anys et al. (1992), Elalaoui et al., (2015), Gourfi et al. (2018), and Benzougagh et al. (2020a, b).

However, it is important to acknowledge that this model does not account for significant types of degradation such as gully expansion and landslides, as pointed out by Renard et al. (1997) and Teng et al. (2019). The primary equation employed in the RUSLE model is shown in Eq. (1), was used to predict the annual soil loss:

$$A = R * K * LS * C * P \qquad (1)$$

where A is the mean annual soil loss (t/ha/an), R is the rainfall erosivity factor (MJ/ha. h. an), K is the soil erodibility factor (t.h/ha. MJ.mm); LS = factor topography (L en m, S en %); C = cover management factor (dimensionless); and P = conservation practice factor (dimensionless).

2.4.1 R- Factor Estimation

The kinetic energy of rain, which strongly contributes to the removal of solid particles related to the intensity of rainfall and depends on the size and speed of falling drops as well as the average intensity of rainfall during 30 min, was used to estimate the climatic aggressiveness factor according to Wischmeier and Smith's (1978) formula. The monthly and yearly average rainfall data accessible at the reference climatic stations (Table 2 and Fig. 3) were used. Thus, Rango and Arnoldus (1987) produced the following applicable equation, Eq. (2):

$$Log(R) = 1.74 log \sum \frac{Pi^2}{P} + 1.29 \qquad (2)$$

where pi is the monthly rainfall, p is the annual rainfall, and R is the rainfall erosivity factor.

Table 2 Coordinates of the reference rainfall stations

ID	Stations	Coordinates X	Y	Altitude (m)
S1	Abadla	200	1295	250
S2	Aghbalou	27,615	8305	1070
S3	Chichaoua	18,153	1112	340
S4	Imin El Hammam	2414	724	770
S5	Nkouris	23,835	55	1100
S6	Sidi Hssain	2291	7017	1030
S7	Sidi Rahal	3031	1178	690
S8	Taferiat	29,125	1075	760
S9	Tahanaout	2559	804	925

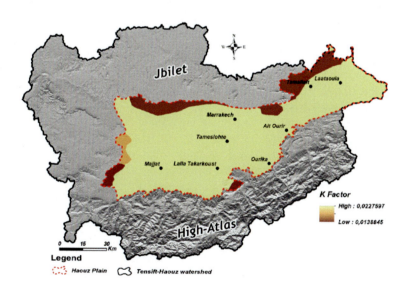

Fig. 3 Map of the spatial distribution of the K-factor

2.4.2 K-factor Estimation

The inherent vulnerability of the soil to erosion, as indicated by the soil erodibility factor (K) determined from profile parameters, was assessed using the soil type map obtained from Fig. 1 of the Digital Soil Map of the World (DSMW), as detailed in Eq. 3. This map was originally published by the Food and Agriculture Organization (FAO) in 2007. Additionally, the K factor (2000) was estimated using the Williams equation (Eq. (4)).

$$K = fcsand * fcl-si * forgc * fhisand * 0.1317 \qquad (3)$$

where, $fcsand$ a factor which lowers the K indicator in soils with high coarse-sand content and higher for soils with little sand, $fcl\text{-}si$ gives low soil erodibility factors for soils with high clay-to-silt ratios, $forgC$ reduces K values in soils with high organic carbon content, while $fhisand$ lowers K values for soils with extremely high sand content:

$$fcsand = 0.2 + 0.3e^{-0.256.ms.\left(1-\frac{msilt}{100}\right)} \qquad (4)$$

$$fcl - si = \left(\frac{msilt}{mc + msilt}\right)^{0.3} \qquad (5)$$

$$forgc = 1 - \frac{0.0256.orgC}{orgC + e^{(3.72-2.95.orgC)}} \qquad (6)$$

$$fhisand = 1 - \frac{0.7.\left(1-\frac{ms}{100}\right)}{\left(1-\frac{ms}{100}\right) + e^{-5.51+22.9.\left(1-\frac{ms}{100}\right)}} \qquad (7)$$

where, ms: : Percent sand content (0.05–2.0 mm diameter particles); $msilt$: Percent silt content (0.002–0.05 mm); mc: Percent clay content (<0.002 m) and $orgC$: Percent organic carbon content of the layer (%).

The generic soil unit information report included in the FAO-published soil type map is where the percentages of sand (ms), silt (msilt), clay (mc), and organic carbon (orgC) are taken from Table 3, which provides information on soil units and computations of csand, cl-si, orgC, and hisand for each type of soil in the research area.

2.4.3 LS-Factor Estimation

The LS factor illustrates the effect of slope steepness and length on soil erosion (Fig. 4). The relative slope length and steepness values, L and S, show how easily the 22.1 m long, 9% slope unit plot of Wischmeier and Smith (1978) is. The slope can be mathematically calculated for the original USLE using a digital elevation model;

Table 3 Soil types, K-values, and the computation of $f$csand, $f$cl-si, $f$orgC and $f$hisand of the Williams (2000) in study area

Soil code	ms (sand) Top soil %	msilt (Silt) Top soil %	mc (clay) Topsoil %	orgC oraganic carbon %	Fcsand	F cl-si	F orgc	F hisand	K usle	K
Xk10-2a	48,7	29,9	21,6	0,64	0,20,005	0,8495	0,9768	0,9993	0,1659	0,021,846
Bk10-2b	81,6	6,8	11,7	0,44	0,20,000	0,7406	0,9906	0,7185	0,1054	0,013,885
Re5-b	68,3	15,1	16,6	0,5	0,20,000	0,8005	0,9874	0,9634	0,1523	0,020,059
Xk4-2a	47,7	30,9	22,6	0,65	0,20,006	0,8482	0,9758	0,9994	0,1655	0,021,795
Jc13-2a	39,6	39,9	20,6	0,65	0,20,068	0,8826	0,9758	0,9999	0,1728	0,022,760
IL-Re-2c	58,9	16,2	24,9	0,97	0,20,000	0,7563	0,9272	0,9942	0,1394	0,018,364

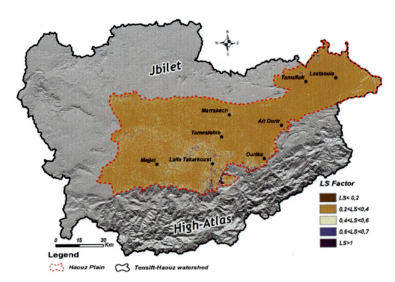

Fig. 4 Map of the topographic factor LS on the perimeter of the Haouz plain

however, the slope length must be measured or estimated. Slope length estimation is not possible owing to the variability, size, and topography of the topography and land use activities. Several formulae can be used to determine the combined LS factor (Wischmeier and Smith 1978). This study calculated the LS factor distributed over the field using the contributing surface unit method. With the aid of a tool from the hydrologic modeling extension of ArcGIS Spatial Analyst tools, the flow accumulation was extracted from the 30 m resolution set of (DEM) data. Moore and Burch (1986) and Panagos et al. (2015) provided methods for computing the LS factor, as shown in Eq. (8):

$$\text{LS} = (\text{Flowaccumulation} \times \frac{\text{cellsize}}{22.13})^{0.4} * (\frac{\sin(\text{slope})}{0.0896})^{1.3} \quad (8)$$

2.4.4 C- Factor Estimation

The C-factor describes how soil loss is affected by vegetation, management, and erosion control techniques (Wischmeier and Smith 1978; Mengistu et al. 2015). The C-factor was most strongly influenced by changes in land cover, which also caused a significant increase in soil loss. Its values vary from 1 for entirely bare land to 0 for entirely covered water or land surface (Mengistu et al. 2015). Because of geographical and temporal changes, numerous studies have classified land cover units using remotely sensed data to estimate C-factor values with extensive ground truth Teferi and Bewket (2009). In this study, we employed the Normalized Difference Vegetation Index (NDVI) (depicted in Figs. 6a–c) and utilized supervised classification techniques to ascertain the distribution of vegetation cover within the study area. This determination was made using a land use map (shown in Figs. 5a–c) derived from Landsat images captured on three different dates: OLI on 2020–09–11, ETM + on September 1, 2002, and TM on September 16, 1992. As detailed in Table 4, each land-use category was assigned a C-factor value based on information from available literature sources.

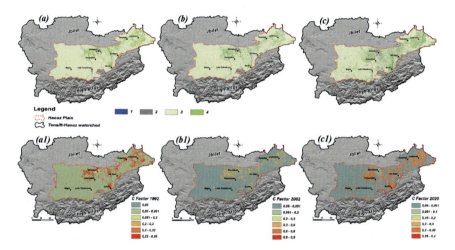

Fig. 5 Land use map of the study area in **a** 1992, **b** 2002, and **c** 2020. And a map of the spatial distribution of the C-factor in: **a1** 1992; **b1** 2002, and **c1** 2020

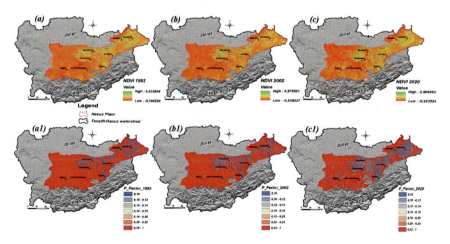

Fig. 6 NDVI-based vegetation extracted from the Landsat scene in **a** date 1992, **b** date 2002, **c** date 2020 and Map of the spatial distribution of the P-factor in **a1** 1992; **b1** 2002 and **c1** 2020

Table 4 Value and area in (Ha) of C factors for the different land cover types considered

Gride	LU/LC	Area (Ha) 1992	2002	2020	Factor C	Source
1	Hydrographic network + Eroded area	5125,70	5933,96	6413,02	0,18	Ganasri and Ramesh (2016), Hurni (1985), Meshesha et al. (2012)
2	Build up-land	4452,12	5749,75	8204,63	0,09	
3	Bare soil	636,095,20	614,697,30	520,042,00	0,05	
4	Agricultural land and vegetation	75,584,27	94,046,26	153,924,30	0,35	

Table 5 Percent area (Ha) and P-factor values suggested by Wischmeier and Smith (1978)

Land-use type	Slope (%)	Area (Ha) 1992	2002	2020	Factor P	Source
Agricultural land	0–5	62,260,7	74,556,7	121,509	0,10	Wischmeier and Smith (1978), Bewket and Teferi (2009) Ganasri and Ramesh, 2016)
	5–10	11,088,9	15,644,9	25,552,2	0,12	
	10–20	2025,82	3423,85	6014,92	0,14	
	20–30	156,488	321,99	647,358	0,19	
	30–50	37,427	72,31	146,564	0,25	
	50–100	0,193,088	1,51,489	3,84,236	0,33	
Another land	All	645,332	626,063	534,414	1	

2.4.5 P-factor Estimation

The P-factor shows the amount of soil lost owing to slope-side straight-line cropping once a conservation technique has been implemented (Wischmeier and Smith 1978; Meshesha et al. 2012). By changing the amount, pattern, slope, or direction of surface runoff, conservation techniques, including contour cropping, strip cropping, and terrace construction, primarily affect water erosion by lowering the volume and rate of runoff (Renard et al. 1997). The P-factor has a value between 0 and 1. Ganasri and Ramesh (2016). The research area was divided into agricultural and other land use groups to calculate the P-factors Wischmeier and Smith (1978) recommended. As stated in Table 5, agricultural land was divided into six different slope classes, each receiving a P-value, whereas all other land uses received a value of 1 (Fig. 6).

2.5 Results Validation

Selected field observations were performed to evaluate the consistency and coherence of model outputs. The validity of the model outputs was compared to numerical data outputs from similar published research spanning the entire Tensift Watershed because there were no prior case studies specific to the study region. The input layers were designed to employ the best available land use, topography, precipitation, and soil data to reduce errors in soil erosion risk calculations.

3 Results and Discussion

3.1 Spatial Distribution of RUSLE Factors

Under the WGS 1984 UTM Zone 29N reference system, all layers of the different model factors (R, K, LS, C, and P) were arranged in a grid with 30 × 30 m cells. By incorporating several variables into the RUSLE model in a GIS context, it was possible to estimate annual soil loss in the Haouz Plain. The geographical distribution of the soil loss in the Haouz Plain is shown in Fig. 7. The estimated soil loss values are presented in Table 6. Soil loss map for 1992 (Fig. 7a), 2002 (Fig. 7b), and 2020 (Fig. 7c) was divided into four categories: very low, moderate, high, and very high. The results show that water erosion in the study area is low to medium; it also shows a decrease for the moderate and extremely high soil loss classes and an increase in the high soil loss class between 1992 and 2020, due to changes in the C-factor. (4) and Figs. (5a1–c1). With the 2020 C-table factor, the soil loss assessment indicated that 88.63% of the Haouz Plain still experienced very low and moderate soil loss (<26 t/ha/yr), whereas 11.37% was subject to high to extremely high soil loss (>35 t/ha/yr). The average annual soil loss is low in the Haouz Plain and high in the High Atlas. The average annual soil losses were 13.03, 12.30, and 14.90 t/ha/year, respectively in 1992, 2002, and 2020, respectively, which is consistent with the results obtained in the Tensift watershed by the author (Meliho et al., 2018). Utilizing precipitation erosivity, soil types, slope, and land use, the spatial pattern obtained by the RUSLE model was used to assess soil erosion by water. Water erosion is accelerated in places with particularly steep bare slopes and high precipitation erosivity.

Furthermore, several studies have found that the soil erosion rate is more responsive to precipitation, supporting the findings of this study. In 1992, the southern part of the Haouz plain (referred to as central Haouz) exhibited an erosivity of 36.71

Fig. 7 Spatial distribution of soil erosion in the Haouz plain in **a** 1992 **b** 2002 **c** 2020

Table 6 Soil erosion severity classes by total land area in 1992, 2002, and 2020

Year	Soil loss t/ha/yr	Severity class	Area (Ha)	Percent %
2020	<12	Very low	383,530,00	53,25
	12 26	Moderate	199,773,00	27,74
	26–35	High	128,229,31	17,80
	>35	Very high	8708,79	1,21
2002	<12	Very low	340,652,33	47,30
	12 26	Moderate	323,859,00	44,97
	26–35	High	51,775,30	7,19
	>35	Very high	3954,47	0,55
1992	<12	Very low	527,501,00	73,24
	12 26	Moderate	171,853,00	23,86
	26–35	High	16,996,80	2,36
	>35	Very high	3890,30	0,54

MJ.mm/ha.hr.year, with a gradual decrease in erosivity as one moved towards the northwest, as indicated in Fig. 8. The lowest recorded R-value in that year was 17.40 MJ.mm/ha.hr.year. In contrast, the erosivity in the eastern Haouz Plain (referred to as the eastern Haouz Plain) was much higher in 2002, as depicted in Fig. 8b, with a value of 85.97 MJ.mm/ha.hr.year. Similar to central Haouz, erosivity tended to diminish as one moved towards the western part of the plain. Finally, in 2020, Fig. 8c shows that the western portion of the Haouz plain (western Haouz) has an erosivity of 48.70 MJ.mm/ha.hr.year and tends to drop towards the eastern area. This is the lowest R-value measured, 30.40 MJ.mm/ha.hr.year. 5.70 MJ.mm/ha.hr.year is the lowest R-value recorded. The significantly higher R-value is linked to a higher sensitivity to erosion and a greater capacity of rainwater to erode the soil from the surface.

The K-factor values range from 0.013 to 0.23 t.h/ha. MJ.mm (Fig. 3). The soil with the highest K-factor value was susceptible to erosion and dominated by silt and clayey silts. The western part of the Haouz Plain shows lower K-factor values and is, therefore, less susceptible to erosion. The LS factor map in Fig. 4 shows values ranging from 0.01 1. More than 83% of the research area is covered by a value between 0.2 and 0.4, which has a gradual slope and favors a low to moderate soil loss. The middle portion of the research region features high terrain that is directly connected to the High Atlas range of the Tensift watershed, increasing the area's potential vulnerability to soil erosion by water. High LS values correlated with this connection. The C-factor indicates the impact of vegetation and other types of soil surface cover on soil erosion and offers insight into land use issues crucial for development planning. The C-factor map showed regional and temporal fluctuations between 1992 (Fig. 5a1), 2002 (Fig. 5b1), and 2020 (Fig. 5c1); the values typically varied from 0.05 to 0.35. The areas that were cultivated, and hence, more susceptible to soil erosion had the highest C-factor. The fluctuations in P-factor values between

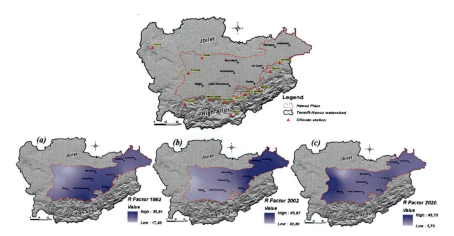

Fig. 8 Location map of weather stations where precipitation data were retrieved for this study and map of the spatial distribution of the R-factor in **a** 1992, **b** 2002, and **c** 2020

1992 (Fig. 6a1), 2002 (Fig. 6b1), and 2020 (Fig. 6) are depicted (Fig. 6c1). It shows the effect of several erosion management techniques on the erosion rate. The croplands used on slope classes larger than 50%, where enhanced sensitivity to soil erosion was noticeable and where high values of the P-factor were found.

3.2 Evaluation of the Evolution of Annual Soil Losses

Equation 1 was employed within ArcGIS 10.5.1 environment to generate potential soil erosion maps for 1992, 2002, and 2020. The estimated soil loss across the Haouz Plain exhibited a range of variations. In 1992, it ranged from 26 t ha^{-1} yr^{-1} in the plain area to 35 t ha^{-1} yr^{-1}, covering 2.90% of the total area. In 2002, it ranged from 26 t ha^{-1} yr^{-1} in the plain to 35 t ha^{-1} yr^{-1} for an area encompassing 7.74%. This pattern persisted in 2020 with a consistent loss rate of 19% across the region. These findings underscore the regional heterogeneity of soil loss within the Haouz Plain. This heterogeneity can primarily be attributed to climate change, land cover variations, and suboptimal land management practices. Specifically, soil loss averaged less than 12 t ha^{-1} yr^{-1} in 1992, whereas it ranged between 12 and 26 t ha^{-1} yr^{-1} between 2002 and 2020.

4 Conclusions

This research provides a broad picture of how soil erosion in the Haouz Plain changed between 1992, 2002, and 2020 due to climatic change and sociospatial development. In the study region, soil erosion caused by water is a significant issue. The findings of this study show that human actions, such as disregarding land conservation and spatial variance in rainfall, are the primary causes of soil erosion in the region. The findings also indicate 4.84 and 11.28% changes between 1992 and 2002, and 2002 and 2020, respectively. It is clear that the Haouz Plain does not need conservation measures to be implemented from the geographical distribution pattern of the erosion risk map for the years 1992, 2002, and 2020. Therefore, a risk map for water erosion can assist direct expenditures to prevent soil losses and set priorities for efficient repair measures. The soil erosion risk assessment also indicated places with a high risk of erosion, where efforts and resources should be focused to prevent future erosion. The approaches RUSLE, RS, and GIS adopted by this present chapter offer potentially useful approaches for identifying areas likely to be most susceptible to soil water erosion over time and prioritizing areas for effective, sustainable land management planning based on erosion severity classes. We recommend using the same approaches for similar cases in MENA regions.

5 Recommendations

In light of this chapter, several recommendations can be made:

- Raise awareness among stakeholders, policymakers, and local communities about the significant issue of water-induced soil erosion in the Haouz Plain and its detrimental effects on the environment and agricultural productivity. Emphasize the need for collective action to address the causes of soil erosion and promote sustainable land management practices.
- Advocate for the implementation of land conservation measures, considering the identified primary causes of soil erosion in the region. Encourage the adoption of strategies such as terracing, contour plowing, and the use of cover crops to mitigate soil erosion and promote soil health.
- Highlight the importance of considering both climatic change and sociospatial development in soil erosion assessments. Encourage the integration of climate change adaptation measures and land-use planning strategies to minimize erosion risks in the face of changing environmental and socio-economic conditions.
- Promote the use of risk maps for water erosion as a tool for directing expenditures and setting priorities for efficient repair measures. Encourage the allocation of resources to areas with a high risk of erosion, based on the findings of the erosion risk assessment.

- Encourage the adoption of the RUSLE (Revised Universal Soil Loss Equation), remote sensing (RS), and Geographic Information System (GIS) approaches for similar cases in the MENA (Middle East and North Africa) regions. Highlight the potential usefulness of these approaches in identifying areas susceptible to soil water erosion and prioritizing targeted land management planning based on erosion severity classes.
- Facilitate knowledge exchange and collaboration among researchers, practitioners, and policymakers in the MENA regions to share experiences, best practices, and lessons learned in addressing soil erosion challenges. Foster regional cooperation to develop context-specific solutions and strategies that consider local environmental and socio-economic conditions.
- Seek funding and resources to support ongoing monitoring and assessment of soil erosion dynamics in the Haouz Plain and other vulnerable regions. Continuously monitor changes in land cover, climatic conditions, and socio-spatial development to update risk maps and inform adaptive management strategies.

These recommendations aim to address soil erosion issues in the Haouz Plain and similar regions by promoting awareness, implementing conservation measures, integrating climate change adaptation and land-use planning, utilizing risk maps, sharing knowledge, and mobilizing resources for sustainable land management.

Funding: This research received no external funding.

DataAvailability Data Availability Statement The data presented in this study are available on request from the corresponding author.

Conflicts of Interest The authors declare no conflict of interest.

References

Alexakis D, Hadjimitsis D, Agapiou A (2013) Integrated use of remote sensing, GIS and precipitation data for the assessment of soil erosion rate in the catchment area of Yialias in Cyprus. Atmos Res 131:108–124. https://doi.org/10.1016/j.atmosres.2013.02.013

Al-Quraishi AMF (2003) Soil erosion risk prediction with RS and GIS for the Northwestern Part of Hebei Province, China. Pakistan J Appl Sci 3(10–12):659–669

Al-Quraishi AMF, Negm AM (2020) Environmental remote sensing and GIS in Iraq. Springer Water, Springer, Cham. ISBN: 978–3–030–21343–5

Anys H, Bonn F, Merzouk A (1992) Mapping and calculating water erosion using remote sensing and GIS, the case of the Oued Aricha watershed (Settat, Morocco). Géo Observateur 2:37–51. http://hdl.handle.net/11143/7804

Bag R, Mondal I, Dehbozorgi M, Rakhohori BSP, Das DN, Bandyopadhyay J, Pham QB, Al-Quraishi AMF, Nguyen XC (2022) Modelling and mapping of soil erosion susceptibility using machine learning in a tropical hot sub-humid environment. J Clean Prod 364:132428. https://doi.org/10.1016/j.jclepro.2022.132428

Bammou Y, Bouskri I, Brahim B, Kader S, Brahim I, Abdelkrim B, Spalević V (2023b) The contribution of the frequency ratio model and the prediction rate for the analysis of landslide risk in the Tizi N'tichka area on the national road (RN9) linking Marrakech and Ouarzazate. CATENA 232:107464. https://doi.org/10.1016/j.catena.2023.107464

Bammou Y, Benzougagh B, Bensaid A, Igmoullan B, Al-Quraishi AMF (2023a) Mapping of current and future soil erosion risk in a semi-arid context (Haouz plain-Marrakech) based on CMIP6 climate models, the analytical hierarchy process (AHP) and RUSLE. Model Earth Syst Environ 1–14. https://doi.org/10.1007/s40808-023-01845-9

Benmansour M, Mabit L, Nouira A, Moussadek R, Bouksirate H, Duchemin M, Benkdad A (2013) Assessment of soil erosion and deposition rates in a Moroccan agricultural feld using fallout 137Cs and 210Pbex. J Environ Radioact 115:97–106. https://doi.org/10.1016/j.jenvrad.2012.07.013

Benzer N (2010) Using the geographical information system and remote sensing techniques for soil erosion assessment. Pol J of Environ Stud 19(5):881–886

Benzougagh B, Dridri A, Boudad L, Kodad O, Sdkaoui D, Bouikbane H (2017) Evaluation of natural hazard of Inaouene watershed river in Northeast of Morocco: application of morphometric and geographic information system approaches. Int J Innov Appl Stud 19(1):85

Benzougagh B, Baamar B, Dridri A, Boudad L, Sadkaoui D, Mimich K (2020a) Relationship between landslide and morpho-structural analysis: a case study in Northeast of Morocco. Appl Water Sci 10(7):1–10. https://doi.org/10.1007/s13201-020-01258-4

Benzougagh B, Sarita GM, Dridri A, Boudad L, Sadkaoui D, Mimich K, Khaled MK (2020b) Mapping of soil sensitivity to water erosion by RUSLE model: case of the Inaouene watershed (Northeast Morocco). Arab J Geosci 13(21):1–15. https://doi.org/10.1007/s12517-020-06079-y

Benzougagh B, Dridri A, Boudad L, Baamar B, Sadkaoui D, Khedher KM (2022) Identification of critical watershed at risk of soil erosion using morphometric and geographic information system analysis. Appl Water Sci 12:1–20. https://doi.org/10.1007/s13201-021-01532-z

Benzougagh B, Boudad Larbi P, Dridri Abdallah P, Driss S (2016) Utilization of GIS in morphometric analysis and prioritization of sub-watersheds of Oued Inaouene (Northeast Morocco). Europ Sci J 12(6). https://doi.org/10.19044/esj.2016.v12n6p266

Benzougagh B, Meshram SG, Fellah BE, Mastere M, El Basri M, Ouchen I, Turyasingura B (2023) Mapping of land degradation using SAM approach: the case of Inaouene watershed (Northeast Morocco). Model Earth Syst Environ https://doi.org/10.1007/s40808-023-01711-8

Bewket W, Teferi E (2009) Assessment of soil erosion hazard and prioritization for treatment at the watershed level: case study in the Chemoga Watershed, Blue Nile Basin. Ethiopia Land Degrad Dev 20(2009):609–622. https://doi.org/10.1002/ldr.944

Bhattacharya RK, Bhattacharya ND, Chatterjee K (2020) Das Sub-basin prioritization for assessment of soil erosion susceptibility in Kangsabati, a plateau basin: a comparison between MCDM and SWAT models. Sci Total Environ 734:139474. https://doi.org/10.1016/j.scitotenv.2020.139474

Bonn F (1998) Spatialization of soil erosion models using remote sensing and GIS: possibilities, errors and limitations. Drought 9(3):185–192

Damnati B, Radakovitch O, Ibrahimi S (2004) Use of cesium-137 for estimating erosion rates in a watershed in northern Morocco. Drought 15(2):195–199

Demirci A, Karaburun A (2012) Estimation of soil erosion using RUSLE in a GIS framework: a case study in the Buyukcekmece lake watershed, northwest Turkey. Environ Earth Sci 66:903–913. https://doi.org/10.1007/s12665-011-1300-9

Efe R, Ekinci D, Curebal I (2008) Erosion analysis of Sahin Creek watershed (NW of Turkey) using GIS based on Rusle (3d) method. J Appl Sci 8:49–58. https://doi.org/10.3923/jas.2008.49.58

Elalaoui A, Marrakchi C, Fekri A, Maimouni S, Aradi M (2015) Implementation of a qualitative model for mapping areas at risk of water erosion in the Atlas Range: the case of the upstream Tessaoute watershed. Software (High Atlas, Central, Morocco, USA). Eur Sci J 11(29):106–121

Farhan Y, Nawaiseh S (2015) Spatial assessment of soil erosion risk using RUSLE and GIS techniques. Environ Earth Sci 74:4649–4669. https://doi.org/10.1007/s12665-015-4430-7

Ganasri BP, Ramesh H (2016) Assessment of soil erosion by RUSLE model using remote sensing and GIS—a case study of Nethravathi Basin. Geosci Front 7:953–961. https://doi.org/10.1016/j.gsf.2015.10.007

Gourfi AD, Daoudi L, Shi Z (2018) The assessment of soil erosion risk, sediment yield and their controlling factors on a large scale: Example of Morocco. J Afr Earth Sc 147:281–299. https://doi.org/10.1016/j.jafrearsci.2018.06.028

Haregeweyn N, Atsushi T, Jean P, Mitsuru T, Dereg TM, Ayele AF, Jean N, Enyew A (2017) Comprehensive assessment of soil erosion risk for better land use planning in river basins: a case study of the Upper Blue Nile River Sci. Total Environ 574:95–108. https://doi.org/10.1016/j.scitotenv.2016.09.019

Haregeweyn N, Tsunekawa A, Nyssen J, Poesen J, Tsubo M, Meshesha DT, Schütt B, Adgo E, Tegegne F (2015) Soil erosion and conservation in Ethiopia: a review. Prog Phys Geogr 39(6):750–774. https://doi.org/10.1177/0309133315598725

Hossini H, Karimi H, Mustafa YT, Al-Quraishi AMF (2022). Role of effective factors on soil erosion and land degradation: a review. In: Al-Quraishi AMF, Mustafa YT, Negm AM (eds) Environmental degradation in Asia. Earth and environmental sciences library. Springer, Cham. https://doi.org/10.1007/978-3-031-12112-8_11

Hurni H, Hans H, Karl H, Brigitte P, Hanspeter L (2008) Soil erosion and conservation in global agriculture. In: Vlek PLG (ed) Land use and soil resources. Springer, Netherlands, pp 41–71. https://doi.org/10.1007/978-1-4020-6778-5_4

Jiang S, Alves A, Rodrigues F, Ferreira J Jr, Pereira FC (2015) Mining point-of-interest data from social networks for urban land use classification and disaggregation. Comput Environ Urban Syst 53:36–46. https://doi.org/10.1016/j.compenvurbsys.2014.12.001

Jinren RN, Yingkui KL (2003) Approach to soil erosion assessment in terms of land-use structure changes. J Soil Water Conserv 58(3):158–169

Kader S, Novicevic R, Jaufer L (2022) Soil management in sustainable agriculture: analytical approach forthe ammonia removal from the diary manure. Agric For 68(4):69–78. https://doi.org/10.17707/AgricultForest.68.4.06

Kader S, Raimi MO, Spalevic V, Iyingiala AA, Bukola RW, Jaufer L, Butt TE (2023) A concise study on essential parameters for the sustainability of Lagoon waters in terms of scientific literature. Turkish J Agric For 47(3):288–307. https://doi.org/10.55730/1300-011X.3087

Kader S, Jaufer L, Bashir O, Olalekan Raimi M (2023) A comparative study on the stormwater retention of organic waste substrates biochar, sawdust, and wood bark recovered from Psidium Guajava L. Species. Agric For 69(1):105–112. https://doi.org/10.17707/AgricultForest.69.1.09

Kouli M, Soupios P, Vallianatos F (2009) Soil erosion prediction using the revised universal soil loss equation (RUSLE) in a GIS framework, Chania, Northwestern Crete. Greece Environ Geol 57:483–497. https://doi.org/10.1007/s00254-008-1318-9

Lal R (2001) Soil degradation by erosion. Land Degrad Dev 12:519–539. https://doi.org/10.1002/ldr.472

Markhi A, Laftouhi NE, Soulaimani A, Fniguire F (2015). Quantification and evaluation of water erosion using the RUSLE model and integrated deposition in a GIS. Application in the n'Fis watershed in the Marrakech High Atlas (Morocco). Eur Sci J 11(29):340–356

Mehwish M, Nasir MJ, Raziq A, Al-Quraishi AMF (2024) Soil erosion vulnerability and soil loss estimation for Siran River watershed, Pakistan: an integrated GIS and remote sensing approach. Environ Monit Assess 196:104. https://doi.org/10.1007/s10661-023-12262-x

Meliho M, Khattabi A, Mhammdi N (2018) A GIS-based approach for gully erosion susceptibility modelling using bivariate statistics methods in the Ourika watershed, Morocco. Environ Earth Sci 77:1–14. https://doi.org/10.1007/s12665-018-7844-1

Meliho M, Khattabi A, Mhammdi N, Hongming Z (2016) Mapping of water erosion risks using the revised universal soil loss equation, remote sensing, and GIS in the Ourika watershed (High Atlas, Morocco). Europ Sci J 12:227–297. https://doi.org/10.19044/esj.2016.v12n32p277

Meliho M, Khattabi A, Mhammdi N (2020) Spatial assessment of soil erosion risk by integrating remote sensing and GIS techniques: a case of Tensift watershed in Morocco. Environ Earth Sci 79:207 https://doi.org/10.1007/s12665-020-08955-y

Mengistu D, Bewket W, Lal R (2015) Soil erosion hazard under the current and potential climate change induced loss of soil organic matter in the upper Blue Nile (Abay) River Basin. Ethiop ReaearchGate 137–163. https://doi.org/10.1007/978-3-319-09360-4_7

Merzouk A, Fenjiro I, Laouina A (1996) Mapping the evolution of erosion patterns in the Western Rif (Morocco): a multidate study using watershed GIS. In: De Noni G, Lamachère Georges DN, Lamachère JM, Roose E (eds) Surface conditions and runoff and erosion risks. Bulletin–Erosion Network, vol 16, pp 444–456

Merzouki T (1992) Diagnosis of siltation in large Moroccan Dams. Moroccan J Civil Eng 38:46–50

Meshesha DT, Tsunekawa A, Tsubo M, Haregeweyn N (2012) Dynamics and hotspots of soil erosion and management scenarios of the Central Rift Valley of Ethiopia Int. J Sediment Res 27(2012):84–99. https://doi.org/10.1016/S1001-6279(12)60018-3

Molla T, Sisheber B (2017) Estimating soil erosion risk and evaluating erosion control measures for soil conservation planning at Koga watershed in Ethiopia's highlands Solid. Earth 8(2017):13–25. https://doi.org/10.5194/se-8-13-2017

Moore ID, Burch GJ (1986) (1986) Physical basis of the length-slope factor in the universal soil loss equation Soil Science. Soc Am J 50(5):1294–1298. https://doi.org/10.2136/sssaj1986.03615995005000050042x

Neupane B, Mandal U, Al-Quraishi A, Ozdemir, M, Rai R (2023) Environmental threat of soil erosion in the Gwang Khola watershed, Chure Region of Nepal. Iraqi Geol J 56:194–206. https://doi.org/10.46717/igj.56.1e.14ms-2023-5-24

Panagos P, Ballabio C, Borrelli P, Meusburger K, Klik A, Rousseva S, Alewell C (2015) Rainfall erosivity in Europe. Sci Total Environ 511:801–814. https://doi.org/10.1016/j.scitotenv.2015.01.008

Pandey A, Chowdary VM, Mal BC (2007) Identifcation of critical erosion prone areas in the small agricultural watershed using USLE, GIS and remote sensing. Water Resour Manag 21:729–746. https://doi.org/10.5194/10.1007/s11269-9061-z

Piqué A, Soulaimani A, Hoepffner C (2007) Geology of Morocco, Geode. Imprimerie El Watanya, Marrakesh, Morocco

Prasannakumar V, Shiny R, Geetha N, Vijith H (2011) Spatial prediction of soil erosion risk by Remote Sensing, GIS and RUSLE approach: a case study of Siruvani River Watershed in Attapady Valley, Kerala, India. Environ Earth Sci 46:965–972. https://doi.org/10.1017/s12665-011-0913-3

Prasannakumar V, Vijith H, Abinod S, Geetha N (2012) Estimation of soil erosion risk within a small mountainous sub-watershed in Kerala, India, using revised universal soil loss equation (RUSLE) and geo-information technology. Geosci Front 3:209–215. https://doi.org/10.1016/j.gsf.2011.11.003

Rango A, Arnoldus HMJ (1987). Watershed Management. In: FAO Technical Papers, pp 1–11

Renard KG, Foster GR, Weesies GA, McCool DK, Yoder DC (1997) Predicting soil erosion by water: a guide to conservation planning with the revised universal soil loss equation (RUSLE). Agriculture Handbook No. 703, USAD-ARS, Washington

Rosas MA, Gutierrez RR (2019) Assessing soil erosion risk at national scale in developing countries: the technical challenges, a proposed methodology, and a case history. Sci Total Environ. https://doi.org/10.1016/j.scitotenv.2019.135474

Rozos D, Skilodimou HD, Loupasakis C, Bathrellos GD (2013) Application of the revised universal soil loss equation model on landslide prevention. An example from N. Euboea (Evia) Island, Greece. Environ Earth Sci 70:3255–3266. https://doi.org/10.1007/s12665-013-2390-3

Sestras P, Mircea S, Roşca S, Bilaşco Ş, Sălăgean T, Dragomir LO, & Kader S (2023) GIS based soil erosion assessment using the USLE model for efficient land management: a case study in an area with diverse pedo-geomorphological and bioclimatic characteristics. Notulae Bot Horti Agrobotan Cluj-Napoca 51(3):13263–13263. https://doi.org/10.15835/nbha51313263

Tahiri M, Tabyaoui H, El Hammichi F, Tahiri A, El Haddi H (2014) Evaluation and Quantification of Erosion and Sedimentation from RUSLE, MUSLE and Deposition Models Integrated in a GIS. Application to the Oued Sania Subbasin (Tahaddart Basin, North-West Rif, Morocco). Eur J Sci Res 125(2):157–178

Teng M, Huang C, Wanga P, Zeng L, Zhoua Z, Xiao W, Huang Z, Liu C (2019). Impacts of forest restoration on soil erosion in the three gorges reservoir area, China. Sci Total Environ 697:134164. https://doi.org/10.1016/j.scitotenv.2019.134164

Wischmeier WH, Smith DD (1978) Predicting rainfall erosion losses: a guide to conservation planning. Department of Agriculture, Science and Education Administration

Wolka K, Tadesse H, Garedew E, Yimer F (2015) Soil erosion risk assessment in the Chaleleka wetland watershed, Central Rift Valley of Ethiopia Environ. Syst Res 4(5). https://doi.org/10.1186/s40068-015-0030-5

GIS-Based Erosion Potential Method (EPM) for Soil Degradation Evaluation: A Case Study the Northeast of Morocco

Mohammed El Brahimi, Brahim Benzougagh, Mohamed Mastere, Bouchta El Fellah, Ayad M. Fadhil Al-Quraishi, Najia Fartas, and Khaled Mohamed Khedhe

Abstract Soil degradation is a significant environmental concern, posing severe challenges to sustainable agriculture and land management. The Northeast region of Morocco is particularly susceptible to soil erosion, owing to a combination of factors, including topography, geomorphology, geology, climate, and land use practices. This chapter introduces a GIS-based Erosion Potential Method (EPM) designed to assess soil degradation in the Oued Amter watershed in Northeast Morocco. This study developed a comprehensive erosion risk assessment model by integrating various spatial data, such as topography, rainfall, land cover, and soil properties. The results of this study reveal that soil loss varies significantly, ranging from 55.91 to 44,088.33 m^3/km^2/year across the Oued Amter watershed. These findings

---

M. El Brahimi · B. Benzougagh (✉) · M. Mastere · B. E. Fellah · N. Fartas
Geophysics and Natural Hazards Laboratory, Department of Geomorphology and Geomatics (D2G), Scientific Institute, Mohammed V University in Rabat, Avenue Ibn Batouta, Agdal, PO Box 703, 10106 Rabat-City, Morocco
e-mail: brahim.benzougagh@is.um5.ac.ma

M. El Brahimi
e-mail: mohammed_elbrahimi2@um5.ac.ma

M. Mastere
e-mail: mohamed.mastere@is.um5.ac.ma

B. E. Fellah
e-mail: bouchta.elhellah@um5.ac.ma

N. Fartas
e-mail: najia_fartas@um5.ac.ma

A. M. F. Al-Quraishi
Petroleum and Mining Engineering Department, Faculty of Engineering, Tishk International University, Erbil 44001, Kurdistan Region, Iraq
e-mail: ayad.alquraishi@tiu.edu.iq; ayad.alquraishi@gmail.com

K. M. Khedhe
Department of Civil Engineering, College of Engineering, King Khalid University, Abha 61421, Saudi Arabia
e-mail: kkhedher@kku.edu.sa

© The Author(s), under exclusive license to Springer Nature Switzerland AG 2024
A. M. F. Al-Quraishi and Y. T. Mustafa (eds.), *Natural Resources Deterioration in MENA Region*, Earth and Environmental Sciences Library,
https://doi.org/10.1007/978-3-031-58315-5_8

underscore the urgent need to address the region's soil conservation and land use planning. Policymakers, land managers, and researchers can use EPM outcomes to make informed decisions and implement effective measures to safeguard the soil and promote sustainable land use practices.

Keywords Water erosion · Geographical information system · Remote sensing · EPM module · Oued Amter watershed

1 Introduction

Water erosion is a major cause of soil degradation worldwide and has been a significant environmental problem for centuries (Eswaran et al. 2019; Bag et al. 2022; Benzougagh et al. 2022; Bammou et al. 2023). This is defined as the detachment and transportation of soil particles by water. Water erosion occurs when rainfall, snowmelt, or irrigation water hits the soil surface, causing soil particles to detach, transport, and deposit elsewhere (Morgan 2009; Benzougagh et al. 2016; Ougougdal et al. 2020; Elliot and Flanagan 2023; Mehwish et al. 2024). The impacts of water erosion are significant, and include soil nutrient depletion, reduced crop yields, and increased sedimentation in water bodies. Water erosion significantly threatens agricultural productivity and environmental sustainability (Al-Quraishi 2003; Benzougagh et al. 2017, 2020a, b; Bazhenova et al. 2022). Several factors influence the rate and extent of water erosion, including the rainfall intensity, soil type, slope gradient, vegetation cover, and land use. The rainfall intensity determines how much water hits the soil surface and affects the erosion rate (Zhao and Hou 2019; Benzougagh et al. 2020a, b; Neupane et al. 2023; Chen et al. 2023).

Soils with lower infiltration rates are more susceptible to water erosion because they cannot absorb large amounts of water, which leads to surface runoff and soil detachment. The slope gradient affects the velocity of surface runoff, with steeper slopes leading to faster water flow, which increases erosion rates. Vegetation cover plays a crucial role in protecting the soil surface from water erosion by reducing the impact of rainfall and increasing the infiltration rates (Liu et al. 2020; El Brahimi et al. 2022; Hossini et al. 2022; Singh et al. 2023). Finally, land use and management practices such as tillage, crop rotation, and grazing can exacerbate or mitigate water erosion (Wassie 2020; Ghosh et al. 2022; Van Huyssteen and du Preez 2023).

Various management practices, including conservation tillage, terracing, contour farming, cover crops, and agroforestry, can be implemented to prevent or mitigate water erosion (Guisan and Thuiller 2005; Akhtar and Riaz 2017). These practices aim to reduce surface runoff and soil detachment, increase infiltration rates, and protect the soil surface from the effects of rainfall (Alghamdi 2018). The effectiveness of these practices varies depending on local soil and climatic conditions, but they have been demonstrated to reduce water erosion rates significantly. Water erosion risk assessment is essential for identifying areas vulnerable to water erosion and for implementing appropriate management practices to prevent or mitigate erosion (Al-Quraishi 2004; Mihi et al. 2020; Kebede et al. 2023; Hagras 2023). Several methods have been developed to assess water erosion risk, ranging from empirical to modeling approaches.

– Empirical Methods:

Empirical methods involve using simple field-based techniques to assess the water erosion risk based on the presence or absence of specific factors that influence erosion rates. One of the most common empirical methods is the Universal Soil Loss Equation (USLE) (Uddin et al. 2018; Talchabhadel et al. 2020; Kader et al. 2022; Kumar et al. 2023), which estimates the average annual soil loss due to water erosion based on six key factors: rainfall, soil erodibility, slope length and steepness, vegetation cover, and management practices. Other empirical methods include the Revised Universal Soil Loss Equation (RUSLE), Morgan-Morgan-Finney (MMF) model, and Water Erosion Prediction Project (WEPP) (Zema et al. 2020; Svoray 2022; Halder 2023).

– Modeling Methods:

Modeling methods use computer-based models to simulate erosion rates and identify areas at high risk of erosion (Girma and Gebre 2020; Bagheri et al. 2023). These methods require extensive data input, including rainfall data, soil properties, topography, vegetation cover, and management practices. One of the most widely used modeling methods is the Soil and Water Assessment Tool (SWAT), which simulates water erosion processes at the watershed scale by integrating hydrological, agricultural, and environmental processes (Wang et al. 2019; Akoko et al. 2021; Aloui et al. 2023). Other modeling methods include the Revised SWAT (SWAT +) method (Yen et al. 2019; Senent-Aparicio et al. 2021; Llanos-Paez et al. 2023), and Soil and Water Assessment Tool with Geospatial Information (SWAT-G) (Meshesha et al. 2021; Föeger et al. 2022; El Ghoul et al. 2023), and Water Erosion Prediction Project model for arid and semiarid lands (WEPPAL) (Chen et al. 2021; Li et al. 2022; Gao et al. 2023).

– Remote Sensing Methods:

Remote sensing methods involve the use of satellite or aerial imagery to identify areas that are at a high risk of water erosion based on specific indicators (Schaefer and Thinh 2019; Cerbelaud et al. 2022; Danchenkov et al. 2023), including vegetation cover, soil moisture, and topography. These methods provide a rapid and cost-effective approach to assessing erosion risk over large areas but require calibration and validation against field-based data. Some of the most commonly used remote sensing methods for water erosion risk assessment include the Normalized Difference Vegetation Index (NDVI) (Almagro et al. 2019; Sharma et al. 2022; De Oliveira Oliveira et al. 2023), Soil Adjusted Vegetation Index (SAVI), and Topographic Wetness Index (TWI) (Berhanu and Bisrat 2018; Zhi et al. 2021; Das et al. 2023).

Soil erosion is a pressing issue in many parts of the world, and Rif in Morocco is no exception. Rif is a mountainous region in northern Morocco, where the study area is located, and is known for its rich culture and diverse landscapes (El Brahimi et al. 2022; Kader et al. 2023b). However, this region faces the threat of soil erosion due to various natural and anthropogenic factors. Soil erosion is a process in which the top layer of soil is removed by wind, water, or other agents, resulting in the loss of valuable nutrients and degradation of land (Fartas et al. 2022; Kader et al. 2023a). In extreme cases, this process can lead to reduced agricultural productivity, biodiversity loss, and desertification. In RIF, soil erosion is a major concern because of the region's unique topography, fragile ecosystems, and the overuse of land for agriculture and forestry (Benzougagh et al. 2022; Lamane et al. 2022). Therefore, it is essential to understand the causes and impacts of soil erosion in RIF and to implement effective strategies to mitigate the problem and ensure sustainable land use practices for future generations.

The objective of this study was to use the Erosion Potential Method (EPM) in the Oued Amter watershed, which is located in the Rif of Morocco. EPM is an empirical method used to assess water erosion risk based on the presence or absence of specific factors that influence erosion rates (Pandey et al. 2021; Micić Ponjiger et al. 2023). It is a relatively simple and cost-effective method that can be used to evaluate erosion potential across a range of scales, from small catchments to larger watersheds. EPM is based on criteria that assess soil erodibility, vegetation cover, slope length and steepness, and climate. These criteria are scored on a scale of 0 to 10, with higher scores indicating a greater erosion potential (Pourghasemi et al. 2021; Abbate et al. 2023). The scores for each criterion are then multiplied to obtain a single erosion potential index, which can be used to identify areas with high, moderate, and low erosion potential.

The EPM results can be used to identify areas of high, moderate, and low erosion potential, and can be overlaid onto maps to provide visual representations of erosion risk. The results can also be used to prioritize areas for erosion control measures such as planting vegetation, constructing terraces or berms, or implementing conservation tillage practices. In addition to identifying erosion-risk areas, EPM can also be used to track changes in erosion potential over time. By using the EPM to evaluate the erosion potential before and after the implementation of erosion control practices, land managers can assess the effectiveness of these practices and make adjustments as needed.

However, it is essential to note that the EPM only estimates the erosion potential and does not consider the complex interactions between various factors that influence erosion rates, such as land-use practices, soil management, and climate variability. Therefore, it should be combined with other erosion risk assessment methods, local knowledge, and experience. Overall, the Erosion Potential Method is a valuable tool for assessing the water erosion risk and identifying areas that require management practices to prevent or mitigate erosion. It is a relatively simple and cost-effective method that land managers and policymakers can use to prioritize erosion control efforts cost-effectively.

2 Materials and Methods

2.1 Study Area

The Oued Amter watershed, situated in the province of Chefchaouen, northwest Morocco (Fig. 1), spans an area of 300 km² and has a perimeter of almost 100 km. It lies southeast of Tetouan and is positioned between Oued Laou and Jebha in the Tanger-Tetouan-Al Houceima region. Its northern boundary meets the Mediterranean Sea at an outlet with coordinates X = 555,115.06 m and Y = 516,611.278 m, whereas the Bab Berred Mountain chain marks the southern boundary. The watershed is located between latitudes 35° and 35°15' N and longitudes 4°40' and 4°50' E. The region boasts mountainous terrain, with elevations ranging from 0 to 2100 m (Jebel Tizirane), exhibiting various landforms such as structural features, depressions, ravines, and more (El Brahimi et al. 2022).

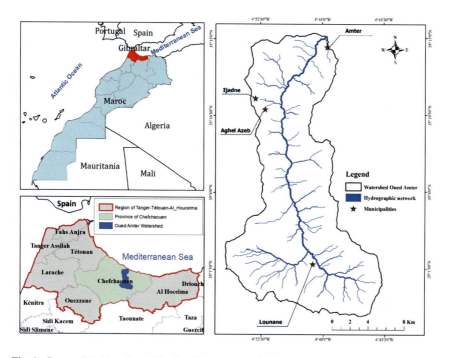

Fig. 1 Geographical location of the Oued Amter watershed

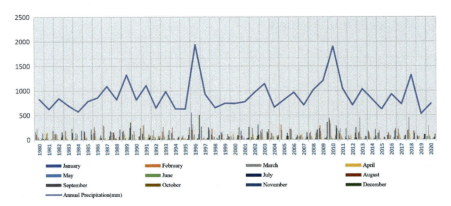

Fig. 2 Annual precipitation of the station of Chefchaouen (1980–2020) (El brahimi et al. 2022)

2.2 Precipitation

A semi-arid Mediterranean climate with seasonal solid contrasts and irregularities in precipitation characterizes this basin. This basin is marked by variable precipitation from one month to another, with an average annual rainfall of approximately 600 mm/year. The rainy season in the study area extends from November to April, peaking in December and January. The driest months were July and August (Fig. 2).

2.3 Geological Context

The Rifan chain is part of the Betic-Rifan arc, located in northwest Morocco at the southern margin of the western Mediterranean. This geological complex is relatively recent, dating back to the early Tertiary (Durand-Delga et al. 1999; Chalouan et al. 2008; Michard et al. 2020). The Rifan chain is structured by various tectonic movements resulting from the collision of the Eurasian and African plates, the collapse of the Alboran Sea, and the great thrusts that extend over several kilometers along the coast. Very complex geological structures form the RIF domain. It is divided into three groups of paleogeographic units: outer units, flysch sheets, and inner units (Vitale et al. 2015; El Hilali et al. 2023). The external Rif is constituted by a belt of folds and thrusts composed of beds of evaporite rocks from the Upper Triassic to Miocene, originating in the North African continental crust. The current structures result from upper Miocene tectonic phases (Azdimousa et al. 2019; Gimeno-Vives et al. 2020).

The Internal Rif (study area) is part of the Alboran paleogeographic plate, formed mainly by Paleozoic rocks covered by Mesozoic layers interposed between the two Eurasia-Africa plates. The Internal Rif region comprises three major units: Sebtides, the Ghomarides, and limestone ridges. The Sebtides are mainly composed of metamorphic rocks, such as schist and gneiss, whereas the Ghomarides are composed of sedimentary rocks, such as sandstone and limestone. The limestone ridge is made up of well-layered limestone formations (Chalouan and Michard 2004; Perri et al. 2022). Formations can also be distinguished from the pre-dorsal at the boundaries of the limestone ridge. These formations are composed of sandstones, marls, and limestones, and are believed to have been deposited during the Lower Cretaceous (Hernández-Molina et al. 2016). The Internal Rif region is also known for its tectonic activity, which has resulted in the formation of folds, faults, and thrusts. These tectonic features have created a complex and diverse geological landscape that has attracted the attention of geologists worldwide (Vitale et al. 2015; Sestras et al., 2023). The study area is characterized by high rainfall, which can contribute to the soil saturation and destabilization of steep slopes. This, combined with the geological structure, creates a high risk of landslides, rockfalls, and soil erosion (Fig. 3). Additionally, human activities such as deforestation, overgrazing, and construction can exacerbate this problem, as they can destabilize the soil and increase the risk of landslides and erosion.

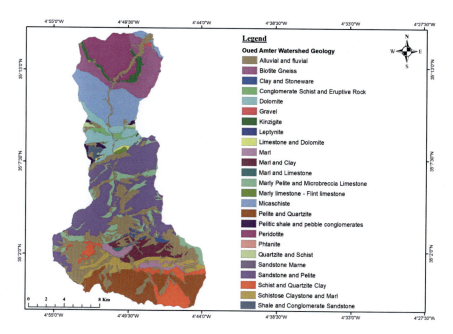

Fig. 3 Geological map of the Oued Amter watershed (Based on the Bab Berret Geological Map:1/250000)

2.4 The Relief

The relief of a watershed plays a crucial role in determining its erosion risk (Pandey et al. 2021; Benzougagh et al. 2022). Relief refers to the shape and elevation of the land surface, including the hills, valleys, and slopes. High-relief watersheds with steep slopes and rugged terrain are more susceptible to erosion than are low-relief watersheds with gentle slopes (Benzougagh et al. 2016, 2017; Kadam et al. 2018). Figure 4b and Table 1 show the altitude distribution in the Oued Amter Watershed. The analysis of this distribution showed that the average altitude of the watershed was 700 m. Land use practices, such as deforestation, urbanization, and agriculture, can increase erosion risk by removing vegetation cover and disturbing the soil surface. This increases the potential for surface runoff and soil erosion, leading to sedimentation in downstream water bodies and the loss of productive land (Fig. 4a).

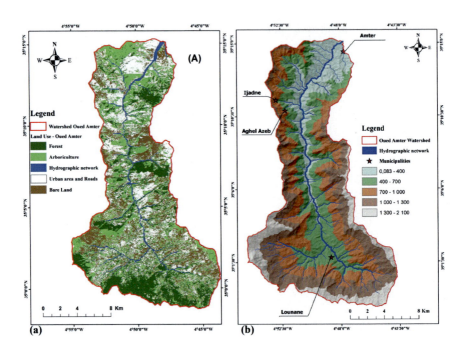

Fig. 4 **a** Land use map **b** Altitude map of the Oued Amter watershed

Table 1 Morphological characteristics of Oued Amter watershed

Parameters	Value
Surface	300 km^2
Perimeter	100 km
GRAVELUS compactness index	1,61
Length of main stream	35 km
Length of equivalent rectangle	42,78 km
Width of equivalent rectangle	7,01 km
Maximum elevation	2100 m
Minimum elevation	0 m (Outlet point)
Average elevation	700 m
Average slope	28,05 m/km–0,28%

3 Methods

3.1 Justification for the Choice of the EPM Model

Several models are available for estimating soil erosion potential, and the choice of which model to use depends on various factors, such as the research question, data availability, and the characteristics of the study area. One possible justification for using EPM is that it is a widely recognized and well-established model extensively used in the scientific literature (Boote et al. 1996; Kefi et al. 2012; Vatandaşlar and Yavuz 2023). The EPM has been applied in various settings, from small watersheds to entire regions, and has been validated using field observations in various environments (Spalevic et al. 2020; Baharvand and Pradhan 2022).

Additionally, the EPM has been shown to incorporate a range of relevant variables, such as topography, soil properties, and land use, into its predictions. This can help provide a comprehensive understanding of the factors contributing to erosion potential and can be useful for identifying at-risk areas and developing appropriate management strategies (Benzougagh et al. 2020a, b, 2022). Moreover, the EPM model is relatively straightforward, and many software tools are available for its implementation. This can make it an attractive option for researchers and practitioners who may not have extensive modeling expertise or access to specialized software. Although several models are available for estimating soil erosion potential, the Erosion Potential Method is a well-established and widely used approach that offers a comprehensive and accessible tool for understanding the factors contributing to erosion in a given area.

3.2 Methodology of the Model Application

EPM can be applied using a simple methodology that involves the following steps (Fig. 5), the details of which are shown in Fig. 6.

- **Collecting soil data**: Soil data such as texture, organic matter content, and permeability are necessary to determine the soil erodibility factor. These data can be obtained from soil surveys or by analyzing soil samples in a laboratory.
- **Measure the slope length and gradient**: The length and gradient of the slope can be measured using a surveyor's level or other tools. These values can be obtained by using topographic or digital elevation models.
- **The erosion index** was calculated using the soil erodibility factor, slope length, and slope gradient, and the erosion index for the field was calculated using a standardized equation. The resulting erosion index indicates the potential for erosion in the field.
- **Interpretation of the results**: The erosion index value can be interpreted using a classification system that assigns a risk level to the field based on the index value. This can help prioritize fields for erosion control measures and evaluate the effectiveness of existing erosion control practices.
- **Erosion control measures**: Based on the results of the EPM assessment, appropriate erosion control measures can be implemented to reduce the risk of erosion. These may include practices, such as conservation tillage, cover cropping, and terracing.

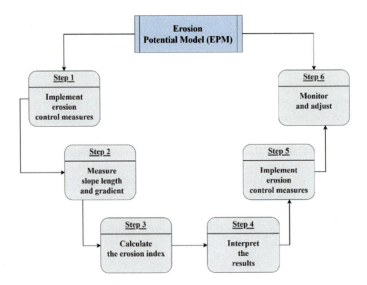

Fig. 5 Steps of the EPM model methodology

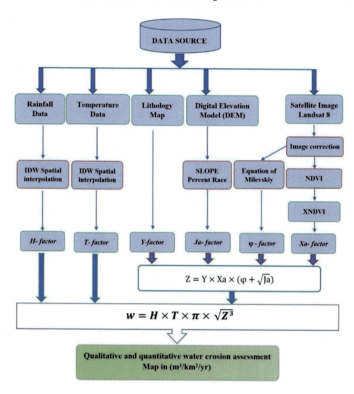

Fig. 6 Details of the EPM methodology in the study area

– **Monitor and adjust**: The effectiveness of erosion control measures should be monitored over time, and adjustments should be made as necessary to maintain their effectiveness.

The application of this model requires a large amount of data related to the nature of rocks and soils, slope, precipitation, temperature, and vegetation cover, in addition to information that can be obtained during field visits. Therefore, we are supported by various sources of information emphasizing the use of remote sensing and geographic information systems to produce thematic maps of various factors and indicators of the equation. Then, the integration and installation of these maps, networking of their databases, and realization of a set of spatial analysis functions. This model integrates several factors that influence erosion dynamics. The erosion coefficient Z is considered as the base value for all calculations (Gavrilovic 1972; Gavrilovic et al. 2008; Al Ghamdi 2018) and allows the determination of the erosive potential in the watersheds. It combines four factors responsible for erosion: slopes (Ja), the sensitivity of lithological formations to erosion (Y), vegetation cover protection (Xa), and erosive state (φ), which are then combined with temperature (T) and rainfall (H). The overlap of these factors allowed for the elaboration of the soil loss map (Fig. 14). The combination type is defined using the following formula:

$$w = H \times T \times \pi \times \sqrt[3]{Z} \qquad (1)$$

where:

W: Average annual sediment production (m3/ km2/yr).

T: Temperature coefficient, calculated as follows:

$$T = \sqrt{\frac{T}{10} + 0.1} \qquad (2)$$

T: Average annual temperature in °C.

H: Average annual precipitation (mm/year).

Z: Erosion coefficient calculated as follows

$$Z = Y \times Xa \times (\varphi + \sqrt{Ja}) \qquad (3)$$

where:

Y: the value of the coefficient of soil resistance due to erosion. It depends on the parent rock; the type of soil and climate varies between 0.05 and 1.

Xa: Watershed regularization coefficient, relating to soil protection, atmospheric phenomena influence, and erosive forces related to natural conditions.

φ: Coefficient expressing the type and degree of visible erosion processes in the watershed.

Ja: Average watershed slope index.

4 Results and Discussion

4.1 The Qualitative Assessment of Hydrological Erosion in the Amter Oued Basin Through the Gavrilovic Model (EPM).

In the following subsections, the results related to the factors soil erodibility coefficient, slope index, soil protection coefficient, erosion rate coefficient, and erosion potential index.

4.1.1 Soil Erodibility Coefficient (Y)

The soil erodibility coefficient (Y) is a key parameter in the EPM that is used to estimate soil erosion (Dragičević et al. 2018; Sakuno et al. 2020). The EPM is a widely used approach for estimating soil erosion rates in agricultural areas, and is based on the concept of the universal soil loss equation (USLE) (Panagos et al. 2014; Pandey et al. 2021). The soil erodibility coefficient (Y) represents the inherent susceptibility of soil to erosion. The soil texture, organic matter content, structure, permeability, and slope influence it. The higher the Y value, the more efficient the soil erosion (Phinzi and Ngetar 2019; Chen et al. 2023). The Y value was determined by conducting field experiments to measure the soil loss under controlled rainfall, slope, and soil management practices (El Kateb et al. 2013; Ebabu et al. 2019). The data collected from these experiments were used to calculate the Y value using the following equation:

$$Y = (A/S)^{(1/n)} \tag{4}$$

where:

- **Y**: Soil erodibility coefficient
- **A**: Average annual soil loss (ton/acre/year)
- **S**: Average slope length (feet)
- **n**: Slope exponent (dimensionless).

The Y value is specific to a particular soil type and is typically reported in tons/acre/year. An essential parameter in soil erosion modeling is the estimation of soil loss and sediment yield under various land-use and management scenarios. This factor shows the sensitivity of the soil to water erosion and depends on a set of laboratory analyses regarding texture, structure, organic matter percentage, and permeability. It is determined for a given soil using the following relationship (Wischmeier and Smith 1978):

$$K = \frac{(2.1 \times M1.14 \times 10 - 4 \times (12 - a) + 3.25 \times (b - 2) + 2.5(c - 3))}{100} \tag{5}$$

where:

- **M**: calculated using the formula M = (% of silt) × (100–% clay).
- **a**: Percentage of organic matter.
- **b**: Code for permeability.
- **c**: Code for structure.

In the study area, obtaining a precise soil map may not be feasible because of resource and time constraints. Therefore, alternative sources of information will be used to assess soil characteristics. For this purpose, we will rely on the geological maps of Bab Berret and Rif, which are available at scales of 1/25,000 and 1/500,000, respectively. Although these maps do not provide detailed information on the soil properties, they offer valuable insights into the geological and landform characteristics of the area. By integrating this information with other data sources, such as field observations and remote sensing, we can better understand soil variability and its impact on various environmental processes.

The lithological formations in the study basin significantly affected the erosion intensity. Marly rocks are the predominant geological features and are highly susceptible to erosive processes, particularly water-related ones. More than half of the basin's total area is classified as highly sensitive (0.21 to 0.43) because of the prevalence of these rocks. As a result, erosion is a major concern in the study area, and understanding the underlying factors that contribute to it is crucial for effectively managing the basin's natural resources. By identifying and analyzing the distribution of susceptible areas, we can develop targeted strategies to mitigate erosion and its impact on soil quality, water resources, and biodiversity (Fig. 7).

Based on Fig. 7, the Soil Erodibility Coefficient (Y) values in the Oued Amter Basin range from 0.14 to 0.43, indicating a range of rock and soil resistance to water erosion. Rocks with high erosion capacity cover more than half of the basin, primarily in the middle section and some narrow zones in the south and north. In contrast, low-capacity rocks were scattered in the northern section and narrow areas at the bottom and south of the basin. The varying sensitivity and resistance of the rocks in the study area contribute to different forms of water erosion and varying degrees of soil degradation. By analyzing these patterns and understanding the underlying factors

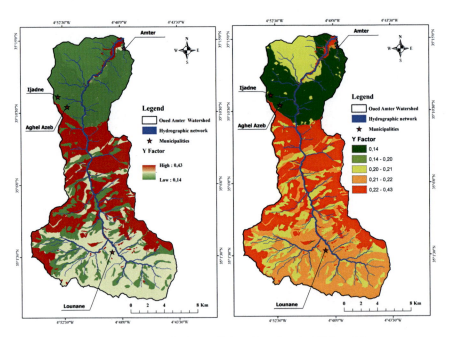

Fig. 7 Map of soil and rock susceptibility coefficients based on the EPM model

contributing to erosion susceptibility, we can develop targeted strategies to mitigate the impacts of erosion on soil quality, water resources, and biodiversity.

4.1.2 Slope Index (Ja)

The Slope Index (Ja) is a crucial parameter of the EPM, which estimates the potential for water-induced erosion in a given area. The Ja parameter represents the slope length and steepness, as well as the topographic characteristics of the landscape, which can affect the rate and intensity of water runoff and soil erosion (Wischmeier and Smith 1978; Tariq and Akhtar 2022). High Ja values indicate steep slopes and longer slopes, which are more susceptible to erosion, while low Ja values indicate flatter terrain and less susceptibility to erosion. The Ja parameter is used in conjunction with other EPM parameters, such as the Soil Erodibility Coefficient (Y) and Rainfall Erosivity Factor (R), to estimate the amount of soil loss and sediment yield in a given area. By assessing the Ja value across a study area, we can identify areas that are particularly vulnerable to water-induced erosion and prioritize management interventions to reduce erosion and its impacts on soil quality and water resources (Renard et al. 1997; Wang et al. 2016).

The watershed of the Amter River is characterized by a strong presence of steep to very steep slopes (Fig. 8). This terrain configuration favors runoff, causes rapid water displacement from precipitation, and increases the amount of water flowing downstream. Thus, these slopes provide strong erosive power from water and favor an increased rate of eroded materials from upstream to downstream in the basin. A Digital Elevation Model (DEM) with a 30-m resolution. Figure 8 illustrates the range of the gradient values in the study area, which varied from 0 to 163.75%. The DEM provided valuable information on the length and steepness of slopes and the topographic features that influence water runoff and soil erosion. High values of Ja indicate steeper and longer slopes that are more prone to erosion, while lower values suggest flatter terrain that is less susceptible to erosion. By analyzing the Ja values across the study area, we can identify locations particularly vulnerable to water-induced erosion and prioritize management measures to mitigate soil loss and sediment yield.

4.1.3 Soil Protection Coefficient (Xa)

The Soil Protection Coefficient (Xa) is a key parameter in EPM for assessing soil erosion potential in a given area (Solaimani et al. 2009; Elaloui et al. 2022). The Xa parameter represents the degree of protection provided by the vegetation cover and soil management practices against water-induced erosion. High Xa values indicate that the soil is well protected, whereas low Xa values suggest that the soil is more vulnerable to erosion (Rahman et al. 2015; Barakat et al. 2023). The Xa parameter is combined with other EPM parameters, such as the Soil Erodibility Coefficient (Y) and Rainfall Erosivity Factor (R), to estimate the amount of soil loss and sediment

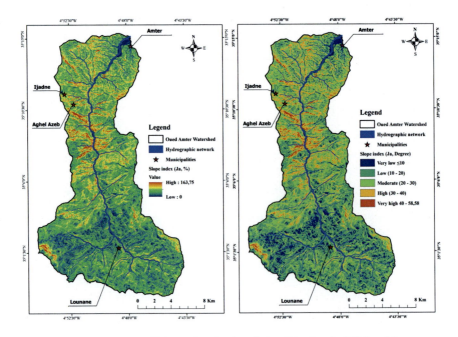

Fig. 8 Map of regression coefficient values for slope index (Ja) using the EPM model

yield in a given area (Fig. 8). By analyzing the Xa value across a study area, we can identify areas that require additional measures to improve soil protection and reduce the impact of water-induced erosion on soil quality and water resources (Hosseini and Ashraf, 2015; Chalise et al. 2019).

In the study area, the degree of soil protection varied depending on the vegetation cover index, as indicated by the range of values for the Soil Protection Coefficient (Xa) in our study basin. Xa values range from 0.11 in areas with significant vegetation cover (e.g. mature forests, thick vegetation) to 0.85 in areas with low protection due to lack of vegetation cover (e.g. seasonal and degraded mature croplands). These areas were scattered throughout the basin in narrow zones. By analyzing the Xa values in the study area, we can identify areas that require additional measures to improve soil protection, such as implementing vegetation restoration or soil management practices, to reduce the impact of water-induced erosion on soil quality and water resources (Fig. 9).

4.1.4 Erosion Rate Coefficient (Φ)

The Erosion Rate Coefficient (φ) is a significant parameter used in the EPM to estimate the amount of soil loss due to water-induced erosion in a given area (Dragičević et al. 2018; Zhou et al. 2023). The φ value represents the rate of soil erosion caused by water and is influenced by various factors such as topography, vegetation cover, and

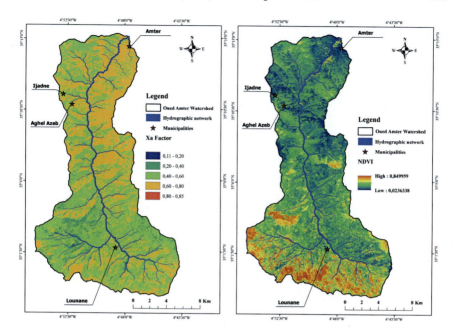

Fig. 9 Map of soil protection factor (Xa) values based on EPM model

soil erodibility. High φ values indicate higher rates of soil loss, while low φ values indicate lower rates of soil loss (Istanbulluoglu and Bras, 2005; Maliqi and Singh 2019). The φ value is combined with other EPM parameters, such as the Soil Erodibility Coefficient (Y) and the Rainfall Erosivity Factor (R), to estimate the amount of soil loss and sediment yield in a given area. By assessing the φ value across a study area, we can identify areas particularly vulnerable to water-induced erosion and prioritize management interventions to reduce erosion and its impacts on soil quality and water resources (Tangestani 2006; Efthimiou et al. 2017).

Remote Sensing (RS) and Geographic Information Systems (GIS) provide valuable tools for calculating erosion indicators from Landsat images by utilizing the square root of the third band (TM3) divided by the maximum radiation value (Qmax) to generate erosion indices (Fig. 10). In simpler terms, this approach allows the identification of areas with high erosion rates by producing high erosion indices and vice versa (Mesrar et al. 2015). For Landsat 8, the type and extent of erosion coefficient φ can be accurately determined using the following formula:

$$\varphi = \sqrt{\frac{R}{Qmax}} \qquad (6)$$

where:

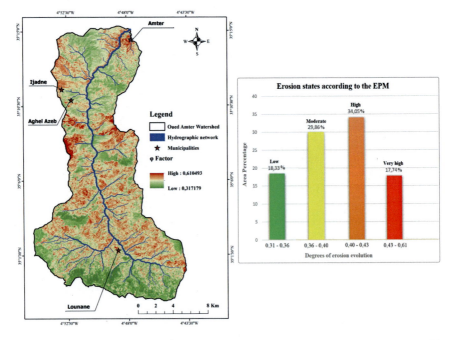

Fig. 10 Mapping erosion evolution coefficients (ϕ), using the EPM model: a spatial representation of erosion dynamics

- **R**: band 4 of Landsat image 8,
- **Qmax**: radiance maximum of band 4.

Various factors, including vegetative cover, lithological formations, slope, climate, and land use practices, determine the erosive state of a given area. In this study, the intensity of erosion varies depending on the distribution of these agents, resulting in a range of erosive state values from weak (0.31) to strong (0.61), as depicted in Fig. 10. Interestingly, more than half of the basin area is classified as having a very strong erosive state, with values falling within the 0.40–0.61 range. These findings underscore the critical need for effective erosion management strategies to mitigate erosion's environmental and societal impacts in this region.

4.1.5 Erosion Potential Index (Z)

A key component of the EPM is the Erosion Potential Index (Z), which estimates the relative erosion potential of different areas within the study area (Maliqi and Singh 2019; Kabo-Bah et al. 2021). The Erosion Potential Index was calculated by combining various factors contributing to soil erosion, including soil texture, slope angle, and rainfall intensity. Each of these factors was assigned a weighting factor based on its relative importance in contributing to erosion, and these factors were

Table 2 Potential erosion coefficients in the study area (Gavrilovic et al. 2006)

Degrees of Erosion (Z)	Potential erosion coefficient	Area (km²)	Percentage of area (%)
0,01–0,19	Very low	1,0629	0,36
0,20–0,40	Low	11,8233	4,11
0,41–0,70	Moderate	106,726	37,11
0,71–1	High	109,808	38,18
1,01–3,129	Very high	58,1193	20,21

then combined to produce a single index value for each location within the study area (Ouyang et al. 2018; Benzougagh et al. 2023). This parameter was derived by combining several coefficient maps in a GIS system, as described by the EPM equation. Specifically, it integrates the Soil Erodibility Coefficient map (Y), the soil protection coefficient (Xa), the average slope of the study area as a percentage (Ja), and the coefficient of erosion processes visible in the watershed (ϕ). By merging these layers, we obtained a comprehensive measure of the erosion probability that accounts for factors such as soil properties, terrain characteristics, and erosion patterns.

$$Z = Xa \times Y \times (\varphi + \sqrt{Ja}) \qquad (7)$$

The erosion coefficient (Z) is typically categorized into several levels by using the Gavrilovic model. This model classified the erosion coefficient on a scale ranging from 0.01 to 3.129, with different categories corresponding to different levels of erosion risk (Table 2).

Based on the classification results presented in Table 1, it is evident that a significant portion of the watershed has a high or very high risk of erosion, with a value of (Z) greater than 0.71. Specifically, 58.39% of the total area fell into this category. Conversely, only a small fraction of the watershed (4.47%) exhibited low to very low erosion intensities (Z varies between 0.01 and 0.4). Most of the watershed (37.11%) falls into the moderate erosion intensity category, indicating moderate erosion risk in these areas. The analysis revealed that over 50% of the watershed exhibited a strong or very strong potential for erosion, with a corresponding Z value greater than 0.71. These areas are characterized by steep to very steep slopes and moderate slopes with reduced vegetative cover. These high-risk zones were distributed throughout the watershed as narrow strips, encompassing various parts of the basin (Fig. 11).

4.2 Quantitative Assessment of Hydraulic Erosion Using EPM Model.

The EPM model was used to quantify hydraulic erosion in the Oued Amter Basin. A hydric erosion map (Fig. 15) was generated by multiplying the different thematic layers of the EPM equation, revealing a clear spatial variability of erosion risk across

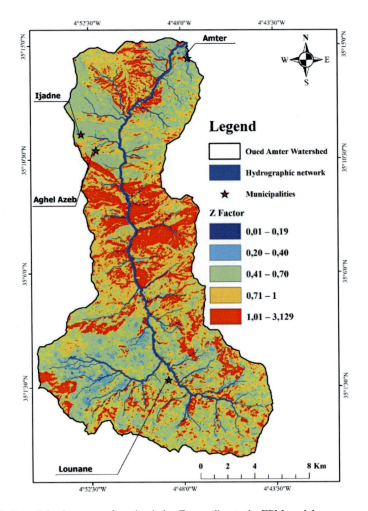

Fig. 11 Potential values map of erosion index Z according to the EPM model

the study area. The entire watershed appears to be affected by erosion dynamics, particularly in higher areas with steep slopes and areas devoid of vegetation. These findings provide valuable information for assessing the potential impact of hydraulic erosion in the basin and developing appropriate management strategies.

4.2.1 Temperature Coefficient (Heat Index) (T)

The Temperature Coefficient (T) is a crucial parameter in the EPM model that measures the potential for soil erosion in a given area. The T coefficient is also known as the Heat Index and represents the effect of temperature on soil erosion.

The (T) coefficient is calculated by integrating several climatic and meteorological factors that affect the ability of the soil to resist erosion due to increased temperatures. These factors included air temperature, relative humidity, solar radiation, wind speed, and rainfall. (Ghobadi et al. 2011; Lense et al. 2023). To calculate this coefficient, we used the following equation:

$$T = \sqrt{\frac{t}{10} + 0.1} \tag{8}$$

where:

- **T**: Temperature Coefficient (T).
- **t**: Average annual temperature (in °C).

Higher values of the T coefficient indicate a higher potential for soil erosion owing to the effect of heat on soil moisture and structure. Soil moisture is reduced as temperature increases, leading to increased soil particle detachment and erosion. In addition, higher temperatures can lead to changes in soil structure and reduce the stability of the soil, making it more susceptible to erosion (Dang et al. 2015; Zeghmar et al. 2022). The Landsat 8 satellite images used to calculate the Temperature Coefficient (T) indicate substantial spatiotemporal variations at the watershed scale, adding an extra dimension to the spatial variability of water erosion in the Oued Amter watershed. As a result, the collective characteristics of the generative, regulatory, and accelerative erosion factors in the watershed contribute significantly to its fragility and heightened vulnerability to the initiation and expansion of water erosion (Fig. 12).

4.2.2 Precipitation Coefficient (Precipitation Index) (H)

The Precipitation Coefficient, or Precipitation Index (H), is an essential component of the EPM used to assess the risk of soil erosion used to assess soil erosion in a given area. EPM considers various factors, such as topography, land use, soil type, and rainfall patterns, to estimate the risk of erosion in a particular area (Wang et al. 2017; Marko et al. 2023). The Precipitation Coefficient, represented by variable H, measures the amount and intensity of rainfall in a particular location. It was calculated by dividing the annual precipitation by the number of rainy days per year. The resulting value indicates the frequency and intensity of rainfall in the area, which is a critical determinant of the erosion risk (Pereira et al. 2023). In the study area, the upstream–downstream variation in precipitation (between 1637 and 1937 mm) and their spatiotemporal intensities favor the removal of considerable quantities of materials (Fig. 13). Usually, higher values of H indicate a greater risk of soil erosion, as more intense and frequent rainfall events increase the likelihood of soil loss. Conversely, lower values of H suggest a lower risk of erosion because rainfall events are less intense and less frequent.

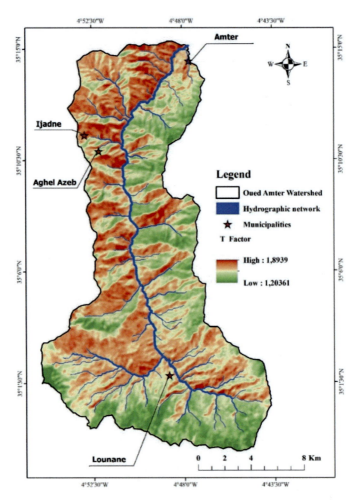

Fig. 12 Map of temperature coefficient (T) according to the EPM

4.2.3 Estimation of Soil Losses According to the EPM Model of Gavrilovic

The EPM is a widely used technique for predicting and assessing soil loss due to erosion. It is based on the evaluation of various soil properties including erodibility, slope, vegetation cover, and erosive forces (Sharma et al. 2018; Marko et al. 2023). Soil loss in EPM refers to the amount of soil removed from a given area over a specified period. It can occur due to different erosion types, including sheet erosion, rill erosion, gully erosion, and wind erosion. The amount of soil loss depends on various factors, such as the severity of erosion, soil properties, and erosive forces acting on the soil (Sharma et al. 2018). According to Zachard's classification (1982),

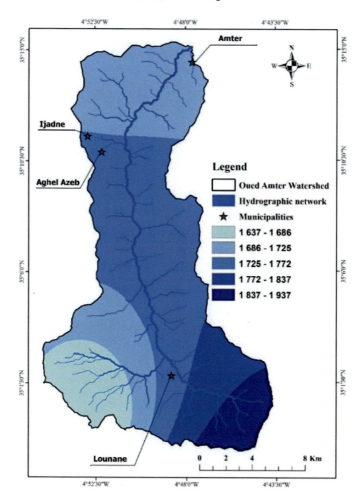

Fig. 13 Map of precipitation values (H) in the study area according to the EPM

the estimation and spatial mapping of soil losses using the erosion potential method (EPM), model, and identification of erosion classes (Table 3) through the integration of all EPM model factors results in the generation of an average annual volume map (W), as depicted in the following equation:

$$w = H \times T \times \pi \times \sqrt[3]{Z} \qquad (9)$$

where:

- **T:** temperature coefficient (°C); **H:** annual precipitation (mm).; **π:** 3, 14; **Z:** erosion coefficient.

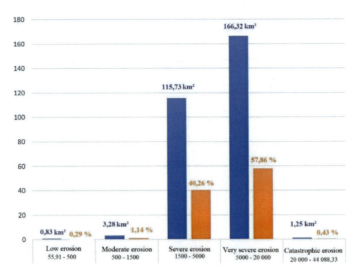

Fig. 14 Areas (in % and km^2) with erosion rates (m^3/km^2/yr) in the study area

To estimate the total quantities (masses) of eroded sediments G (in ton/Km2/year), we use the equation: G = W. ρ (Eq.)

Where:

- **W:** Average annual volume of eroded sediment (m3/km2/an); **ρ:** Density (ton/m3)

The results of soil loss in the Oued Amter watershed are consistent with both the findings of the study and visual observations (Table.4). Figures 14 and 15 further illustrate that nearly 90% of the total study area is subject to catastrophic erosion, while the remaining 10% is strong to very strongly eroded.

5 Discussion

RS and GIS technology have revolutionized erosion mapping, which has previously required significant field work and expert analysis. Professionals evaluated terrain slopes, observed erosion patterns, and vegetation types and determined the erosion coefficient based on geomorphological and geological knowledge. However, this process has become more efficient and accurate when used by RS and GIS. This study utilized cartographic materials, satellite images, GIS technology, and an EPM model to create an erosion map for the Oued Amter watershed in RIF, Morocco. Therefore, to create an erosion map for the study area, it is necessary to have an existing land use map and a relief slope map. The EPM analysis estimated the sediment production in the entire river catchment area. Hence, applying the EPM model requires a series of geoprocessing operations on the vector and raster layers before

GIS-Based Erosion Potential Method (EPM) for Soil Degradation ...

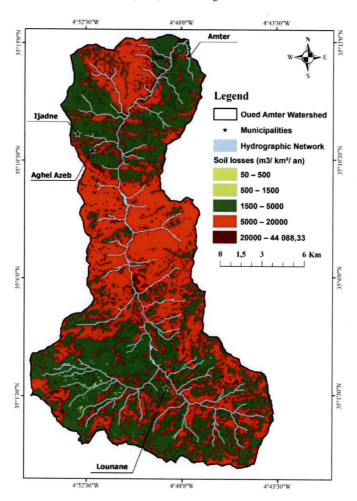

Fig. 15 Map of soil losses in the Amter watershed

Table 3 Loss classes in soil (Zachard 1982)

Class	Soil losses (m³/ha/year)	Soil losses (m3/km²/an)	Appreciation
1	<0.5	<50	Insignificant erosion
2	0.5–5	50–500	Low erosion
3	5–15	500–1500	Moderate erosion
4	15–50	1500–5000	Severe erosion
5	50–200	5000–20,000	Severe erosion
6	>200	>20,000	Catastrophic erosion

Table 4 Degree of erosion in the Oued Amter watershed

Classes	Soil loss (m³/km²/yr)	Degree of erosion	Area (km²)	Area (%)
1	55,91–500	Low erosion	0,837,515	0,291,369
2	500–1500	Moderate erosion	3,28,842	1,14,403
3	1500–5000	Severe erosion	115,739	40,2652
4	5000–20,000	Very severe erosion	166,327	57,8646
5	20,000–44 088,33	Catastrophic erosion	1,25,017	0,43,493

arriving at the final calculation using GIS software for this utility. The research findings indicated that the erosion coefficient obtained from the visually unverified erosion map was statistically reliable. The map's accuracy is increased using proportional and instrumentally determined land use boundaries and clearly defined slope areas. Factors such as soil erodibility, soil protection, slope, temperature, and precipitation are essential for erosion control.

Mapping using cartographic materials and satellite images offers several significant advantages. First, it reduces the time and cost of fieldwork required for erosion map production. Second, it enables a more comprehensive preparation for field trips. Third, it facilitates project implementation by providing a better understanding of erosion hazards and investment risks. Fourth, this makes it easier to define proposals for anti-erosion measures. Finally, field checks can focus on critical and unclear erosion areas verified on the terrain, allowing for the proposed anti-erosion measures. The study's findings indicate that soil loss varied significantly, ranging from 55.91 to 44,088.33 m³/km²/year. Using the EPM method to map erosion has proved valuable for environmental monitoring and water resource management. Additionally, combining these techniques yields satisfactory results for accurately assessing soil erosion. After comparing our study's methodology with other studies conducted in different regions of Morocco, we found that the erosion rate in the RIF region, particularly in the Oued Amter watershed, was significantly higher than that in other areas of the country (Table.4 and Fig. 16). This can be attributed to several variables, including soil erodibility, soil protection, slope, temperature, rainfall, land use, and land cover (Table 5).

6 Conclusions

Water erosion is a significant environmental problem that affects agricultural productivity and sustainability. Several factors influence the rate and extent of water erosion, including the rainfall intensity, soil type, slope gradient, vegetation cover, and land use. Preventing or mitigating water erosion requires appropriate management practices tailored to local soil and climatic conditions. Understanding the factors influencing water erosion and implementing effective management practices are crucial for achieving sustainable agricultural production and environmental conservation.

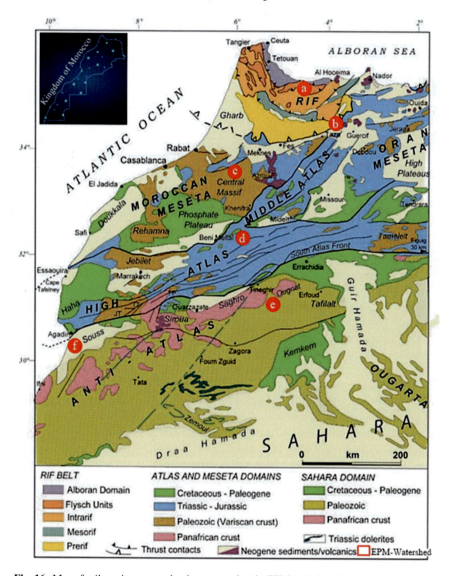

Fig. 16 Map of soil erosion rate evaluation cases using the EPM model in Morocco: **a**: Amter watershed, **b**: Inaouene watershed, **c**: Beht watershed, **d**: Ouaoumana watershed, **e**: Toudgha watershed, **f**: Souss basin

The main objective of this study was to apply EPM to the Oued Amter watershed to estimate soil erosion and investigate its spatial distribution. This study demonstrates that the EPM, coupled with RS and GIS techniques, is valuable for identifying and analyzing soil loss in areas such as the Amter Basin. Soil erodibility, soil protection, slope, temperature, and rainfall are critical factors for erosion control. The study results indicated that the soil loss varied from 55.91 to 44 088.33 m^3/km^2/

Table 5 Results of the estimation of sediment production comparing the EPM model in some river catchments within the Moracin territory

Watershed	Region (Country)	Soil losses m^3/km^2/yr	References
Oued Amter (a)	North of Morocco (Rif)	55.91 to 44 088.33	Present study
Oued Inaouene (b)	Northeast of Morocco (Pre-Rif)	0.181 to 97.642	Marouane et al. (2021)
Oued Beht (c)	Central Massif (Morocco)	0.12 to 20.400	Lakhili et al. (2021)
Ouaoumana (d)	Middle Atlas (Morocco)	23.47 to 39 007	Ennaji et al. (2022)
Toudgha (e)	Southeast of Morocco	1 to 4000	El Badaoui et al. (2021)
Souss (f)	Middle western of Morocco,	0.02 to 181	Argaz et al. (2019)

year. Erosion mapping using the EPM method proved effective for environmental monitoring and water resource management, providing satisfactory combined results.

This study found that over 90% of the basins experienced significant sedimentation. Only 2% of the total basin area was relatively preserved, with low-to-moderate erosion. This dynamic can have negative consequences, particularly on the soil fertility and hydraulic infrastructure. The high level of sedimentation in the majority of the basins suggests that erosion control measures are necessary to prevent further soil degradation and to protect the hydraulic infrastructure. It is recommended that conservation efforts should be prioritized in preserved areas with low-to-moderate erosion to maintain the current soil quality and prevent erosion from increasing. Such conservation measures may include land management practices, vegetation cover, and other erosion control strategies, which would help mitigate the negative impacts of erosion on soil fertility and hydraulic infrastructure. This approach adopted in this study can be used to develop appropriate land management and conservation strategies to prevent and reduce soil erosion in the Amter Basin area and all other similar basins.

Funding: This research received no external funding.

DataAvailability Data Availability Statement The data presented in this study are available on request from the corresponding author.

Conflicts of Interest The authors declare no conflict of interest.

References

Abbate A, Mancusi L, Frigerio A, Papini M, Longoni L (2023) CRHyME (Climatic rainfall hydrogeological model experiment): a new model for geo-hydrological hazard assessment at the basin scale. Nat Haz Earth Syst Sci Discus 1–51. https://doi.org/10.5194/nhess-2023-15

Akhtar MS, Riaz M (2017) Comparative evaluation of RUSLE, MUSLE and EPM models for prediction of soil erosion. Environ Monit Assess 189(6):283

Akoko G, Le TH, Gomi T, Kato T (2021) A review of SWAT model application in Africa. Water 13(9):1313. https://doi.org/10.3390/w13091313

Alghamdi AG (2018) Biochar as a potential soil additive for improving soil physical properties—a review. Arab J Geosci 11(24):766. https://doi.org/10.1007/s12517-018-4056-7

Almagro A, Thomé TC, Colman CB, Pereira RB, Junior JM, Rodrigues DBB, Oliveira PTS (2019) Improving cover and management factor (C-factor) estimation using remote sensing approaches for tropical regions. Int Soil Water Conserv Res 7(4):325–334. https://doi.org/10.1016/j.iswcr.2019.08.005

Aloui S, Mazzoni A, Elomri A, Aouissi J, Boufekane A, Zghibi A (2023) A review of soil and water assessment tool (SWAT) studies of Mediterranean catchments: applications, feasibility, and future directions. J Environ Manage 326:116799. https://doi.org/10.1016/j.jenvman.2022.116799

Al-Quraishi AMF (2003) Soil erosion risk prediction with RS and GIS for the Northwestern Part of Hebei Province, China. J Appl Sci 3(10–12):659–669

Al-Quraishi AMF (2004) Assessment of soil erosion risk using RUSLE and geoinformation technology for North Shaanxi Province, China. J China Univ Geosci 15:31–39

Argaz, A., Ouahman, B., Darkaoui, A., Bikhtar, H., Ayouch, E., & Lazaar, R. (2019). Flood hazard mapping using remote sensing and GIS Tools: A case study of souss watershed. *Journal of Materials and Environmental Science, 10*(2), 170–181.

Azdimousa A, Jabaloy-Sánchez A, Talavera C, Asebriy L, González-Lodeiro F, Evans NJ (2019) Detrital zircon U-Pb ages in the Rif Belt (northern Morocco): paleogeographic implications. Gondwana Res 70:133–150. https://doi.org/10.1016/j.gr.2018.12.008

Bag R, Mondal I, Dehbozorgi M, Bank SP, Das DN, Bandyopadhyay J, Al-Quraishi AMF, Nguyen XC (2022) Modelling and mapping of soil erosion susceptibility using machine learning in a tropical hot sub-humid environment. J Clean Prod 364:132428. https://doi.org/10.1016/j.jclepro.2022.132428

Bagheri M, Ibrahim ZZ, Mansor S, Abd Manaf L, Akhir MF, Talaat WIAW, Wolf ID (2023) Hazard assessment and modeling of erosion and sea level rise under global climate change conditions for coastal city management. Nat Hazard Rev 24(1):04022038. https://doi.org/10.1061/(ASCE)NH.1527-6996.0000593

Baharvand S, Pradhan B (2022) Erosion and flood susceptibility evaluation in a catchment of Kopet-Dagh mountains using EPM and RFM in GIS. Environ Earth Sci 81:490. https://doi.org/10.1007/s12665-022-10598-0

Bammou Y, Benzougagh B, Bensaid A, Igmoullan B, Al-Quraishi AMF (2023) Mapping of current and future soil erosion risk in a semi-arid context (haouz plain-Marrakech) based on CMIP6 climate models, the analytical hierarchy process (AHP) and RUSLE. Model Earth Syst Environ 1–14. https://doi.org/10.1007/s40808-023-01845-9

Barakat A, Rafai M, Mosaid H, Islam MS, Saeed S (2023) Mapping of water-induced soil erosion using machine learning models: a case study of Oum Er Rbia Basin (Morocco). Earth Syst Environ 7(1):151–170. https://doi.org/10.1007/s41748-022-00317-x

Bazhenova OI, Tyumentseva EM, Golubtsov VA (2022) Soil erosion on the agricultural lands of the Asian Part of Russia (Siberia): processes, intensity and areal distribution. In Global degradation of soil and water resources: regional assessment and strategies. Springer Nature Singapore, Singapore, pp 475–497. https://doi.org/10.1007/978-981-16-7916-2_30

Benzougagh B, Dridri A, Boudad L, Kodad O, Sdkaoui D, Bouikbane H (2017) Evaluation of natural hazard of Inaouene Watershed River in Northeast of Morocco: application of morphometric and geographic information system approaches. Int J Innov Appl Stud 19(1):85

Benzougagh B, Baamar B, Dridri A, Boudad L, Sadkaoui D, Mimich K (2020a) Relationship between landslide and morpho-structural analysis: a case study in Northeast of Morocco. Appl Water Sci 10(7):1–10. https://doi.org/10.1007/s13201-020-01258-4

Benzougagh B, Sarita GM, Dridri A, Boudad L, Sadkaoui D, Mimich K, Khaled MK (2020b) Mapping of soil sensitivity to water erosion by RUSLE model: case of the Inaouene watershed (Northeast Morocco). Arab J Geosci 13(21):1–15. https://doi.org/10.1007/s12517-020-06079-y

Benzougagh B, Dridri A, Boudad L, Baamar B, Sadkaoui D, Khedher KM (2022) Identification of critical watershed at risk of soil erosion using morphometric and geographic information system analysis. Appl Water Sci 12:1–20. https://doi.org/10.1007/s13201-021-01532-z

Benzougagh B, Boudad Larbi P, Dridri Abdallah P, Driss S (2016) Translation: "utilization of GIS in morphometric analysis and prioritization of sub-watersheds of Oued Inaouene (Northeast Morocco). Europ Sci J 12(6). https://doi.org/10.19044/esj.2016.v12n6p266

Benzougagh B, Meshram SG, Fellah BE (2023) Mapping of land degradation using spectral angle mapper approach (SAM): the case of Inaouene watershed (Northeast Morocco) Model. Earth Syst Environ https://doi.org/10.1007/s40808-023-01711-8

Berhanu B, Bisrat E (2018) Identification of surface water storing sites using topographic wetness index (TWI) and normalized difference vegetation index (NDVI). J Nat Resourc Dev 8:91–100. https://doi.org/10.5027/jnrd.v8i0.09

Boote KJ, Jones JW, Pickering NB (1996) Potential uses and limitations of crop models. Agron J 88(5):704–716. https://doi.org/10.2134/agronj1996.00021962008800050005x

Cerbelaud A, Lefèvre J, Genthon P, Menkes C (2022) Assessment of the WRF-Hydro uncoupled hydro-meteorological model on flashy watersheds of the Grande Terre tropical island of New Caledonia (South-West Pacific). J Hydrol Region Stud 40:101003. https://doi.org/10.1016/j.ejrh.2022.101003

Chalise D, Kumar L, Spalevic V, Skataric G (2019) Estimation of sediment yield and maximum outflow using the IntErO model in the Sarada river basin of Nepal. Water 11(5):952. https://doi.org/10.3390/w11050952

Chalouan A, Michard A, Kadiri KE, Negro F, Lamotte DFD, Soto JI, Saddiqi O (2008) The rif belt. Continental evolution: the geology of Morocco: structure, stratigraphy, and tectonics of the Africa-Atlantic-Mediterranean triple junction, 203–302. https://doi.org/10.1007/978-3-540-77076-3_5

Chalouan A, Michard A (2004) The Alpine Rif Belt (Morocco): a case of mountain building in a subduction-subduction-transform fault triple junction. Pure Appl Geophys 161:489–519. https://doi.org/10.1007/s00024-003-2460-7

Chen J, Li Z, Xiao H, Ning K, Tang C (2021) Effects of land use and land cover on soil erosion control in southern China: implications from a systematic quantitative review. J Environ Manage 282:111924. https://doi.org/10.1016/j.jenvman.2020.111924

Chen S, Zhang G, Zhu P, Wang C, Wan Y (2023) Impact of land use type on soil erodibility in a small watershed of rolling hill northeast China. Soil Tillage Res 227:105597

Danchenkov A, Belov N, Bubnova E, Myslenkov S (2023) Foredune defending role: vulnerability and potential risk through combined satellite and hydrodynamics approach. Remote Sens Appl Soc Environ 30:100934. https://doi.org/10.1016/j.rsase.2023.100934

Dang DN, Lanarde L, Jeannin M, Sabot R, Refait P (2015) Influence of soil moisture on the residual corrosion rates of buried carbon steel structures under cathodic protection. Electrochim Acta 176:1410–1419

Das J, Saha P, Mitra R, Alam A, Kamruzzaman M (2023) GIS-based data-driven bivariate statistical models for landslide susceptibility prediction in Upper Tista Basin, India. Heliyon 9(5). https://doi.org/10.1016/j.heliyon.2023.e16186

De Oliveira Oliveira VFR, Vick EPV, Bacani VMB (2023) Analysis of seasonal environmental fragility using the normalized difference vegetation index (NDVI) and soil loss estimation in the Urutu watershed, Brazil. https://doi.org/10.21203/rs.3.rs-2557676/v1

Dragičević N, Karleuša B, Ožanić N (2018) Modification of erosion potential method using climate and land cover parameters. Geomat Nat Haz Risk 9(1):1085–1105. https://doi.org/10.1080/19475705.2018.1496483

Durand-Delga M, Gardin S, Olivier P (1999) Datation des flyschs éocrétacés maurétaniens des Maghrébides: la formation du Jbel Tisirène (Rif, Maroc). Comptes Rendus De L'académie Des Sciences-Series IIA-Earth and Planetary Science 328(10):701–709. https://doi.org/10.1016/S1251-8050(99)80180-3

Ebabu K, Tsunekawa A, Haregeweyn N, Adgo E, Meshesha DT, Aklog D, Yibeltal M (2019) Effects of land use and sustainable land management practices on runoff and soil loss in the Upper Blue Nile basin, Ethiopia. Sci Total Environ 648:1462–1475. https://doi.org/10.1016/j.scitotenv.2018.08.273

Efthimiou N, Lykoudi E, Karavitis C (2017) Comparative analysis of sediment yield estimations using different empirical soil erosion models. Hydrol Sci J 62(16):2674–2694. https://doi.org/10.1080/02626667.2017.1404068

El Badaoui, K., Algouti, A., Algouti, A., & Hadach, F. (2021). Analyse des sédiments de la rivière du Toudgha et l'évolution de l'occupation du sol du bassin versant, (Flanc sud du haut Atlas central. Maroc). *Journal International Sciences et Technique de l'Eau et de l'Environnement*, 6(1), 9–20.

El Brahimi M, El Fellah B, Mastere M, Benzougagh B, El Basri M, Fartas N (2022) Quantification of soil sensitivity to water erosion by the RUSLE model in the Oued Amter Watershed, NorthWestern Morocco. Iraqi Geol J 41–56. https://doi.org/10.46717/igj.55.2C.4ms-2022-08-17

El Ghoul I, Sellami H, Khlifi S, Vanclooster M (2023) Impact of land use land cover changes on flow uncertainty in Siliana watershed of northwestern Tunisia. CATENA 220:106733. https://doi.org/10.1016/j.catena.2022.106733

El Kateb H, Zhang H, Zhang P, Mosandl R (2013) Soil erosion and surface runoff on different vegetation covers and slope gradients: a field experiment in Southern Shaanxi Province, China. CATENA 105:1–10. https://doi.org/10.1016/j.catena.2012.12.012

EL Hilali M, Bounab A, Timoulali Y, El Messari JES, Ahniche M (2023) Seismic site-effects assessment in a fluvial sedimentary environment: case of Oued Martil floodplain Northern Morocco. Nat Haz 1–23. https://doi.org/10.1007/s11069-023-06032-8

Elaloui A, Khalki EME, Namous M, Ziadi K, Eloudi H, Faouzi E, Chehbouni A (2022) Soil erosion under future climate change scenarios in a semi-arid region. Water 15(1):146. https://doi.org/10.3390/w15010146

Elliot WJ, Flanagan DC (2023) Estimating WEPP cropland soil erodibility from soil properties. In Soil erosion research under a changing climate, January 8–13, 2023, Aguadilla, Puerto Rico, USA (p 1). American Society of Agricultural and Biological Engineers. https://doi.org/10.13031/soil.23010

Ennaji, N., Ouakhir, H., Halouan, S., & Abahrour, M. (2022, October). Assessment of soil erosion rate using the EPM model: case of Ouaoumana basin, Middle Atlas, Morocco. In *IOP conference series: earth and environmental science* (Vol. 1090, No. 1, p. 012004). IOP Publishing. https://doi.org/10.1088/1755-1315/1090/1/012004

Eswaran H, Lal R, Reich PF (2019) Land degradation: an overview. Response Land Degrad 20–35. https://doi.org/10.1201/9780429187957

Fartas N, El Fellah B, Mastere M, Benzougagh B, El Brahimi M (2022) Potential soil erosion modeled with RUSLE approach and geospatial techniques (GIS tools and remote sensing) in Oued Joumouaa Watershed (Western Prerif-Morocco). Iraqi Geol J 47–61. https://doi.org/10.46717/igj.55.2B.5Ms-2022-08-21

Föeger LB, Buarque DC, Pontes PRM, de Oliveira Fagundes H, Fan FM (2022) Large-scale sediment modeling with inertial flow routing: assessment of Madeira River basin. Environ Model Softw 149:105332. https://doi.org/10.1016/j.envsoft.2022.105332

Gao J, Zhang Y, Zuo L (2023) The optimal explanatory power of soil erosion and water yield in karst mountainous areas. J Geog Sci 33(10):2077–2093. https://doi.org/10.1007/s11442-023-2166-y

Gavrilovic M (1972) Crimes within the territory of the Socialist Republic of Serbia. Yugoslavian J Crimin Crim L 10:479

Gavrilovic, I. T., Hormigo, A., Yahalom, J., DeAngelis, L. M., & Abrey, L. E. (2006). Long-term follow-up of high-dose methotrexate-based therapy with and without whole brain irradiation for newly diagnosed primary CNS lymphoma. *Journal of Clinical Oncology, 24*(28), 4570–4574.

Gavrilovic Z, Stefanovic M, Milovanovic I, Cotric J, Milojevic M (2008) Torrent classification–base of rational management of erosive regions. In: IOP conference series: earth and environmental science (vol 4(1)). IOP publishing, p 012039. https://doi.org/10.1088/1755-1307/4/1/012039

Ghobadi Y, Pirasteh S, Pradhan B, Ahmad NB, Shafri HZBM, Sayyad GA, Kabiri K (2011) Determine of correlation coefficient between EPM and MPSIAC models and generation of erosion maps by GIS techniques in Baghmalek watershed, Khuzestan, Iran. In Proceedings of the 5th Symposium on Advances in Science and Technology SAStech, Mashhad, Iran (pp 1–12)

Ghosh A, Rakshit S, Tikle S, Das S, Chatterjee U, Pande CB, Mattar MA (2022) Integration of GIS and remote sensing with RUSLE model for estimation of soil erosion. Land 12(1):116. https://doi.org/10.3390/land12010116

Gimeno-Vives O, Mohn G, Bosse V, Haissen F, Zaghloul MN, Atouabat A, de Lamotte DF (2020) Reply to comment by Michard et al. on "The mesozoic margin of the maghrebian tethys in the rif belt (Morocco): evidence for polyphase rifting and related magmatic activity". Tectonics 39. https://doi.org/10.1029/2020TC006165

Girma R, Gebre E (2020) Spatial modeling of erosion hotspots using GIS-RUSLE interface in Omo-Gibe River basin, Southern Ethiopia: implication for soil and water conservation planning. Environ Syst Res 9(1):1–14. https://doi.org/10.1186/s40068-020-00180-7

Guisan A, Thuiller W (2005) Predicting species distribution: offering more than simple habitat models. Ecol Lett 8(9):993–1009. https://doi.org/10.1111/j.1461-0248.2005.00792.x

Hagras A (2023) Estimating water erosion in the EL-Mador Valley Basin, South-West Matrouh City, Egypt, using revised universal soil loss equation (RUSLE) model through GIS. Environ Earth Sci 82(1):47. https://doi.org/10.1007/s12665-022-10722-0

Halder JC (2023) The integration of RUSLE-SDR lumped model with remote sensing and GIS for soil loss and sediment yield estimation. Adv Space Res 71(11):4636–4658. https://doi.org/10.1016/j.asr.2023.01.008

Hernández-Molina FJ, Sierro FJ, Llave E, Roque C, Stow DAV, Williams T, Brackenridge RE (2016) Evolution of the gulf of Cadiz margin and southwest Portugal contourite depositional system: tectonic, sedimentary and paleoceanographic implications from IODP expedition 339. Mar Geol 377:7–39. https://doi.org/10.1016/j.margeo.2015.09.013

Hosseini M, Ashraf MA (2015) Application of the SWAT model for water components separation in Iran. Springer Japan. https://doi.org/10.1007/978-4-431-55564-3

Hossini H, Karimi H, Mustafa YT, Al-Quraishi AMF (2022) Role of effective factors on soil erosion and land degradation: a review. In: Environmental degradation in Asia: land degradation, environmental contamination, and human activities. pp 221–235 https://doi.org/10.1007/978-3-031-12112-8_11

Istanbulluoglu E, Bras RL (2005) Vegetation-modulated landscape evolution: effects of vegetation on landscape processes, drainage density, and topography. J Geophys Res Earth Surf 110(F2). https://doi.org/10.1029/2004JF000249

Kabo-Bah KJ, Guoan T, Yang X, Na J, Xiong L (2021) Erosion potential mapping using analytical hierarchy process (AHP) and fractal dimension. Heliyon 7(6). https://doi.org/10.1016/j.heliyon.2021.e07125

Kadam AK, Umrikar BN, Sankhua RN (2018) Assessment of soil loss using revised universal soil loss equation (RUSLE): a remote sensing and GIS approach. Remote Sensing of Land, 2(1), 65–75. https://doi.org/10.21523/gcj1.18020105

Kader, S., Novicevic, R., Jaufer, L. (2022). Soil management in sustainable agriculture: analytical approach forthe ammonia removal from the diary manure. Agric For 68(4):69–78. https://doi.org/10.17707/AgricultForest.68.4.06

Kader S, Raimi MO, Spalevic V, Iyingiala AA, Bukola RW, Jaufer L, Butt TE (2023) A concise study on essential parameters for the sustainability of Lagoon waters in terms of scientific literature. Turkish J Agric For 47(3):288–307. https://doi.org/10.55730/1300-011X.3087

Kader S, Jaufer L, Bashir O, Olalekan Raimi M (2023) A comparative study on the stormwater retention of organic waste substrates niochar, sawdust, and wood bark recovered from Psidium Guajava L. Species. Agric For 69(1):105–112. https://doi.org/10.17707/AgricultForest.69.1.09

Kebede L, Temesgen M, Fanta A, Kebede A, Rockström J, Melesse AM (2023) Effect of locally adapted conservation tillage on runoff, soil erosion, and agronomic performance in semiarid rain-Fed farming in Ethiopia. Land 12(3):593. https://doi.org/10.3390/land12030593

Kefi M, Yoshino K, Setiawan Y (2012) Assessment and mapping of soil erosion risk by water in Tunisia using time series MODIS data. Paddy Water Environ 10:59–73. https://doi.org/10.1007/s10333-011-0265-3

Kumar G, Kurothe RS, Viswakarma AK, Mandal D, Sena DR, Mandal U, Dinesh D (2023) Assessment of soil vulnerability to erosion in different land surface configurations and management practices under semi-arid monsoon climate. Soil Tillage Res 230:105698. https://doi.org/10.1016/j.still.2023.105698

Lamane H, Moussadek R, Baghdad B, Mouhir L, Briak H, Laghlimi M, Zouahri A (2022) Soil water erosion assessment in Morocco through modeling and fingerprinting applications: a review. Heliyon. https://doi.org/10.1016/j.heliyon.2022.e10209

Lense GHE, Lämmle L, Ayer JEB, Lama GFC, Rubira FG, Mincato RL (2023) Modeling of soil loss by water erosion and its impacts on the Cantareira system, Brazil. Water 15(8):1490. https://doi.org/10.3390/w15081490

Li P, Chen J, Zhao G, Holden J, Liu B, Chan FKS, Mu X (2022) Determining the drivers and rates of soil erosion on the loess plateau since 1901. Sci Total Environ 823:153674. https://doi.org/10.1016/j.scitotenv.2022.153674

Liu B, Xie Y, Li Z, Liang Y, Zhang W, Fu S, Guo Q (2020) The assessment of soil loss by water erosion in China. Int Soil Water Conserv Res 8(4):430–439. https://doi.org/10.1016/j.iswcr.2020.07.002

Llanos-Paez O, Estrada L, Pastén-Zapata E, Boithias L, Jorda-Capdevila D, Sabater S, Acuña V (2023) Spatial and temporal patterns of flow intermittency in a Mediterranean basin using the SWAT+ model. Hydrol Sci J 68(2):276–289. https://doi.org/10.1080/02626667.2022.2155523

Maliqi E, Singh SK (2019) Quantitative estimation of soil erosion using open-access earth observation data sets and erosion potential model. Water Conserv Sci Eng 4:187–200. https://doi.org/10.1007/s41101-019-00078-1

Marko O, Gjoka K, Shkodrani N, Gjipalaj J (2023) Climate change effect on soil erosion in Vjosa River Basin. J Ecol Eng 24(2):92–100. https://doi.org/10.12911/22998993/156831

Marouane, L., Lahcen, B., & Valérie, M. (2021). Assessment and mapping of water erosion by the integration of the Gavrilovic "EPM" model in the Inaouene watershed, Morocco. In *E3S web of conferences* (Vol. 314, p. 03009). EDP Sciences. https://doi.org/10.1051/e3sconf/202131403009

Mehwish M, Nasir MJ, Raziq A, Al-Quraishi AMF (2024) Soil erosion vulnerability and soil loss estimation for Siran River watershed, Pakistan: an integrated GIS and remote sensing approach. Environ Monit Assess 196:104. https://doi.org/10.1007/s10661-023-12262-x

Meshesha TW, Wang J, Melaku ND, McClain CN (2021) Modelling groundwater quality of the Athabasca River Basin in the subarctic region using a modified SWAT model. Sci Rep 11(1):13574. https://doi.org/10.1038/s41598-021-92920-7

Mesrar H, Sadiki A, Navas A, Faleh A, Quijano L, Chaaouan J (2015) Soil erosion and causal factors modeling, case of the Sahla catchment, Central Rif, Morocco. Zeitschrift Für Geomorphologie 59(4):495–514. https://doi.org/10.1127/zfg/2015/0169

Michard A, Saddiqi O, Chalouan A, Chabou MC, Lach P, Rossi P, Youbi N (2020) Comment on "the mesozoic margin of the maghrebian tethys in the rif belt (Morocco): evidence for poly-phase rifting and related magmatic activity" by Gimeno-Vives et al. Tectonics 39(4):e2019TC006004. https://doi.org/10.1029/2019TC006004

Mićić Ponjiger T, Lukić T, Wilby RL, Marković SB, Valjarević A, Dragićević S, Morar C (2023) Evaluation of rainfall erosivity in the western balkans by mapping and clustering ERA5 reanalysis data. Atmosphere 14(1):104. https://doi.org/10.3390/atmos14010104

Mihi A, Benarfa N, Arar A (2020) Assessing and mapping water erosion-prone areas in northeastern Algeria using analytic hierarchy process, USLE/RUSLE equation, GIS, and remote sensing. Appl Geomat 12(2):179–191. https://doi.org/10.1007/s12518-019-00289-0

Morgan RPC (2009) Soil erosion and conservation. John Wiley & Sons

Neupane B, Mandal U, Al-Quraishi A, Ozdemir M (2023) Environmental threat of soil erosion in the Gwang Khola watershed, Chure Region of Nepal. Iraqi Geol J 56:194–206. https://doi.org/10.46717/igj.56.1e.14ms-2023-5-24

Ougougdal HA, Khebiza MY, Messouli M, Bounoua L (2020) Delineation of vulnerable areas to water erosion in a mountain region using SDR-InVEST model: a case study of the Ourika watershed, Morocco. Sci Afric 10:e00646. https://doi.org/10.1016/j.sciaf.2020.e00646

Ouyang W, Wu Y, Hao Z, Zhang Q, Bu Q, Gao X (2018) Combined impacts of land use and soil property changes on soil erosion in a mollisol area under long-term agricultural development. Sci Total Environ 613:798–809. https://doi.org/10.1016/j.scitotenv.2017.09.173

Panagos P, Christos K, Cristiano B, Ioannis G (2014) Seasonal monitoring of soil erosion at regional scale: an application of the G2 model in Crete focusing on agricultural land uses. Int J Appl Earth Obs Geoinf 27:147–155. https://doi.org/10.1016/j.jag.2013.09.012

Pandey S, Kumar P, Zlatic M, Nautiyal R, Panwar VP (2021) Recent advances in assessment of soil erosion vulnerability in a watershed. Int Soil Water Conserv Res 9(3):305–318. https://doi.org/10.1016/j.iswcr.2021.03.001

Pereira FC, Donovan M, Smith CM, Charters S, Maxwell TM, Gregorini P (2023) A geospatial modelling approach to understand the spatio-temporal impacts of grazing on soil susceptibility to erosion. Soil Systems 7(2):30. https://doi.org/10.3390/soilsystems7020030

Perri F, Martín-Martín M, Maaté A, Hlila R, Maaté S, Criniti S, Critelli S (2022) Provenance and paleogeographic implications for the cenozoic sedimentary cover of the ghomaride complex (Internal Rif Belt), Morocco. Mar Petrol Geol 143:105811. https://doi.org/10.1016/j.marpetgeo.2022.105811

Phinzi K, Ngetar NS (2019) The assessment of water-borne erosion at catchment level using GIS-based RUSLE and remote sensing: a review. Int Soil Water Conserv Res 7(1):27–46. https://doi.org/10.1016/j.iswcr.2018.12.002

Pourghasemi HR, Honarmandnejad F, Rezaei M, Tarazkar MH, Sadhasivam N (2021) Prioritization of water erosion–prone sub-watersheds using three ensemble methods in Qareaghaj catchment, southern Iran. Environ Sci Pollut Res 28:37894–37917. https://doi.org/10.1007/s11356-021-13300-2

Rahman MR, Shi ZH, Chongfa C, Dun Z (2015) Assessing soil erosion hazard-a raster-based GIS approach with spatial principal component analysis (SPCA). Earth Sci Inf 8:853–865. https://doi.org/10.1007/s12145-015-0219-1

Renard KG, Foster GR, Weesies GA, McCool DK, Yoder DC (1997) Predicting soil erosion by water: a guide to conservation planning with the Revised Universal Soil Loss Equation (RUSLE). US Department of Agriculture, Agriculture Handbook, p 703

Sakuno NRR, Guiçardi ACF, Spalevic V, Avanzi JC, Silva MLN, Mincato RL (2020) Adaptation and application of the erosion potential method for tropical soils. Rev Ciência Agron 51. https://doi.org/10.5935/1806-6690.20200004

Schaefer M, Thinh NX (2019) Evaluation of land cover change and agricultural protection sites: a GIS and remote sensing approach for Ho Chi Minh City, Vietnam. Heliyon 5(5). https://doi.org/10.1016/j.heliyon.2019.e01773

Senent-Aparicio J, Blanco-Gómez P, López-Ballesteros A, Jimeno-Sáez P, Pérez-Sánchez J (2021) Evaluating the potential of Glofas-era5 river discharge reanalysis data for calibrating the SWAT model in the Grande San Miguel River Basin (El Salvador). Remote Sens 13(16):3299. https://doi.org/10.3390/rs13163299

Sestras P, Mircea S, Roşca S, Bilaşco Ş, Sălăgean T, Dragomir LO, Kader S (2023) GIS based soil erosion assessment using the USLE model for efficient land management: a case study in an area with diverse pedo-geomorphological and bioclimatic characteristics. Notulae Botanicae Horti Agrobotanici Cluj-Napoca 51(3):13263–13263. https://doi.org/10.15835/nbha51313263

Sharma N, Acharya S, Kumar K, Singh N, Chaurasia OP (2018) Hydroponics as an advanced technique for vegetable production: an overview. J Soil Water Conserv 17(4):364–371. https://doi.org/10.5958/2455-7145.2018.00056.5

Sharma M, Bangotra P, Gautam AS, Gautam S (2022) Sensitivity of normalized difference vegetation index (NDVI) to land surface temperature, soil moisture and precipitation over district Gautam Buddh Nagar, UP, India. Stoch Environ Res Risk Assess 1–11. https://doi.org/10.1007/s00477-021-02066-1

Singh MC, Sur K, Al-Ansari N, Arya PK, Verma VK, Malik A (2023) GIS integrated RUSLE model-based soil loss estimation and watershed prioritization for land and water conservation aspects. Front Environ Sci 11:1136243. https://doi.org/10.3389/fenvs.2023.1136243

Solaimani K, Modallaldoust S, Lotfi S (2009) Investigation of land use changes on soil erosion process using geographical information system. Int J Environ Sci Technol 6:415–424. https://doi.org/10.1007/BF03326080

Spalevic V, Barovic G, Vujacic D, Curovic M, Behzadfar M, Djurovic N, Billi P (2020) The impact of land use changes on soil erosion in the river basin of Miocki Potok, Montenegro. Water 12(11):2973. https://doi.org/10.3390/w12112973

Svoray T (2022) Assessments of erosion risk. In: A geoinformatics approach to water erosion: soil loss and beyond. Springer International Publishing, Cham, pp 205–263. https://doi.org/10.1007/978-3-030-91536-0_6

Talchabhadel R, Nakagawa H, Kawaike K, Prajapati R (2020) Evaluating the rainfall erosivity (R-factor) from daily rainfall data: an application for assessing climate change impact on soil loss in Westrapti River basin, Nepal. Model Earth Syst Environ 6:1741–1762. https://doi.org/10.1007/s40808-020-00787-w

Tangestani MH (2006) Comparison of EPM and PSIAC models in GIS for erosion and sediment yield assessment in a semiarid environment: Afzar Catchment, Fars Province, Iran. J Asian Earth Sci 27(5):585–597. https://doi.org/10.1016/j.jseaes.2005.06.002

Tariq M, Akhtar K (2022) Mulching is an approach for a significant decrease in soil erosion. In: Mulching in agroecosystems: plants, soil & environment. Springer Nature Singapore, Singapore, pp 59–70. https://doi.org/10.1007/978-981-19-6410-7_4

Uddin K, Abdul Matin M, Maharjan S (2018) Assessment of land cover change and its impact on changes in soil erosion risk in Nepal. Sustainability 10(12):4715. https://doi.org/10.3390/su10124715

Van Huyssteen CW, du Preez CC (2023) Agricultural land degradation in South. Africa. https://doi.org/10.1007/698_2022_922

Vatandaşlar C, Yavuz M (2023) Useful indicators and models for assessing erosion control ecosystem service in a semi-arid forest landscape. Environ Monit Assess 195(1):208. https://doi.org/10.1007/s10661-022-10814-1

Vitale S, Zaghloul MN, El Ouaragli B, Tramparulo FDA, Ciarcia S (2015) Polyphase deformation of the Dorsale Calcaire complex and the maghrebian flysch basin units in the Jebha area (Central Rif, Morocco): new insights into the Miocene tectonic evolution of the Central Rif belt. J Geodyn 90:14–31. https://doi.org/10.1016/j.jog.2015.07.002

Wang ZJ, Jiao JY, Rayburg S, Wang QL, Su Y (2016) Soil erosion resistance of "Grain for Green" vegetation types under extreme rainfall conditions on the Loess Plateau, China. CATENA 141:109–116. https://doi.org/10.1016/j.catena.2016.02.025

Wang Y, Zhang J, Chen W, Sun G (2017) Assessment of soil erosion risk and its spatial distribution in the Loess Plateau region of China using the revised universal soil loss equation (RUSLE) and geo-detector. Sustainability 9(7):1179

Wang J, Zhuan R, Chu L (2019) The occurrence, distribution and degradation of antibiotics by ionizing radiation: an overview. Sci Total Environ 646:1385–1397. https://doi.org/10.1016/j.scitotenv.2018.07.415

Wassie SB (2020) Natural resource degradation tendencies in Ethiopia: a review. Environ Syst Res 9(1):1–29. https://doi.org/10.1186/s40068-020-00194-1

Wischmeier WH, Smith DD (1978) Predicting rainfall erosion losses—a guide to conservation planning. In: Agriculture handbook, p 537

Yen H, Park S, Arnold JG, Srinivasan R, Chawanda CJ, Wang R, Zhang X (2019) IPEAT+: A built-in optimization and automatic calibration tool of SWAT+. Water 11(8):1681. https://doi.org/10.3390/w11081681

Zachar, D. (1982). Erosion factors and conditions governing soil erosion and erosion processes. In *Soil erosion* (pp. 205–387).

Zeghmar A, Marouf N, Mokhtari E (2022) Assessment of soil erosion using the GIS-based erosion potential method in the Kebir Rhumel Watershed, Northeast Algeria. J Water Land Dev (52). https://doi.org/10.24425/jwld.2022.140383

Zema DA, Nunes JP, Lucas-Borja ME (2020) Improvement of seasonal runoff and soil loss predictions by the MMF (Morgan-Morgan-Finney) model after wildfire and soil treatment in Mediterranean forest ecosystems. CATENA 188:104415. https://doi.org/10.1016/j.catena.2019.104415

Zhao L, Hou R (2019) Human causes of soil loss in rural karst environments: a case study of Guizhou, China. Sci Rep 9(1):3225. https://doi.org/10.1038/s41598-018-35808-3

Zhi J, Zhou Z, Cao X (2021) Exploring the determinants and distribution patterns of soil mattic horizon thickness in a typical alpine environment using boosted regression trees. Ecol Ind 133:108373. https://doi.org/10.1016/j.ecolind.2021.108373

Zhou CB, Chen YF, Hu R, Yang Z (2023) Groundwater flow through fractured rocks and seepage control in geotechnical engineering: theories and practices. J Rock Mech Geotech Eng 15(1):1–36. https://doi.org/10.1016/j.jrmge.2022.10.001

Predicting Soil Erosion Using RUSLE Model in Duhok Governorate, Kurdistan Region of Iraq

Azade Mehri, Hazhir Karimi, Yaseen T. Mustafa, Ayad M. Fadhil Al-Quraishi[ID], and Saman Galalizadeh

Abstract Soil erosion is a main cause of land degradation, adversely impacting soil health, soil fertility, and soil carbon. This study used the revised universal soil loss equation (RUSLE) model and geoinformatics tools to predict soil erosion rates in the Duhok governorate, located in the Kurdistan Region of Iraq (KRI). The RUSLE model integrates several to predict annual soil loss: soil erodibility factor (K), rainfall erosivity factor (R), slope (L), slope steepness factor (S), land use/cover factor (C), and conservation methods (P). Information layers for these variables were created using ArcGIS 10.3, showing the extent of soil erosion. Additionally, we identified and prioritized erosion hotspots to implement conservation practices. The results showed that approximately 78% of the study area experienced very low and low levels of soil erosion, whereas only 4 and 7% were subject to high and very high erosion rates, respectively. The findings also emphasized the significance of slope and land-use interactions in accelerating soil erosion, particularly in agricultural areas characterized by steep slopes.

A. Mehri
Young Researchers and Elite Club, Gorgan Branch, Islamic Azad University, Gorgan, Iran
e-mail: az.mehri@yahoo.com

H. Karimi (✉)
Department of Biological Science, University of Alabama, Tuscaloosa, USA
e-mail: hazhir.karimi25@gmail.com

Y. T. Mustafa
Department of Environmental Science, Faculty of Science, University of Zakho, Zakho, Kurdistan Region, Iraq
e-mail: yaseen.mustafa@uoz.edu.krd

A. M. F. Al-Quraishi
Petroleum and Mining Engineering Department, Faculty of Engineering, Tishk International University, Erbil 44001, Kurdistan Region, Iraq
e-mail: ayad.alquraishi@tiu.edu.iq; ayad.alquraishi@gmail.com

S. Galalizadeh
School of Science, Edith Cowan University, Joondalup, WA, Australia
e-mail: s.galalizadeh@ecu.edu.au

Keywords Land degradation · RUSLE · GIS · Land use and land cover · Duhok

1 Introduction

Soil erosion is a serious issue affecting ecosystems and food security (FAO 2019). Soil erosion can also lead to the loss of important topsoil, a drop in nutrients and organic matter, and a reduction in rooting depth, all of which can reduce crop yield (Sun et al. 2014; Al-Quraishi 2003). Moreover, erosion significantly threatens the functions of soil ecosystems, such as food, fiber, purification, and energy provision (Hossini et al. 2022; Karimi et al. 2018). Sediments from erosion affect water usage for different purposes such as irrigation, groundwater resilience, power generation, and fishing (Sun et al. 2014; Moghaddasi et al. 2022). Common types of erosion include water, rill, wind, scalding, and tunnel erosion (Tadesse et al. 2017; Wang et al. 2024). The most prevalent form of soil erosion is water erosion, caused by floods, precipitation, and surface runoff (Bag et al. 2022; Oldeman 1994). Wind erosion is another prevalent type of soil erosion that occurs predominantly in arid and semi-arid regions (FAO 2019; Fadhil 2009).

Both anthropogenic and natural factors drive soil erosion. Agricultural practices, deforestation, urbanization, and overgrazing are some of the most essential human causes of soil erosion (KouroshNiya et al. 2020; Pirsaheb et al. 2020; Fadhil 2013). In addition, natural factors, such as climate and weather, land use patterns, physiography, and soil characteristics, affect soil erosion rates (Neamat and Karimi 2020; Karimi et al. 2020; Sadeghi et al. 2021). Estimating and assessing erosion rates is crucial for developing preventative and restoration methods because they can help understand the causes of erosion and make more accurate predictions under various scenarios (Teng et al. 2018; Tsunekawa et al. 2017; Belayneh et al. 2019). Duhok, located in the north of Iran, is expected to experience soil erosion due to land use/cover changes, rapid population growth, and forest fires. However, few have evaluated the soil erosion rate in this region. Thus, this study aims to fill this gap and explore the erosion rate in Duhok to implement suitable practices.

1.1 Soil Erosion Models

If they provide accurate estimates, trends, and future erosion scenarios, soil erosion models can be used to assess soil loss (Ganasri and Ramesh 2016; Koirala et al. 2019). Several quantitative and reliable soil erosion models have been developed. The models differ in complexity, procedures, data, outputs, and scales (Igwe et al. 2017). Les complex erosion models at the watershed scale produce less accurate forecasts than the relatively basic ones (Al-Quraishi 2004). The application of a model depends on the soil type, climate, and geomorphology of a region (Igwe et al. 2017; Fereshtehpour et al. 2024). Generally, there are three types of soil erosion models:

Table 1 Soil erosion averages for various land use classes

Land use/cover type	Average soil erosion (t ha^{-1} yr^{-1})
Built-up	1.29
Agriculture	178.04
Forest	4.55
Rock	0.70
Soil	107.70

empirical, statistical, conceptual, and physics-based models. The soil erosion models listed in Table 1 are widely utilized.

1.1.1 Empirical Models

Empirical models are the simplest and mainly developed based on statistical observations. Empirical models can estimate and measure the soil erosion rate when limited data are available and are particularly useful in identifying sediment sources and nutrient generation (Merritt et al. 2003; Igwe et al. 2017). Empirical models are more straightforward than the other models.

- **Universal Soil Loss Equation (USLE)**

Wischmeier and Smith (1978) created the USLE, which has been predominantly employed in croplands or gently sloping terrains (Igwe et al. 2017). It provides a statistical relationship between soil erosion rates and variables such as vegetation, soil, terrain, climate, and human activity. The availability of parameter values makes the USLE rather simple to employ; however, this model has several shortcomings, such as disregarding the impact of rainfall and runoff processes on soil erosion (Merritt et al. 2003).

- **Revised Universal Soil Loss Equation (RUSLE)**

Renard et al. (1997) introduced RUSLE, an improved and computerized version of the USLE RUSLE developed by integrating further research and practical expertise. RUSLE projects the average yearly rate of soil loss under numerous management approaches and erosion control scenarios (Igwe et al. 2017). This model suits various land-use and cover patterns (Al-Quraishi 2003; Sun et al. 2014).

1.1.2 Conceptual Models

Conceptual models integrate physically based and empirical models (Beck 1987). The main outputs of these models are the sediment yield, and rainfall and runoff are the main inputs (Igwe et al. 2017). Using these models, the quantitative and qualitative effects of land use changes can be observed in a watershed basin. Because the value of each parameter is obtained from stream discharge and sediment concentration, simple conceptual models have fewer problems than other complex models. Therefore, the number of parameters was assessed to reduce model identification problems.

- **Agricultural Non-Point Source Pollution Model (AGNPS)**

AGNPS is one of the most frequently employed conceptual models. AGNPS models runoff, sediment, and nutrient transportation in an agricultural watershed system (Igwe et al. 2017). This approach addresses the difficulty of identifying non-point pollution sources. It requires several fundamental inputs, including elevation, land cover, soil, hydrology, sediment, and chemistry, whereas the output parameters include sediment and hydrology.

1.1.3 Physically-Based Models

Physically based models have been designed to detect non-point source pollution and off-site erosion consequences and effectively predict the geographical distribution of runoff and sediment (Igwe et al. 2017). The Water Erosion Prediction Project (WEPP) is a commonly physically based model developed to evaluate soil erosion in large areas. The four input files required for WEPP are soil, climate, topography, and management (Igwe et al. 2017). The outputs contain the monthly and annual storm runoff (Merritt et al. 2003).

2 Materials and Methods

2.1 Case Study

The Duhok governorate, located in the Kurdistan region has latitudes and longitudes 36°18′12.64" and 37°20′33.55" N and 42°20′25.36" and 44°17′40.50" E, respectively. Its altitude varies from 430 to 2500 m above sea level. The governorate is approximately 11,066 km^2 in size and has seven districts (Fig. 1). There are approximately 1,602,624 residents (Directorate of Statistics, Duhok). Duhok has warm, dry summers, and pleasant winters. The annual average rainfall is about 616 mm, with the highest rainfall rate occurring during the winter. The annual temperature range is between 3°C and 42°C.

2.2 Methods

This study combined RUSLE and GIS to determine annual soil loss in Duhok. The RUSLE model employs Eq. 1 to calculate the average annual erosion.

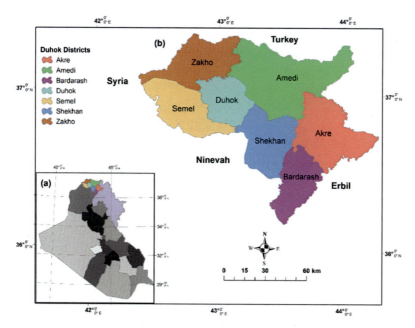

Fig. 1 a Map of Iraq that includes the Kurdistan Region, a A map of the study area (Duhok)

$$A = R \times K \times L \times S \times C \times P \tag{1}$$

A: annual soil erosion (t ha^{-1} yr^{-1})
R: rainfall erosivity
K is the soil erodibility
L: slope length
S: slope steepness
C: land cover and land use
P: conservation practices

The following sections describe the parameter collection, preparation procedure, and factor calculation based on the RUSLE model.

2.2.1 Rainfall Erosivity Factor (R)

The R factor was determined using storm energy (E, MJ m^{-2}) and peak 30-min rainfall intensity (Meshesha et al. 2012). Rainfall intensity represents the erosivity of the precipitation. The spatial distribution of average precipitation was estimated using a geostatistical method utilizing monthly precipitation data from 18 sites from 2002 to 2014 (Fig. 2).

R factor was calculated using the Modified Fournier Index (MFI) (Renard and Freimund 1994) according to Eq. 2.

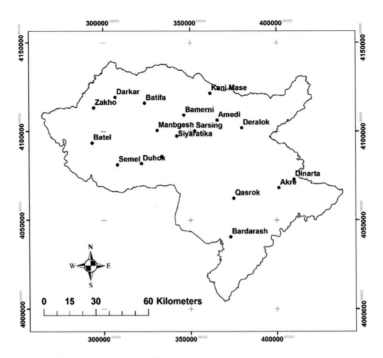

Fig. 2 Location of weather stations in the study area

$$\text{MFI} = \frac{1}{N} \sum_{j=1}^{N} \sum_{i=1}^{12} \frac{P_{ij}^2}{P_j} \quad (2)$$

where Pij is the mean monthly rainfall (mm) in the month i of year j, Pj is the mean annual rainfall (mm) in year j, and N is the number of years. Finally, Eq. (3) was used to estimate the R-factor.

$$\text{IF MFI} < 55 \quad R = (0.07397\text{MFI}1.847)/17.02 \quad (3)$$

$$\text{IF MFI} \geq 55 \quad R = (95.77 - 6.081\text{MFI} + 0.477\text{MFI}2)/17.02$$

2.2.2 Soil Erodibility Factor (K)

The soil erodibility factor (K) is associated with soil permeability, organic matter, and other soil parameters (Sun et al. 2014). This information was acquired from Duhok University. Fifteen sampling locations were included in the study region. The parameter values were entered into ArcGIS, and the point sources were converted to polygon areas using geostatistical techniques. K was determined using the following formula (Sun et al. 2014):

$$K = \{0.2 + 0.3\exp[-0.0256\text{SAN}\frac{(1-SIL)}{100}]\}(\frac{SIL}{CLA+SIL})0.3$$
$$\times (1.0 - \frac{0.25C}{C+\exp(3.72-2.95C)})(1.0 - \frac{0.7SNI}{SNI+\exp(-5.51+22.9SNI)})$$

(4)

SAN, sand fraction (%); SIL, silt fraction (%); CLA, clay fraction (%); C, soil organic carbon content (%); SNI:1-SAN / 100.

2.2.3 Land Cover and Land Use Factor (C)

This factor indicates how different cropping and management practices, crop sequences and yields, soil cover, and subsurface biomass could potentially influence soil erosion rates (Alexakis et al. 2013; Renard et al. 1997). The land use/land cover (LULC) map generated a C-factor map for the research region. The data used to generate the LULC map were obtained from the Copernicus Scientific Data Hub. The map was generated using data collected by Sentinel-2 at level 1-C on August 14, 2019. After being orthorectified, the data were spatially registered on a global reference system and scaled to the level of the uppermost part of the atmosphere (TOA). User Guide for Sentinel-2, published by the Sentinel-2 Team and European Space Agency in 2015. In order to generate a bottom-of-atmosphere (BOA-) corrected reflectance image from Sentinel-2 level 1-C data, the ATCOR method contained within the Sen2Cor plugin of the Sentinel Application Platform (SNAP) software was used to transform the data. This resulted in the data being elevated to level 2-A. (Louis et al. 2016).

Subsequently, the image was resampled to a spatial resolution of 10 m using the nearest neighbor approach, which changes the size of the sample plots. Sub-setting was ultimately used to determine where the study would take place. Subsequently, the image was classified using K-Means Cluster Analysis, which is an unsupervised land classification method. When creating the false-color composite map, K-Means Cluster Analysis was applied to each band and the resulting image was divided into six LULC classes. It comprises towns, cities, wooded areas, agricultural land, water bodies, soil, and rocks. An accuracy assessment of the detailed image was performed to determine the value of the information constructed from the raw data. Consequently, a confusion matrix was built to implement the classification accuracy method. A LULC map was produced when the classified raster image was converted into a vector format. The LULC map had to be classified to use the C-factor value to produce the C-factor map.

2.2.4 Topographic Factor (LS)

Soil erosion is more likely to occur in regions with steep slopes (Alexakis et al. 2013). A DEM layer was employed to calculate the L and S factors. A DEM map was created using Advanced Spaceborne Thermal Emission and Reflection Radiometer (ASTER) data (ASTER-GDEM 2013) downloaded from the Global Data Explorer (USGS 2005). The following equation was used to determine the LS factor (Renard et al. 1997).

$$L = \left(\frac{\lambda}{22.13}\right)^m \tag{5}$$

where λ is the slope length (in meters) and m is the slope length exponent that can be adjusted.

$$m = \frac{\beta}{1+\beta} \quad \beta = \frac{\sin\theta/0.0896}{3.0(\sin\theta)^{0.8}+0.56} \tag{6}$$

where θ denotes the slope angle.

$$S = 10.8 \sin\theta + 0.03 \quad s < 9\% \tag{7}$$

$$S = 16.8 \sin\theta - 0.50 \quad s \geq 9\%$$

where θ is the slope angle and s is the percent slope.

2.2.5 Conservation Practices Factor (P)

The conservation practice factor (P) quantifies the impact of soil erosion management measures, actions, and practices (Alexakis et al. 2013). c 2013). The P factor was assigned a value of 1 because no substantial conservation practices are currently in place in the Duhok Governorate.

2.2.6 Assessing Impacts of Land Use and Slope on Erosion Rate

Evaluating LULC types and slope classes influencing soil erosion rates could help identify hotspots to place appropriate solutions (Meshesha et al. 2012). The current study evaluated the relevance of LULC and slope on soil erosion loss using cross-tabulation in GIS.

2.2.7 Estimation of Soil Erosion

The amount of soil erosion lost was predicted using LULC, soil properties, rainfall intensity, topographic features, and various conservation techniques. GIS is used to convert raster grid layers into the same reference systems and the same spatial resolution (UTM 38 N) (30 m grid). The maps were layered on top of one another to calculate the rate of soil erosion loss and locate the vulnerable areas in the study area that were being studied. The most vulnerable regions were assigned the highest values, while the least vulnerable regions received the lowest values.

3 Results and Discussion

The factor maps produced are shown in Fig. 3. The R value ranged from 209 to 1110 MJ mm ha^{-1} h^{-1} yr^{-1}, with an average value of 525 MJ mm ha^{-1} h^{-1} yr^{-1}. These findings indicate that the Duhok Governorate receives a significant amount of precipitation on an annual basis. The soil erodibility factor, denoted by the letter K, ranged from 0.25 to 0.42 t ha h ha^{-1} MJ^{-1} mm^{-1}, with a mean value of 0.33 t ha h ha^{-1} MJ^{-1} mm^{-1}. The topographic factor (LS) averaged at 4.15 on average, ranging from 0 in level areas to 26.07 in hilly and steep sections. The level areas had topographic factors of 0. The land-cover factor (C) ranged from 0 to 0.36, with a mean value of 0.08.

Seasonal patterns of soil erosion across the region are shown in Fig. 4. The soil erosion was classified into the following five categories: very minor (up to five tons ha-1 per year), slight (5 to 10), moderate (10 to 50), severe (50 to 100), and extremely severe (above 100) (Fig. 5 and Table 2). In the Duhok governorate, the average soil loss was 26.87 ha^{-1} h^{-1} yr^{-1}. Comparatively, the very severe class covers only 6.9% of the study area, whereas the very gentle class covers approximately 68% of the region. The percentage of people who experienced mild, moderate, and severe symptoms was 9.52, 11.1, and 3.68%, respectively.

Table 1 illustrates the average amount of soil lost due to erosion across the various land-use categories. The table shows that agricultural and soil lands face the greatest amount of soil erosion, as indicated by Karimi et al. (2022), who conducted a land degradation analysis for the same study area. Soil erosion rates were the lowest in rock land use, followed by urban and forest areas. Tables 2 and 3 display the proportion of area affected by each erosion severity level and the cross-tabulation outcomes involving slope, land use classification, and average soil erosion. These findings shed light on the significance of the interaction between land use and slope in increasing soil erosion rates in the soil and agricultural classes on steep slopes. The soil and agriculture classes were responsible for the greatest amount of soil loss on an otherwise similar slope. The results of the correlation between the slope and LULC with soil are shown in Fig. 6. For a given type of land use, an increase in slope percentage led to a proportional increase in the rate of soil erosion. The graphs created for the soil and agriculture classes show that the rate of soil erosion on low

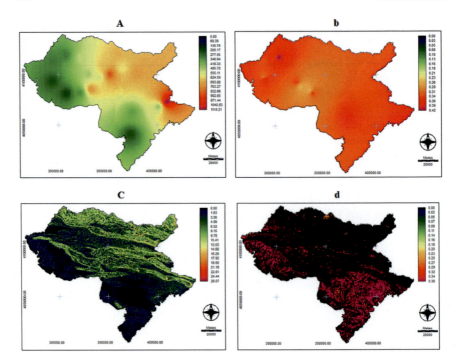

Fig. 3 Maps of RUSLE factors: **a** R facto, **b** K factor, **c** LS factor, **d** C factor

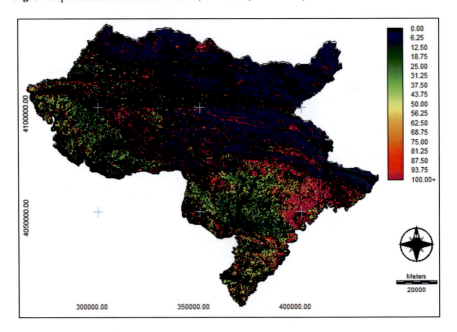

Fig. 4 A map depicting the annual potential for soil erosion in the area under study (t ha-1 yr-1)

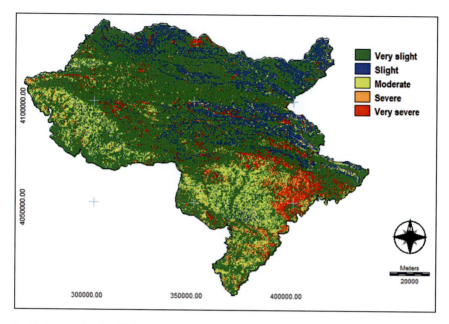

Fig. 5 A map of reclassified annual potential soil erosion

Table 2 The proportion of area affected by each level of erosion severity

No	Erosion status	Soil erosion (t ha^{-1} yr^{-1})	Area (Hectares)	Area (% of total)
1	Very slight	0–5	756,194.18	68.66
2	Slight	5–10	104,835.91	9.52
3	Moderate	10–50	123,664.88	11.23
4	Severe	50–100	40,565.21	3.68
5	Very severe	> 100	76,107.41	6.91

slopes is dramatically increasing. These findings highlight the importance of slopes in the process of soil erosion. Restoring these fragile ecosystems and maintaining constant vigilance in and around them should be a top priority.

The findings illustrated that the study area experience an average soil loss of 26.87 t ha^{-1} yr^{-1}. Soil erosion in some areas has led to the removal of valuable topsoil. The findings highlight the importance of taking prompt action to prevent and reduce soil erosion, particularly in regions with severe and extremely severe erosion, which accounted for approximately 11% of the total governorate. The most sensitive areas are farmlands, whereas forested areas have much less vulnerability. These results also indicate a strong dependence on soil erosion and land-use types. The rate of soil erosion varied within different land use/land cover types, and the maximum rate occurred in agricultural lands, especially on steep slopes.

Table 3 Land use classes, average soil erosion and cross-tabulation of slope (t ha^{-1} yr^{-1})

Slope classes (%) land use	0–10	10–20	20–30	30–40	40–50	50–60	60–70	70–80	80–90	90–100	>100
Built-up	0.26	1.13	2.38	3.55	4.61	5.41	6.20	6.95	7.72	8.53	10.26
Agriculture	23.06	121.96	245.47	356.04	452.72	532.09	603.03	665.24	732.66	787.01	895.55
Forest	0.57	1.69	3.00	4.26	5.45	6.53	7.51	8.38	9.17	9.82	11.20
Rock	0.11	0.43	0.86	1.27	1.67	2.04	2.39	2.70	2.97	3.21	3.76
Soil	33.89	144.51	302.58	436.69	561.67	682.24	792.08	882.70	959.92	1038.83	1150.09

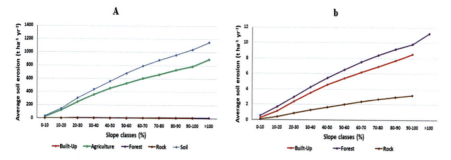

Fig. 6 Average soil erosion changes for land use/cover and slope

Rainfall significantly affected soil erosion loss in the Duhok Governorate. The annual rainfall at some stations exceeded 1000 mm; however, the average annual rainfall across the study area was approximately 710 mm. This rainfall increases soil loss, especially in areas with high slopes and barren lands, with or without partial distribution of vegetation cover. Thus, preserving vegetation cover and forests should be considered in soil conservation because the decline in erosive precipitation and soil loss are strongly affected by vegetation.

Several studies that used RUSLE to estimate the soil erosion loss rate have validated the accuracy of the model (e.g., Meshesha et al. 2012; Sun et al. 2014; Mehri et al. 2018; Fang et al. 2019; Mohammed et al. 2020). Meshesha et al. (2012) used RUSLE to estimate the rate of soil erosion in the Central Rift Valley of Ethiopia. They demonstrated that the farmland erosion rate increased by 80% over the past 30 years, and the accuracy of their assessment was 71.1%. They believed that the fundamental cause of soil erosion was clearing land for agricultural use, which involved cutting down trees. RUSLE was applied by Sun et al. (2014) to analyze soil erosion in the Loess Plateau in China and to establish the role of topography in causing erosion. This study found that agricultural practices caused the highest soil loss. A statistically significant correlation (P = 0.01) existed between the calculated rate and field data (P 0.01). In addition, Al-Abadi et al. used RUSLE to predict annual soil erosion in the Kirkuk governorate in northern Iraq. The yearly erosion rate in the research region is expected to be 2 t ha^{-1} yr^{-1}; however, the rate we measured in our case study was 26.87 t ha^{-1} yr^{-1}.

Previous studies have highlighted the significance of conservation strategies and programs for soil erosion rates. In 2017, Tadesse et al. conducted an assessment to determine Ethiopia's rate of land use/land cover change as well as soil erosion. According to the findings, the rate of soil erosion was reduced by half, from 7.7 tons per hectare per year to 4.8 t ha^{-1} yr^{-1}, as a direct result of implementing management strategies, which included expanding plant life. The RUSLE was utilized by Haregeweyn et al. (2017) to determine the rate of soil erosion and compute the effect of soil erosion both before and after the C and P factors were modified. According to the data, implementing conservation practices resulted in an overall reduction of 52% in the amount of soil lost in the river basin. RUSLE and GIS were utilized in a study conducted by Mehri et al. (2018) in the Gharsoo River Basin of Iran to examine the potential for land use planning to reduce soil erosion rates. New land use was developed because of the implementation of land use planning within the river basin. On average, the risk of soil erosion was reduced to 25.6%, as indicated by the created land use map.

4 Conclusions

The purpose of this study is to utilize the RUSLE model and GIS to assess soil erosion loss and the impacts of various forms of land use, land cover, and slope on soil erosion. While the findings suggest a minimal probability of soil erosion loss in the research area, conservation plans and programs should prioritize specific high-risk sites. Future studies should investigate whether land management techniques, such as terracing, contour farming, and agroforestry, can effectively reduce soil loss in the Duhok governorate. The predicted rate of soil erosion provided by this study has the potential to serve as the foundation for soil conservation efforts in Duhok. The methodology of this study is flexible and can be applied to other areas of Iraq where data and indicators are readily available. We strongly recommend applying this approach in future studies across different regions of Iraq, especially in deteriorated areas, as it holds the potential to enhance land management and the conservation of natural resources.

5 Recommendations

The estimation of soil erosion loss in the Duhok Governorate showed that a significant portion had a slight level. However, severe soil erosion rates have been observed in agricultural classes with steep slopes, and reclamation measures and implementation plans should be considered in these areas. The main recommended plans for farmlands are crop rotation, conservation tillage, contour binding, contour plowing, and cover crops. In addition, some parts of the forest areas within the governorate have degraded due to political conflicts in the last decade. Deforestation and forest fires have significantly increased the rate of soil erosion in these areas. Managers should be concerned about these degradations and fuel accumulation from forest fires; therefore, it is essential to overcome the threats and reclamation of degraded forests. For erosion-sensitive sites, management factors need to be carefully evaluated.

The soil erosion loss rate estimation in the Duhok Governorate indicates that a significant portion of the region experiences a slight level of erosion. However, areas with steep slopes primarily used for agriculture show a severe soil erosion rate. Addressing this issue by implementing reclamation measures in vulnerable areas is crucial. The recommended plans for farmlands include crop rotation, conservation tillage, contour binding, contour plowing, and cover crops. Furthermore, some parts of the forest areas in the governorate have suffered degradation because of political conflicts in the recent past. Deforestation and forest fires have significantly accelerated the rate of soil erosion in these regions. Managers must be attentive to these degradations and fuel accumulations from forest fires. Therefore, addressing these threats and focusing on reclaiming degraded forests are crucial. Management factors must be carefully evaluated for erosion-sensitive sites to ensure the implementation of adequate measures.

Acknowledgements The authors thank the University of Zakho's assistance and the Department of Agriculture at the University of Duhok for providing us with the necessary data and information.

References

Alexakis DD, Hadjimitsis DG, Agapiou A (2013) Integrated use of remote sensing, GIS and precipitation data for the assessment of soil erosion rate in the catchment area of Yialias in Cyprus. Atmos Res 131:108–124

Al-Quraishi AMF (2003) Soil erosion risk prediction with RS and GIS for the Northwestern part of Hebei Province, China. J Appl Sci 3:659–666

Al-Quraishi AMF (2004) Assessment of soil erosion risk using RUSLE and geoinformation technology for North Shaanxi Province, China. J China Univ Geosci 15:31–39

ASTER-GDEM (2013) ASTER global digital elevation model (GDEM). http://gdem.ersdac.jspace systems.or.jp/. Accessed on 15 Aug 2018

Bag R, Mondal I, Dehbozorgi M, Bank SP, Das DN, Bandyopadhyay J, Pham QB, Al-Quraishi AMF, Nguyen XC (2022) Modelling and mapping of soil erosion susceptibility using machine learning in a tropical hot sub-humid environment. J Clean Prod 364:132428

Beck MB (1987) Water quality modeling: a review of the analysis of uncertainty. Water Resour Res 23:1393–1442

Belayneh M, Yirgu T, Tsegaye D (2019) Potential soil erosion estimation and area prioritization for better conservation planning in Gumara watershed using RUSLE and GIS techniques. Environ Syst Res 8:20

Fadhil AM (2009) Land degradation detection using geo-information technology for some sites in Iraq. J Al-Nahrain Univ Sci 12:94–108

Fadhil AM (2013) Sand dunes monitoring using remote sensing and GIS techniques for some sites in Iraq. In: Tan H (ed) PIAGENG 2013: intelligent information, control, and communication technology for agricultural engineering

Fang G, Yuan T, Zhang Y, Wen X, Lin R (2019) Integrated study on soil erosion using RUSLE and GIS in Yangtze river basin of Jiangsu Province (China). Arab J Geosci 12. https://doi.org/10.1007/s12517-019-4331-2

FAO (2019) Soil erosion: the greatest challenge to sustainable soil management. Rome, p 100

Fereshtehpour M, Esmaeilzadeh M, Saleh Alipour R, Burian S (2024) Impacts of DEM type and resolution on deep learning-based flood inundation mapping. Earth Sci Inf 1–21. https://doi.org/10.1007/s12145-024-01239-0

Ganasri BP, Ramesh H (2016) Assessment of soil erosion by RUSLE model using remote sensing and GIS–a case study of Nethravathi Basin. Geosci Front 7:953–961

Haregeweyn N, Tsunekawa A, Poesen J, Tsubo M, Meshesha DT, Fenta AA, Nyssen J, Adgo E (2017) Comprehensive assessment of soil erosion risk for better land use planning in river basins: case study of the Upper Blue Nile River. Sci Total Environ 1(574):95–108. https://doi.org/10.1016/j.scitotenv.2016.09.019

Hossini H, Karimi H, Mustafa YT, Al-Quraishi AMF (2022) Role of effective factors on soil erosion and land degradation: a review. In: Al-Quraishi AMF, Mustafa YT, Negm AM (eds) Environmental degradation in Asia. Earth and environmental sciences library. Springer, Cham

Igwe PU, Onuigbo AA, Chinedu OC, Ezeaku II, Muoneke MM (2017) Soil erosion: a review of models and applications. Int J Adv Eng Res Sci 4(12)

Karimi H, Jafarnezhad J, Kakhani A (2020) Landsat time-series for land use change detection using support vector machine: case study of Javanrud District, Iran. Int Conf Comp Sci Softw Eng (CSASE) 2020:128–131

Karimi H, Jafarnezhad J, Khaledi J, Ahmadi P (2018) Monitoring and prediction of land use/land cover changes using CA-Markov model: a case study of Ravansar County in Iran. Arab J Geosci 11(592)

Karimi H, Mustafa YT, Hossini H, Al-Quraishi AMF (2022) Assessment of land degradation vulnerability using GIS-based multicriteria decision analysis in Zakho District, Kurdistan Region of Iraq. In: Al-Quraishi AMF, Mustafa YT, Negm AM (eds) Environmental degradation in Asia. Earth and environmental sciences library. Springer, Cham

Koirala P, Thakuri S, Joshi S, Chauhan R (2019) Estimation of soil erosion in Nepal using a RUSLE modeling and geospatial tool. Geosciences 9:147

KouroshNiya A, Huang J, Kazemzadeh-Zow A, Karimi HN (2020) Comparison of three hybrid models to simulate land use changes: a case study in Qeshm Island, Iran. Environ Monit Assess 192:302

Louis J, Debaecker V, Pflugetal B (2016) Sentinel-2Sen2Cor:L2A processor for users. In: Living planet symposium, p 91. Prague, Czech Republic

Mehri A, Salman Mahiny A, Mikaeili Tabrizi A, Mirkarimi H, Sadoddin A (2018) Investigation of likely effects of land use planning on reduction of soil erosion rate in river basins: case study of the Gharesoo River Basin. CATENA 167:116–129

Merritt WS, Letcher RA, Jakeman AJ (2003) A review of erosion and sediment transport model. Environ Model Softw 18:761–799

Meshesha DT, Tsunekawa A, Tsubo M, Haregeweyn N (2012) Dynamics and hotspots of soil erosion and management scenarios of the Central Rift Valley of Ethiopia. Int J Sediment Res 27:84–99

Moghaddasi P, Kerachian R, Sharghi S (2022) A stakeholder-based framework for improving the resilience of groundwater resources in arid regions. J Hydrol 609:127737

Mohammed S, Alsafadi K, Talukdar S, Kiwan S, Hennawi S, Alshihabi O, Sharaf M, Harsanyie E (2020) Estimation of soil erosion risk in southern part of Syria by using RUSLE integrating geoinformatics approach. In: Remote sensing applications, p 20

Neamat S, Karimi H (2020) A systematic review of GIS-based landslide hazard mapping on determinant factors from international databases. Int Conf Adv Sci Eng (ICOASE) 2020:180–183

Oldeman LR (1994) The global extent of land degradation. In: Greenland DJ Szabolcs I (eds) Land resilience and sustainable land use. CABI, Wallingford, UK

Pirsaheb M, Nouri M, Karimi H, Mustafa YT, Hossini H, Naderi Z (2020) Occurrence of residual organophosphorus pesticides in soil of some Asian countries, Australia and Nigeria. IOP Conf Ser Mater Sci Eng 737:012175

Renard KG, Foster GR, Weesies GA, McCool DK, Yoder DC (1997) Predicting soil erosion by water: a guide to conservation planning with the revised universal soil loss equation (RUSLE). US Department of Agriculture-Agriculture Handbook No. 703. p 384

Renard KG, Freimund JR (1994) Using monthly precipitation data to estimate the R-factor in the revised USLE. J Hydrol 157:287–306

Sadeghi A, Galalizadeh S, Zehtabian G et al (2021) Assessing the change of groundwater quality compared with land-use change and precipitation rate (Zrebar Lake's Basin). Appl Water Sci 11:170. https://doi.org/10.1007/s13201-021-01508-z

Sun W, Shao Q, Liu J, Zhai J (2014) Assessing the effects of land use and topography on soil erosion on the Loess Plateau in China. CATENA 121:151–163

Tadesse L, Suryabhagavan KV, Sridhar G, Legesse G (2017) Land use and land cover changes and Soil erosion in Yezat Watershed, North Western Ethiopia. Int Soil Water Conserv Res 5:85–94

Teng H, Liang Z, Chen S, Liu Y, Viscarra Rossel RA, Chappell A, Yu W, Shi Z (2018) Current and future assessments of soil erosion by water on the Tibetan Plateau based on RUSLE and CMIP5 climate models. Sci Total Environ 635:673–686

Tsunekawa A, Poesen J, Tsubo M, Meshesha DT, Fenta AA, Nyssen J, Adgo E (2017) Comprehensive assessment of soil erosion risk for better land use planning in river basins: Case study of the Upper Blue Nile River. Sci Total Environ 574:95–108

USGS (2005) The shuttle radar topography mission. http://gdex.cr.usgs.gov/gdex/. Accessed on 10 May 2013

Wang L, Li Y, Gan Y, Zhao L, Qin W, Ding L (2024) Rainfall erosivity index for monitoring global soil erosion. CATENA 234

Wischmeier WH, Smith DD (1978) Predicting rainfall erosion losses. Agriculture Handbook No. 537, USDA-Science and Education Administration, p 58

Drifting Sands, Sand/Dust Storms, and Desertification

Spatiotemporal Variability of Aerosol Optical Depth Over the Arabian Peninsula Using MODIS Data

Abdelgadir Abuelgasim and Ashraf Farahat

Abstract The Arabian Peninsula, located at the western end of the Asian continent, is one of the driest places on Earth, similar to the Sahara Desert in northern Africa. It is predominantly composed of two large extended desert regions, the Empty Quarter and Nufud Desert, which combine to create one of the largest desert regions in the world. The Arabian Peninsula, owing to its massive size and abundance of sand and dust, is considered one of the major atmospheric dust sources in the world. The atmospheric dust emitted by the deserts of this region is transported locally, regionally, and globally by wind. The transported dust significantly contributed to the overall aerosol optical depth (AOD) and region of the area. In-depth studies related to the temporal and spatial variability of dust transport are of great importance because of its adverse effects on human health, climate, ocean conditions, and marine life. In this study, multi-temporal AOD products from the Moderate Resolution Imaging Spectroradiometer (MODIS) were effectively used to identify the seasonal and inter-annual variability and spatial distribution of AOD over the study area. The preliminary findings of the AOD data analysis show that the desert surface of the empty quarter consistently shows high AOD levels along with coastal areas in the region during summer. In contrast, mountainous areas south of Yemen, northeast of Oman, and northwest of Saudi Arabia show the lowest AOD levels due to the absence of significant sand and dust particles. Furthermore, the analysis shows distinctive monthly, seasonal, and annual variability that is strongly linked to the local climatology of the area. It was found that the spring and summer AOD were usually the highest owing to the intensive wind pattern, increased dust emissions, and high humidity. The inter-annual variability of AOD over the study area was not significantly different, mostly related to sporadic dust storm events occurring in particular years. The annual AOD

A. Abuelgasim (✉)
Rabdan Academy, 65 Alinshirah Street, Alsa'adah, Abu Dhabi 22401, United Arab Emirates
e-mail: aabuelgasim@ra.ac.ae

A. Farahat
Department of Physics, College of Engineering and Physics, King Fahd University of Petroleum and Minerals Dhahran, Dhahran 31261, Saudi Arabia
e-mail: farahata@kfupm.edu.sa

© The Author(s), under exclusive license to Springer Nature Switzerland AG 2024
A. M. F. Al-Quraishi and Y. T. Mustafa (eds.), *Natural Resources Deterioration in MENA Region*, Earth and Environmental Sciences Library,
https://doi.org/10.1007/978-3-031-58315-5_10

concentration levels increased at a rate of 0.5% annually during the study period 2003–2019.

Keywords Aerosol optical depth · Dust storms · Haboobs · Arabian Peninsula · Moderate Resolution Imaging Spectroradiometer (MODIS) · Dust emissions

1 Introduction

Terrestrial atmospheric remote sensing refers to the application of remote sensing science and technology for studying the global earth atmosphere. Examples of such applications include studies on the identification, temporal and spatial distribution of air pollutants, and aerosol optical depth (Abuelgasim and Farahat 2020a). The aerosol optical depth (AOD) is a dimensionless quantity that quantitatively measures the degree to which atmospheric aerosols block incident solar radiation (Kaufman et al. 1997). The higher the value, the higher the presence of aerosols and, as such, the greater the attenuation of the incident solar radiation. AOD is defined as the extinction coefficient of solar light in the atmosphere due to aerosols. It represents the sum of both absorption and scattering coefficients. A very low value <0.1 suggests a clean atmosphere with excellent visibility and clear skies. Atmospheric aerosol sources are quite diverse in their sizes and sources (Kaufman et al. 1994).

Aerosols can result from natural processes such as sandstorms suspending dust particles in the atmosphere or wild fire burning (Omari et al. 2019; Csiszar et al. 2003). Aerosols can also be anthropogenic due to human activities resulting from industrial pollution, hydrocarbon industries, urban pollution or urban development, leading to emitting dust particles into the air. Atmospheric aerosols could have a large impact on local weather and climate change. Variability in aerosols' physical characteristics can directly affect Earth's surface temperature as they absorb and scatter solar radiation (Ramanathan et al. 2001). Aerosols also affect cloud formation and properties. Studying aerosols' formation and spatiotemporal distribution in the atmosphere is fundamental to atmospheric science. It is generally believed that the concentrations of aerosols in the atmosphere are strongly influenced by the underlying local land cover, local land-use practices, and the meteorological conditions of the atmosphere in an area, leading to great variability in their spatiotemporal distributions (Abuelgasim and Farahat 2020b). Furthermore, transboundary aerosol transport can strongly influence aerosol concentrations in an area. Identifying the spatiotemporal variability of aerosol concentrations requires regular measurements of their optical properties such as AOD.

AOD can be measured from ground-based stations using spectrometers (Holben et al. 1998). Most ground-based measuring locations are either within or close to large urban centers, and are sometimes located to monitor particular atmospheric events. However, ground-based measurements can accurately measure aerosols' atmospheric concentrations at different spectral ranges much more frequently during the day. Ground-based measurements provide local point measurements of aerosols in the atmosphere's local vertical column and cannot be extended over a broader spatial domain. On the other hand, satellite remote sensing is the only viable method for estimating AOD over a large spatial extent on a continuous and temporal basis with great accuracy and reliability. Advancement in satellite remote-sensing techniques has opened new corridors for the monitoring and mapping of aerosol concentrations over large regions globally.

The primary objectives of this research study are to examine and analyze the spatial and temporal variations of aerosol concentration levels using AOD satellite-based estimates over the Arabian Peninsula (AP). The choice of the AP region as the region for analysis was mainly motivated by its large contribution to the global sand dust aerosols (Abuelgasim and Farahat 2019). The region is also known to experience high levels of aerosol pollution due to a combination of natural and human-made sources, such as dust storms, industrial activities, and transportation, which can have a significant impact on regional and global climate patterns (Ali et al. 2017). Dust storms are prevalent throughout the AP in which the Arabian Desert is a significant contributor to natural dust, with over half of the world's annual average dust emissions originating from this region (Butt et al. 2017; Ali and Assiri 2019). These storms can be generated locally, transported over large distances, or a combination of both. Typically, instability in the atmosphere on a large scale and strong winds at ground level over the AP lead to the onset of dust storms in different regions. During the spring and summer months, these storms tend to happen more often in the eastern and southern areas of the AP (Abuelgasim et al. 2021).

Previous studies have shown that the AP is particularly vulnerable to the effects of aerosols, which can have significant impacts on regional weather patterns, climate change, and public health (Farahat and Abuelgasim 2022; Chowdhury et al. 2022). Additionally, due to the importance of the AP as a hub of oil production and industrial development, aerosol levels above the region may be higher than in other parts of the world (Kumar et al. 2018). Studies conducted earlier have indicated a strong correlation between the fluctuation of aerosol levels in the AP and the frequency and strength of dust storms. Specifically, when there is a rise in either the number or intensity of dust storms, it is anticipated that there will be a corresponding increase in atmospheric aerosol concentration (Alkaabi and Abuelgasim 2021).

The study area is approximately 3,237,500 km^2 with diverse landscapes and major hydrocarbon industries. The landscape is mainly composed of massive-large deserts, large mountain ranges, and vibrant metropolitan cities. The major assumption here is that land cover and meteorological and seasonal conditions in the area are the primary factors influencing the variability of aerosols concentrations at different times and geographic locations within the Arabian Peninsula. AOD levels are used throughout the study as indicator for aerosol concentrations in the atmosphere.

2 Materials and Methods

2.1 Study Area

The Arabian Peninsula is located in the western part of the Asian continent. It is made up of seven countries, namely, Kuwait, Bahrain, Qatar, United Arab Emirates, Sultanate of Oman, Yemen, and Saudi Arabia. It is the largest Peninsula in the world with an approximate area of 3,237,500 km^2 with Saudi Arabia being the largest in land size and population. The Arabian Peninsula is the home of two extended desert regions: the Empty Quarter (Rub Al Khali, about 600,000 km^2) and An Nufud (about 65,000 km^2), connected by the Ad Dahna desert (about 40,000 km^2), which combine to create the massive desert regions in the AP stretching from Yemen to Iraq and from the west coast of Saudi Arabia to Oman, making it one of the major dust sources in the world (Goudie and Middleton 2001).

Major water bodies surround the Peninsula, the Red Sea to the west, the Gulf of Aden, Arabian Sea and Sea of Oman to the south, and the Arabian Gulf on the eastern side (Fig. 1). These water bodies are connected where the Strait of Hormuz connects the Arabian Gulf waters with the Sea of Oman and Bab Elmandab, connecting the Red Sea with the Gulf of Aden. The Peninsula is predominantly a large desert stretching from the northern parts at the border with Iraq and Jordan south to the north of Yemen and Oman. The coastal areas in southern Oman are occupied by massive series of mountains stretching from the Strait of Hormuz to Yemen. The climate is generally hot to extremely hot (Abuelgasim and Farahat 2020a; Albaloushi et al. 2016; Böer 1997), reaching more than 48 degrees during the summer. Rainfall is scarce except for summer rainfall in both Oman and Yemen.

2.2 MODIS Data

In this work, mean daily Aerosol Optical Depth (AOD) data from the Moderate Resolution Imaging Spectroradiometer (MODIS/MAIAC product) (Lyapustin et al. 2011, 2012, 2018) was used for the period 2003–2019 for the Arabian Peninsula. MODIS is a principle instrument onboard of the Terra and Aqua satellites launched on 18 December 1999 and 4 May 2002 respectively, to provide global geophysical data. MODIS uses different algorithms to retrieve data over water and land. For example, the Deep Blue (DB) algorithm (Hsu et al. 2013; Riffler et al. 2010; Sayer et al. 2012) is used to retrieve AOD data over deserts, urban and vegetation regions (Bilal et al. 2013), while the Dark Target (DT) algorithm (Levy et al. 2007) is used to retrieve data over oceans. MODIS DT algorithm also provides AOD at 3 km resolution (Levy et al. 2013), but this product has shown a large uncertainty over urban areas (Nichol and Bilal 2016; Bilal et al. 2019). All data were acquired from the Level-1 and Atmosphere Archive and Distribution System (LAADS) Distributed Active Archive Center (DAAC).

Spatiotemporal Variability of Aerosol Optical Depth Over the Arabian … 195

Fig. 1 Study area map (https://www.nationsonline.org/oneworld/map/Arabia-Map.htm)

2.3 Image Processing and Descriptive Statistics

MODIS AOD data was downloaded in its original format in titles covering the whole study area and neighboring regions. The satellite data processing consisted of mosaicking the different titles covering the study and further cropping areas outside the region's boundaries. The mosaicking process was designed to use the maximum AOD in the case of the presence of overlap between the different MODIS tiles. The resulting images consisted of daily AOD image data, with each pixel representing the estimated AOD at the pixel location (Fig. 2).

Descriptive statistics were generated from the daily AOD image data. This consisted of the average daily, monthly, seasonal, and annual AOD concentrations levels from the satellite data from January 1st 2003 to December 31st 2019. In addition to the mean AOD value, the minimum, maximum and standard deviations measures were calculated to assess the level of AOD variability. The measures from the descriptive statistics were later used to generate multi-temporal AOD level graphs

Fig. 2 MODIS AOD image March 4th 2018

and statistically analyzed to identify multi-temporal trends daily, monthly, and annually. Analysis of the spatial distribution of AOD levels was based on the average monthly and annual images generated from the MODIS data.

3 Results and Discussions

3.1 Monthly Temporal Variability

Figure 3 shows that in the Arabian Peninsula AOD levels are low during January and February of each year. AOD levels start to rise by late March and early April. AOD levels peak during July where June, July and August represent the highest levels. By September AOD levels start to drop, reaching their lowest levels during December.

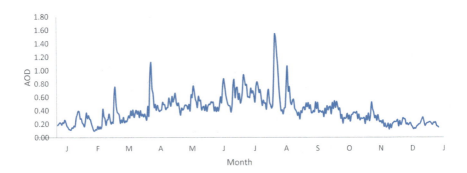

Fig. 3 Daily AOD levels in the Arabian Peninsula 2018

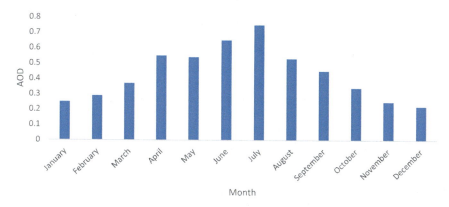

Fig. 4 Monthly average AOD in the Arabian Peninsula 2018

The rise of AOD levels during the summer months is associated with high temperatures leading to rise in local and regional dust storms resulting in higher levels of suspended dust particles. Furthermore, along the coastal areas of the Arabian Peninsula, there is a rise in the amount of evaporation, leading to significant amounts of water vapour levels, increasing the overall AOD. Note that for brevity, only the results of the year 2018 are presented in Fig. 4. Similar years during the study period exhibit identical patterns. Table 1 shows the descriptive statistics for AOD levels during the study period. The statistics confirm that the AOD levels pattern observed during 2018 is the same throughout the study period.

3.1.1 Monthly Variability

Within each month for 2018 data AOD levels show quite a diverse high variability that is believed to be strongly correlated to the meteorological and atmospheric conditions in the study area. This study uses the monthly standard deviation to indicate daily AOD levels variability (Fig. 5).

The highest (standard deviation higher than 0.1) variability in AOD levels is during the summer months due to the extensive dust storms in the area and higher evaporation levels leading to higher levels of AOD. The lowest (standard deviation lower than 0.1) was during the late autumn and early winter season. During the late autumn, less wind activity and lower water evaporation levels result in lower overall AOD levels and variability.

3.1.2 Seasonal Variability

As previously stated, the overall AOD levels in the Arabian Peninsula are higher in the summer and spring seasons and lowest during the autumn season (Fig. 6).

Table 1 Descriptive statistics of AOD level 2003–2019

		2003	2004	2005	2006	2007	2008	2009	2010	2011	2012	2013	2014	2015	2016	2017	2018	2019
January	Mean	0.19	0.18	0.16	0.25	0.17	0.12	0.18	0.24	0.17	0.21	0.19	0.18	0.21	0.22	0.19	0.20	0.18
	Min	0.10	0.11	0.09	0.15	0.11	0.08	0.10	0.08	0.12	0.13	0.12	0.14	0.14	0.12	0.11	0.13	0.12
	Max	0.36	0.36	0.28	0.36	0.28	0.17	0.27	0.65	0.25	0.31	0.25	0.23	0.30	0.34	0.33	0.30	0.24
February	Mean	0.24	0.27	0.22	0.23	0.23	0.30	0.35	0.26	0.28	0.38	0.22	0.22	0.27	0.20	0.30	0.26	0.28
	Min	0.16	0.14	0.14	0.13	0.14	0.13	0.15	0.15	0.19	0.19	0.13	0.16	0.12	0.08	0.20	0.10	0.15
	Max	0.37	0.68	0.44	0.54	0.35	0.96	1.14	0.47	0.41	0.75	0.39	0.30	0.55	0.37	0.48	0.75	0.38
March	Mean	0.39	0.26	0.27	0.31	0.34	0.34	0.43	0.33	0.42	0.62	0.37	0.35	0.35	0.35	0.43	0.34	0.38
	Min	0.17	0.16	0.16	0.12	0.20	0.19	0.19	0.15	0.17	0.32	0.22	0.21	0.19	0.23	0.21	0.22	0.25
	Max	0.76	0.39	0.48	0.56	0.60	0.57	0.71	0.69	1.19	1.75	0.65	0.66	0.59	0.52	1.17	0.92	0.69
April	Mean	0.34	0.31	0.32	0.33	0.39	0.51	0.38	0.37	0.47	0.43	0.41	0.34	0.71	0.28	0.40	0.52	0.39
	Min	0.20	0.17	0.24	0.20	0.29	0.28	0.24	0.23	0.29	0.25	0.22	0.23	0.34	0.19	0.20	0.34	0.23
	Max	0.72	0.58	0.48	0.49	0.57	1.02	0.57	0.52	1.00	0.75	0.78	0.54	1.43	0.43	0.64	1.12	0.80
May	Mean	0.50	0.36	0.35	0.39	0.37	0.39	0.46	0.46	0.44	0.47	0.48	0.34	0.42	0.38	0.39	0.50	0.39
	Min	0.37	0.23	0.24	0.28	0.26	0.23	0.24	0.28	0.30	0.26	0.31	0.22	0.31	0.21	0.19	0.39	0.28
	Max	0.70	0.50	0.48	0.64	0.51	0.64	0.71	0.73	0.64	0.75	0.71	0.45	0.68	0.69	0.73	0.77	0.53
June	Mean	0.36	0.32	0.42	0.42	0.46	0.58	0.49	0.48	0.71	0.56	0.54	0.40	0.48	0.45	0.49	0.62	0.43
	Min	0.26	0.25	0.22	0.27	0.31	0.26	0.33	0.34	0.33	0.36	0.29	0.29	0.30	0.31	0.26	0.38	0.32
	Max	0.49	0.51	0.59	0.77	0.71	1.15	0.79	0.79	1.05	1.15	1.00	0.69	0.93	0.55	1.10	0.93	0.67
July	Mean	0.45	0.32	0.46	0.47	0.37	0.54	0.56	0.55	0.69	0.52	0.49	0.41	0.52	0.45	0.63	0.72	0.50
	Min	0.25	0.24	0.35	0.28	0.28	0.35	0.31	0.38	0.34	0.36	0.33	0.25	0.35	0.30	0.37	0.42	0.34
	Max	0.79	0.47	0.73	0.94	0.49	0.85	1.17	0.84	1.33	0.94	0.83	0.65	0.89	0.59	0.86	1.54	0.86
	Mean	0.41	0.32	0.34	0.39	0.34	0.40	0.48	0.45	0.41	0.48	0.37	0.38	0.39	0.44	0.44	0.50	0.45

(continued)

Spatiotemporal Variability of Aerosol Optical Depth Over the Arabian ... 199

Table 1 (continued)

		2003	2004	2005	2006	2007	2008	2009	2010	2011	2012	2013	2014	2015	2016	2017	2018	2019
August	Min	0.26	0.20	0.22	0.28	0.23	0.24	0.31	0.29	0.30	0.23	0.26	0.29	0.26	0.29	0.28	0.28	0.30
	Max	0.66	0.43	0.64	0.50	0.41	0.64	0.62	0.71	0.61	1.19	0.56	0.56	0.61	0.80	0.61	1.07	0.72
	Mean	0.29	0.27	0.26	0.25	0.34	0.43	0.27	0.32	0.27	0.32	0.28	0.32	0.38	0.35	0.34	0.42	0.38
September	Min	0.19	0.19	0.19	0.19	0.22	0.23	0.19	0.20	0.19	0.21	0.19	0.21	0.24	0.23	0.20	0.29	0.28
	Max	0.36	0.42	0.37	0.34	0.58	0.68	0.37	0.48	0.40	0.58	0.45	0.42	0.54	0.60	0.47	0.55	0.54
	Mean	0.20	0.22	0.22	0.21	0.23	0.27	0.24	0.25	0.26	0.23	0.23	0.24	0.33	0.24	0.27	0.32	0.30
October	Min	0.14	0.11	0.16	0.15	0.17	0.17	0.14	0.13	0.19	0.15	0.17	0.17	0.24	0.17	0.18	0.21	0.23
	Max	0.30	0.32	0.33	0.30	0.32	0.42	0.45	0.39	0.38	0.30	0.44	0.30	0.49	0.32	0.39	0.53	0.41
	Mean	0.20	0.23	0.17	0.21	0.17	0.20	0.18	0.19	0.22	0.20	0.21	0.21	0.23	0.22	0.24	0.22	0.24
November	Min	0.14	0.12	0.11	0.17	0.12	0.13	0.13	0.12	0.15	0.15	0.12	0.13	0.16	0.14	0.16	0.12	0.13
	Max	0.26	0.57	0.23	0.28	0.23	0.26	0.26	0.24	0.34	0.31	0.30	0.30	0.38	0.33	0.33	0.34	0.39
	Mean	0.18	0.19	0.15	0.25	0.17	0.12	0.18	0.21	0.17	0.21	0.18	0.18	0.21	0.21	0.19	0.20	0.18
December	Min	0.10	0.11	0.09	0.15	0.11	0.08	0.10	0.08	0.12	0.13	0.12	0.14	0.14	0.12	0.11	0.13	0.12
	Max	0.36	0.36	0.28	0.36	0.28	0.17	0.27	0.65	0.25	0.31	0.25	0.23	0.30	0.34	0.33	0.30	0.24

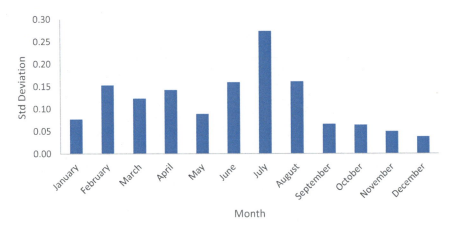

Fig. 5 Monthly standard deviation of AOD in 2018

The meteorological conditions driving this variability suggest that AOD levels in the Arabian Peninsula is driven by natural processes rather than anthropogenic processes.

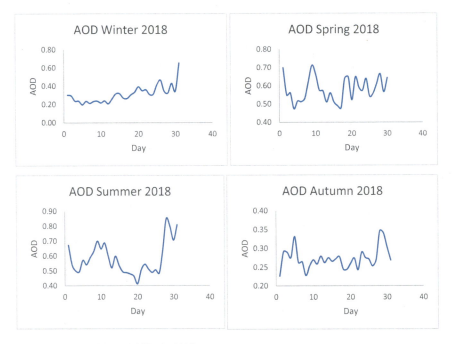

Fig. 6 Seasonal AOD variability in 2018

3.1.3 Inter-Annual Variability

Figure 7 shows the daily AOD concentrations for the entire study period 2003–2019. The figure portrays the previously discussed results, with AOD levels starting low in the winter at the beginning of each year and peaking during the summer time while starting to drop by the autumn season. While the overall monthly temporal variability is similar yearly, the annual mean absolute AOD levels are different. There is a consistent trend of rising annual AOD during the study period 2003–2019 of approximately 0.5% (Fig. 8).

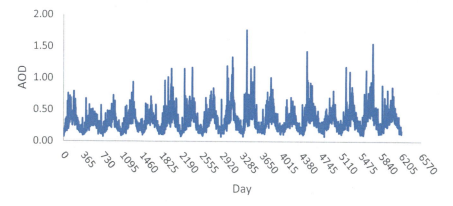

Fig. 7 Daily AOD levels 2003–2019

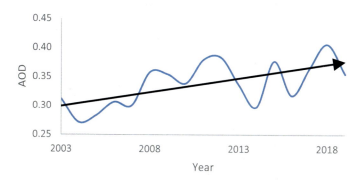

Fig. 8 Mean annual AOD 2003–2019 trends

3.2 Spatial Variability

3.2.1 Monthly Variability

Figure 9 shows the spatial distribution of the average monthly AOD levels over the Arabian Peninsula. Mostly from January to July the areas with higher levels of AOD are located in the central and southern parts of the Arabian Peninsula. These areas comprise the massive desert known as Al Rubi Al Khali, or the empty quarter, and an area of 650,000 km^2 stretches between four countries Saudi Arabia, Oman, UAE, and Yemen (Clark and Amdt 1989). Further south of the empty quarter is Jiddat Alharasis in Oman which is showing extremely higher levels of AOD all year round. Jiddat Aharasis is a stony desert of about 27,000 km^2 in southern central Oman stretching between the two Omani states of Alwusta and Dhufar (Edgell 2006).

The area has a significant presence of fog and humidity due to the monsoon season and coastal evaporation during the summer, while high moisture, fog, and dew during the winter lead to sporadic rainfall (Edgell 2006). This likely explains the higher AOD levels in the area year-round. Furthermore, during the summer, evaporation from nearby water bodies adds to the overall AOD levels in the study area. Mountainous regions in northern and eastern Oman, southern Yemen, and western Saudi Arabia portray lower levels of AOD year-round due to the absence of sand and dust.

3.2.2 Annual Variability

Figure 10 shows the mean annual AOD for all the years during the study period. The general pattern of the spatial distribution of AOD concentrations shows that low AOD levels generally characterize the northern and mountainous areas. The central and southern regions of the study area where the empty quarter are located usually have higher levels along with coastal areas. The Jiddat Alharasis portrays higher levels during all years.

There is quite a variability in the annual mean AOD levels for 2003–2019. This variability is likely strongly related to the study area's wind patterns and dust storms. For example, the years 2008, 2012 and 2018 were characterized by significant outbreaks of dust storms and significantly higher levels of AOD. As can be seen in Fig. 10, there is a consistent pattern of increase in the overall AOD levels within the study area. This can be observed by careful visual analysis of the annual mean AOD concentrations in Fig. 10. The increase is due to changes (higher) in the AOD levels in the central and southern areas of the study area.

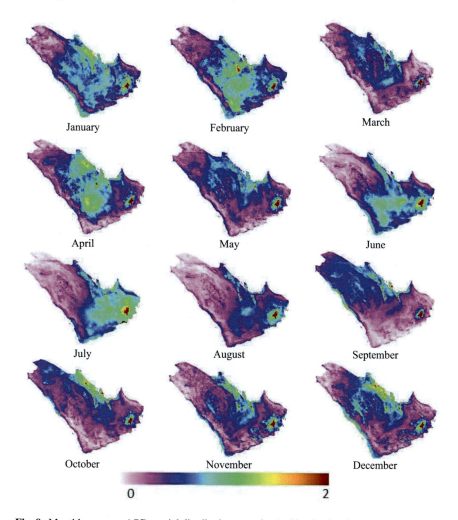

Fig. 9 Monthly average AOD spatial distribution over the Arabian Peninsula

4 Conclusions

Satellite-derived products of AOD show that the occurrence and intensity of AOD varies substantially at daily, seasonal, annual, and perhaps decadal timescales over the Arabian Peninsula. It has been shown that within the study area, AOD levels are low in the winter at the early months of each year (January–March). AOD levels start to increase by early April with the onset of the spring season to reach their peak levels by the summer. The spring and summer seasons have the highest concentration levels of AOD. AOD levels are also varying interannually. It has been observed that there was an approximate annual increase of 0.5% in AOD during the study period

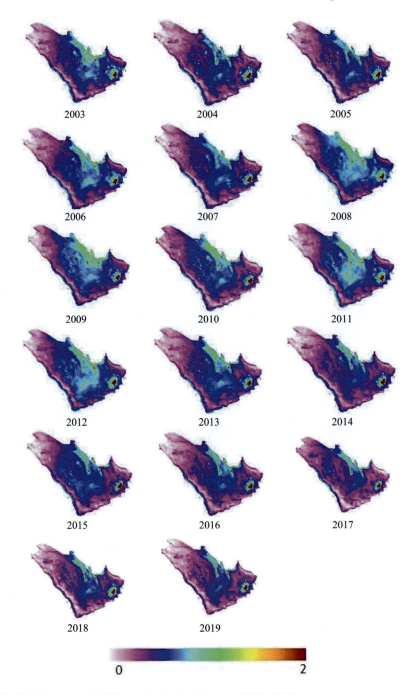

Fig. 10 Mean annual AOD over the Arabian Peninsula 2003–2019

2003–2019. Most of the increase is strongly related to an increase in AOD during the summer months each year. This is likely due to the significant land cover changes in the study area. Land cover change is expected to continue due to the extensive urbanization in the region and other anthropogenic increases, which will lead to further increases in AOD concentrations. Geographically, the massive desert of the empty quarter stretching between Saudi Arabia, Yemen, Oman, and UAE, along with the Stoney desert of Jiddat Alharasis are the areas with the highest AOD levels. Coastal areas in the eastern part of the Arabian Peninsula show higher AOD levels than the western coastal areas.

Due to their large economic impacts, high aerosol loadings and dust storms should be further investigated in the Arabian Peninsula countries. Aerosol loading in the atmosphere could lead to immediate and long-term impacts with varying levels of severity. Aerosol loading has negative impact on the economy including air and seaports shut down, traffic accidents, and soil degradation. Dust storms could lead to serious problems in marine traffic including precluding ships from maneuvering in seaports. Countries with frequent dust storms experience a larger number of flight delays compared to countries with less dust activity. Meanwhile, high aerosol loading in the atmosphere affects the radiation budget by scattering and absorbing solar radiation and by influencing surface temperature (Labban and Farahat 2023). This research suggests the development of an early warning system for large dust events. Further, it is recommended to increase the number of ground monitoring stations that can be envisioned to help in the development of artificial intelligence and neural networking algorithms (Alwadei et al. 2022) to predict aerosol loading and its effects on solar irradiance and dust activities over the Arabian Peninsula.

References

Abuelgasim A, Farahat A (2020a) Effect of dust loadings, meteorological conditions, and local emissions on aerosol mixing and loading variability over highly urbanized semiarid countries: United Arab Emirates case study. J Atmos Solar Terr Phys 199:105215. https://doi.org/10.1016/j.jastp.2020.105215

Abuelgasim A, Farahat A (2020b) Investigations on PM10, PM2.5, and their ratio over the Emirate of Abu Dhabi, United Arab Emirates. Earth Syst Environ. https://doi.org/10.1007/s41748-020-00186-2

Abuelgasim A, Bilal M, Alfaki IA, Spatiotemporal variations and long term trends analysis of aerosol optical depth over the United Arab Emirates. Remote Sens Appl: Soc Environ. https://doi.org/10.1016/j.rsase.2021.100532

Albaloushi R, Alghafri A, Ghazal S, Aljaberi A, Abuelgasim A (2016) Investigations on the seasonal and inter-annual variations of the atmospheric aerosol optical depth in the United Arab Emirates using MODIS satellite data. In: 37th Asian conference on remote sensing: fostering resilient growth in Asia, ACRS 2016; Colombo, Sri Lanka October 16–21, 2016

Alkaabi K, Abuelgasim A (2021) Comparative analysis of pollutant levels during lockdowns across different land-use over the Emirate of Abu Dhabi, United Arab Emirates. The Arab World Geographer 1 September 2021, 24(3): 205–220. https://doi.org/10.5555/1480-6800.24.3.205

Ali MA, Assiri M, Dambul R (2017) Seasonal aerosol optical depth (AOD) variability using satellite data and its comparison over Saudi Arabia for the period 2002–2013. Aerosol Air Qual Res 17(5):1267–1280. https://doi.org/10.4209/aaqr.2016.11.0492

Ali MA, Assiri M (2019) Analysis of AOD from MODIS-merged DT–DB products over the Arabian Peninsula. Earth Syst Environ 3(3):625–636. https://doi.org/10.1007/s41748-019-00108-x

Alwadei S, Farahat A, Ahmed M, Kambezidis HD (2022) Prediction of solar irradiance over the Arabian Peninsula: satellite data, radiative transfer model, and machine learning integration approach. Appl Sci 12:717

Bilal M, Nichol JE, Bleiweiss MP, Dubois D, Rse J (2013) A simplified high resolution modis aerosol retrieval algorithm (sara) for use over mixed surfaces. Remote Sens Environ 136:135–145

Bilal M, Nazeer M, Nichol J, Qiu Z, Wang L, Bleiweiss MP, Shen X, Campbell JR, Lolli S (2019) Evaluation of terra-MODIS C6 and C6.1 aerosol products against Beijing, XiangHe, and Xinglong AERONET sites in China during 2004–2014. Remote Sens 11:486. https://doi.org/10.3390/rs11050486

Böer B (1997) An introduction to the climate of the United Arab Emirates. J Arid Environ 35:3–16. https://doi.org/10.1006/jare.1996.0162

Butt MJ, Ali MA (2017) Assessment of AOD variability over Saudi Arabia using MODIS deep blue products. Environ Pollut 231:143–153. https://doi.org/10.1016/j.envpol.2017.07.104

Chowdhury S, Pozzer A, Haines A, Klingmüller K, Münzel T, Paasonen P, Sharma A, Venkataraman C, Lelieveld J (2022) Global health burden of ambient PM2.5 and the contribution of anthropogenic black carbon and organic aerosols. Environ Int 159:107020. https://doi.org/10.1016/j.envint.2021.107020

Clark A (1989) Amdt R (ed) Lakes of the Rub' al-Khali. Saudi Aramco World. 40(3):28–33. ISSN 0003-7567

Csiszar I, Abuelgasim A, Li Z, Jin J, Fraser R, Hao W-M (2003) Interannual changes of active fire detectability in North America from long-term records of the advanced very high resolution radiometer. J Geophys Res 108(D2):4075. https://doi.org/10.1029/2001JD001373

Edgell HS (2006) Arabian deserts: nature, origin and evolution. Springer. ISBN 978-1-4020-3970-6

Farahat A, Abuelgasim A (2022) Effect of cloud seeding on aerosol properties and particulate matter variability in the United Arab Emirates. Int J Environ Sci Technol 19:951–968. https://doi.org/10.1007/s13762-020-03057-5

Goudie AS, Middleton NJ (2001) Saharan dust storms: nature and consequences. Earth-Sci Rev 56(1–4): 179–204. ISSN 0012-8252,https://doi.org/10.1016/S0012-8252(01)00067-8

Holben BN, Eck TF, Slutsker I, Tanré D, Buis JP, Setzer A, Vermote E, Reagan JA, Kaufman YJ, Nakajima T, Lavenu F, Jankowiak I, Smirnov A (1998) AERONET—a federated instrument network and data archive for aerosol characterization. Remote Sens Environ 66(1):1–16. https://doi.org/10.1016/S0034-4257(98)00031-5

Hsu NC, Jeong M-J, Bettenhausen C, Sayer AM, Hansell R, Seftor CS, Huang J, Tsay S- C (2013) Enhanced deep blue aerosol retrieval algorithm: the second generation. J Geophys Res: Atmos 118(16):9296–9315. https://doi.org/10.1002/jgrd.50712

Kaufman YJ, Gitelson A, Karnieli A, Ganor E, Fraser RS, Nakajima T, Mattoo S, Holben BN (1994) Size distribution and scattering phase function of aerosol particles retrieved from sky brightness measurements. J Geophys Res 99(D5):10341–10356. https://doi.org/10.1029/94JD00229

Kaufman YJ, Tanr D, Remer LA, Vermote EF, Chu A (1997) Operational remote sensing of tropospheric aerosol over land from EOS moderate resolution imaging spectroradiometer. J Geophys Res 102(96):51–67

Kumar KR, Attada R, Dasari HP, Vellore RK, Langodan S, Abualnaja YO, Hoteit I (2018) Aerosol optical depth variability over the Arabian Peninsula as inferred from satellite measurements. Atmos Environ 187:346–357. https://doi.org/10.1016/j.atmosenv.2018.06.011

Labban A, Farahat A (2023) Effect of major dust events on ambient temperature and solar irradiance components over Saudi Arabia. Atmosphere 14:408

Levy RC, Remer LA, Dubovik O (2007) Global aerosol optical properties and application to moderate resolution imaging spectroradiometer aerosol retrieval over land. J Geophys Res 112:D13210. https://doi.org/10.1029/2006JD007815

Levy RC, Mattoo S, Munchak LA, Remer LA, Sayer AM, Patadia F, Hsu NC (2013) The collection 6 MODIS aerosol products over land and ocean. Atmos Meas Tech 6:2989–3034. https://doi.org/10.5194/amt-6-2989-2013

Lyapustin A, Wang Y, Laszlo I, Kahn R, Korkin S, Remer L, Levy R, Reid JS (2011) Multiangle implementation of atmospheric correction (MAIAC): 2. Aerosol algorithm. J Geophys Res 116:D03211. https://doi.org/10.1029/2010JD014986

Lyapustin A, Korkin S, Wang Y, Quayle B, Laszlo I (2012) Discrimination of biomass burning smoke and clouds in MAIAC algorithm. Atmos Chem Phys 12:9679–9686. https://doi.org/10.5194/acp-12-9679-2012

Lyapustin A, Wang Y (2018) MCD19A2 MODIS/Terra+Aqua land aerosol optical depth daily L2G global 1km SIN grid V006. NASA EOSDIS Land Processes DAAC. https://doi.org/10.5067/MODIS/MCD19A2.006

Nichol J, Bilal M (2016) Validation of MODIS 3 km resolution aerosol optical depth retrievals over Asia. Remote Sens 8:328

Omari K, Abuelgasim A, Alhebsi K (2019) Aerosol optical depth retrieval over the city of Abu Dhabi, United Arab Emirates (UAE) using Landsat-8 OLI images. Atmos Pollut Res

Ramanathan V, Crutzen PJ, Kiehl T, Rosenfeld D (2001) Aerosols, climate and the hydrological cycle. Science 294:2119–2124

Riffler M, Popp C, Hauser A, Fontana F, Wunderle S (2010) Validation of a modified AVHRR aerosol optical depth retrieval algorithm over Central Europe. Atmos Meas Tech 3(5):1255–1270. https://doi.org/10.5194/amt-3-1255-2010

Sayer AM, Hsu NC, Bettenhausen C, Ahmad Z, Holben BN, Smirnov A, Thomas GE, Zhang J (2012) SeaWiFS ocean aerosol retrieval (SOAR): algorithm, validation, and comparison with other data sets. J Geophys Res 117:D03206. https://doi.org/10.1029/2011JD016599

The Carrying Loads Composition of Storms Over Iraq

Moutaz A. Al-Dabbas

Abstract One of the worst environmental issues in the Middle East is the occurrence of sand and dust storms. This study examines the amount of load in Iraq's dust storms over four years, from 2007 to 2010. This study aimed to develop a monitoring methodology and investigate SDS's inorganic and organic compositions of the present work covering central and southern Iraq. Samples from the settled dust were collected and analyzed using well-known analytical procedures. The results revealed that the average texture was sandy silty clay; the light minerals were mainly quartz and carbonate with feldspar and gypsum, as well as clay and heavy minerals. Heavy metals, such as Fe, Pb, Zn, Ni, Co, Cd, and Cu. The average uranium activity was lower than the critical dose level. The existing pollens are Pine, Chenopodiaceae, Palmae, Graminea, Olea, Typha, and Artemisia, with some Algae, Fungi, Cuticle, and spores. Fungi and bacteria, such as the gram-positive bacilli, E. coli, and the fungal Aspergillus species, are also common. Viral etiology analysis did not reveal any viral isolates. Aerosols in Iraq are a mixture of naturally occurring and anthropogenic pollutants, and basic information about their composition is vital for the Environmental Risk Assessment of dust in Iraq.

Keywords Sand-dust storms · Mineral · Pollen · Microorganism · Aerosol pollutants · Environmental risk assessment

1 Introduction

Sand and dust storms (SDS) are common climatic phenomena within Iraq and the surrounding countries. They affect human health, the economy, agriculture, and transportation and cause pollution. Northwesterly winds (locally called shamals) are most frequently effective in producing these storms over Iraq. The dry and fine-sized soil particles are lifted upward from the surface by strong winds, such as alluvial plains

M. A. Al-Dabbas (✉)
Department of Geology, College of Science, University of Baghdad, Baghdad, Iraq
e-mail: moutaz.mohammed@sc.uobaghdad.edu.iq

and dried marshlands near the Tigris and Euphrates Rivers. This process occurred in Iraq after a period of low winter precipitation (<400 mm). SDS is associated with many other favorable dust factors, including lack of vegetation, rising air temperature, low relative humidity, air turbulence, and dry soil with less moisture (Walker 2005; Jacquelyn 2009; Al-Zubaidi et al. 2022). Winds pick up many particulates when blowing over different soils, including minerals, organic compounds, soil particles, salts, fertilizers, and substances that are dangerous contaminants (Shi et al. 2005; Awadh 2023). The sources of these elements are believed to be natural, such as rocks, soils, geologic resources, and particulates carried by storms with some industrial materials as artificial sources that contaminate the atmosphere and air with heavy elements. Mineral dust in the atmosphere can be of anthropogenic or natural origin. These dust particles could rise as high as 3000 m in cases of wind turbulence and instability (Wikipedia 2009; Al-Ansari 2013; UNESCO 2013; UNEP, WMO, UNCCD 2016) (Table 1).

Iraq is affected by dust storms, either by suspended or rising storms that are local to Iraqi districts and by regional storms affecting Iraq from outside its international borders (Beg and Al-Sulttani 2020; Rasheed and Al-Ramahi 2021; Mohammed et al. 2022; Abd El-Aal et al. 2023). Carrying particles transported by storms have

Table 1 Average annual precipitation, temperature, dust storm frequency, and dryness index for the Iraqi governorate (Climatological Atlas of Iraq 2000; Sissakian et al. 2013)

Governorate	Average yearly precipitation (mm)	Average yearly temperature (°C)	Average yearly dust storms (days)	Dryness index
Baghdad	150–200	23	12	20–25
Mosul	300–600	20	1–4	5–10
Basra	75–150	24	8–24	15–20
Kirkuk	200–400	22	2–4	5–10
Salahuddin	100–300	23	4–12	10–20
Diyala	150–450	23	4–12	15–20
Anbar	75–150	22	4–8	20–35
Wasit	150–200	23	2–6	20–25
Misan	150–200	23	2–8	15–25
Babil	100–150	23	10–12	25–30
Karbala	75–100	24	8–12	30–35
Najaf	75–100	24	8–12	30–35
DhiQar	100–150	24	6–12	20–30
Qadisiya	100–150	24	6–24	25–30
Muthanna	75–100	24	12–24	25–35
Erbil	400–800	19	1–4	≤ 5
Dohuk	600–800	18	1	≤ 5
Sulimaniya	500–800	18	1–2	≤ 5

dangerous effects on the living. Particles less than 50 microns in size cause lung diseases, bronchitis, and asthma because they carry different types of pollen, spores, fungi, algae, bacteria, and some viruses (Griffin et al. 2002, 2007; Abdullah and Ahmed 2021). Al-Dabbas (2016) concluded that allergens are one of the adverse effects of storms carrying soil particles, minerals, pollens, algae, fungal spores, bacteria, and organic nanoparticles. Al-Hemoud et al. (2018) concluded that the amounts of PM10 particles were directly related to respiratory, bronchial asthma, and cardiovascular illness. Recent research has indicated that wind loads carry radioactive and trace elements (Al-Dabbas 2016, 2018).

The transported radioactive particulates mostly affect human lungs and respiratory systems by inhaling uranium compounds (Al-Dabbas et al. 2011, 2012; Nasif 2011; Iraqi Meteorological Department 2022). The average concentration of Uranium absorbed dose was found to be vital in dust storm loads as investigated by the number of storms blown over Ramadi and Baghdad governorates (Ali and Shejiri 2019). However, uranium compounds in the air can increase radioactivity and chemical toxicity, affect renal function, and cause renal failure (Morton et al. 2002; IAEA 2003). Al-Dousari et al. (2019), concluded that the aeolian activities in Kuwait are of high economic cost, and Sissakian et al. (2013) mentioned that about 347,310 m^3 of sand is removed yearly. This chapter deals with the SDS monitoring methodology and investigates the inorganic and organic composition of dust storms, such as their geochemical, textural, radioactive, and biological characteristics, that were found covering the central and southern regions of Iraq from March 2007 to July 2010.

1.1 Mechanical Processes of Sand Dust Storms

Wind and thermal dynamics are weather factors that favor the formation of storms. The wind is the main factor for soil erosion, either by the transportation process or by the deposition of larger particles that the wind cannot carry. In weather systems, the dynamic temperature process causes turbulence in the air. This blows up the loss of soil particles at high altitudes. For example, fine particles less than 0.005 mm may reach a height of up to 12 km, while fine grains of 0.05–0.005 mm may blow up to 1.5 km in the atmosphere (Al-Dousari et al. 2017).

1.1.1 Suspension

Grain particles of less than 0.1 mm with light density are transported in suspension and lofting of particles by turbulent winds. The suspension-type storm reflects the ability of the wind to carry very fine and fine sand, silt, and clay particles up to 6 km in height and more than 6000 km in distance. These very fine sand, silt, and clay particles remained in the sky as suspended grains. Larger grains precipitated to form sand dune fields (Alonso et al. 2013).

1.1.2 Saltation

The grain sizes from 0.01 to 0.5 mm will erode from the soil but move in continuous skips or jump forward to distinguish the saltation process (Alonso et al. 2013).

1.1.3 Creep

Larger grain sizes higher than 0.5 mm are moving in creeping, and the particles move along the ground by rolling and sliding large sizes and/or light wind favor creep. Approximately 50–80% of the soils and sediments are expected to travel in creeping for some meters and with heights less than 30 m to construct dune zones of sand (Alonso et al. 2013).

1.2 Types of Sand /Dust Storms in Iraq

Sanddust storms are an important activity of speedy winds that decrease the visibility horizon to less than one km (Alonso-Pérez et al. 2013). There are differences in the meanings of dust storms and other storms. For example, sandstorms refer to wind-carrying sand in the air reaching more than 15 m in height and decreasing visibility. The average grain size of sand ranges from 0.15 to 0.30 mm with a speedy wind that exceeds 12 km/h, while dust storms are similar to sand storms, but have distinct properties where smaller particles are distributed through the air (Goldman 2003). There are four types of storms. These include dust storms, suspended dust, and dust. This classification depends on Iraq's climatological conditions, which have two exciting factors: wind speed and visibility (Alonso et al. 2013; Wahab 2007).

1.2.1 Dust Storms

These storms are accompanied by high wind speeds that affect visibility to less than one kilometer and deteriorate below 0.2 km. Therefore, storms are called severe dust or sandstorms. Sissakian et al. (2013) discussed the relationship between the climatic parameters, dryness index, and the number of dust storm days for each Iraqi governorate and concluded that there are seven governorates (Anbar, Babil, Karbala, Najaf, Dhi-Qar, Qadisiyah, and Muthana) with the worst dryness index, less annual rainfall, and higher mean annual temperature with a maximum number of days of a dust storm (Fig. 1, Table 1). Al-Jumaily and Ibrahim (2013) investigated and monitored dust storms in Iraq and the monthly dust storm frequency in Iraq from 2003 to 2012. They concluded that storm occurrences (as a meteorological phenomenon) have continuously increased for a maximum during the years 2012 and April had the maximum number of storms reaching 16 storms, June with 14 storms, followed by March with 12 storms, and less than the extent in the other

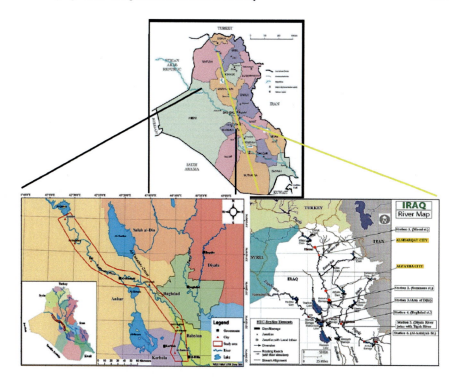

Fig. 1 Iraqi governorates location map, shows the Euphrates river (left) and the Tigris river and its tributaries in Iraq (right), (Wikipedia 2009; Oleiwi and Al-Dabbas 2021; Al-Ali and Al-Dabbas 2022a, b)

months with fewer dust storms (Fig. 2a and b). Jassim et al. (2012) showed a clear increase in the number of dust storm days per year from 2000 to 2012 compared with that from 1985 to 1999 (Fig. 3a and b).

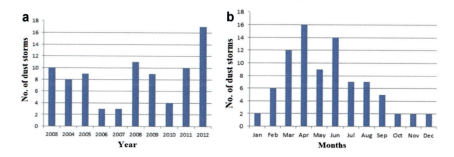

Fig. 2 a The yearly monitored dust storms in Iraq (2003–2012), b month monitored dust storms frequency in Iraq (from 2003 to 2012) (Al-Jumaily and Ibrahim 2013)

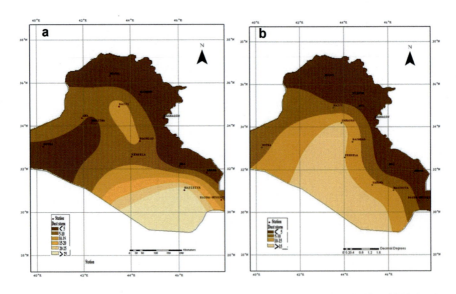

Fig. 3 Annual mean dust storm days in Iraq, **a** from 1985 to 1999), **b** from 2000 to 2012 (Jassim et al. 2012)

1.2.2 Suspended Dust

The horizontal visibility is less than one kilometer with thick suspended dust, but when it is moderate, the visibility range is between 1 and 5 km, and the winds are considered light. Jassim et al. (2012) showed a clear increase in the number of days per year in which dust was suspended from 2000 to 2012 more than from 1985 to 1999 (Fig. 4a and b).

1.2.3 Rising Dust

The horizontal visibility is more than 0.1 km. The windows were considered to be moderate. Jassim et al. (2012) showed a clear increase in the number of days per year when dust increased from 2000 to 2012, more than from 1985 to 1999 (Fig. 5a and b).

1.3 Sand-Dust Storms Hot Points Sources

According to different studies and research projects in Iraq, the areas of source points include deteriorated range lands, irrigated farm fields, bare ground, marshes, and rain-fed agricultural fields. The variable dust source sites of Iraq are effective in creating SDS, but in general, the dust region's source in Iraq may not only be the

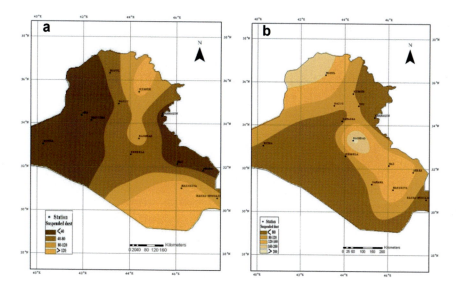

Fig. 4 Annual mean number of days of suspended dust in Iraq, **a** from 1985 to 1999), **b** from 2000 to 2012 (Jassim et al. 2012)

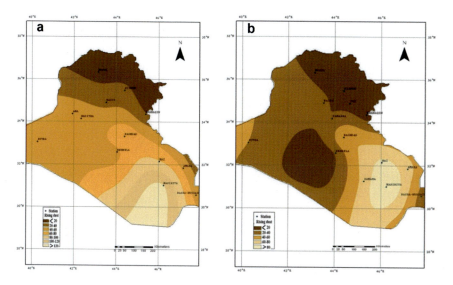

Fig. 5 Annual mean number of days of rising dust in Iraq, **a** from 1985 to 1999), **b** from 2000 to 2012 (Jassim et al. 2012)

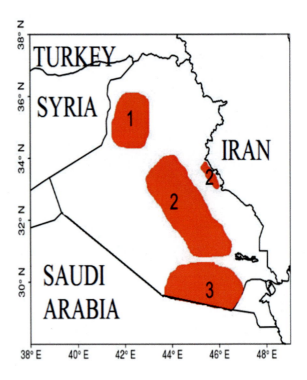

Fig. 6 Sand dust storms (SDS) hotpoint sources (Alonso et al. 2013)

hotspots for SDS or have their local contributions, but other hotspots could also be the origin of the dust located within the countries close by as sources for regional SDS (Awadh 2023). For example, the arid regions in Saudi Arabia's northern half followed the Syrian and Jordanian hotspots. According to Boloorani et al. (2012) and Alonso et al. (2013), dust hotspots have three primary causes (Fig. 6).

1. The Jazera area and Iraqi-Syrian-borders.
2. Central–eastern parts of Iraq.
3. Southern Iraq.

1.4 Evaluation of Blowing Particulates During Storms Passing Iraq with Remote Sensing Technique

Monthly and inter-annual dust climatology with a focus on Iraq was presented to evaluate dust variability during storms. Aerosol optical depth data from MODIS, MISR, SeaWiFS, MERIS, and MODIS's Angström exponent were used for this purpose (Alonso-Pérez et al. 2013). Aerosol Optical Depth (AOD), which measures the vertically integrated extinction of light by aerosols, is wavelength-dependent. AOD, a unitless figure, offers details about the amount of aerosols throughout the entire vertical range of the atmosphere. The number of aerosols in the air column

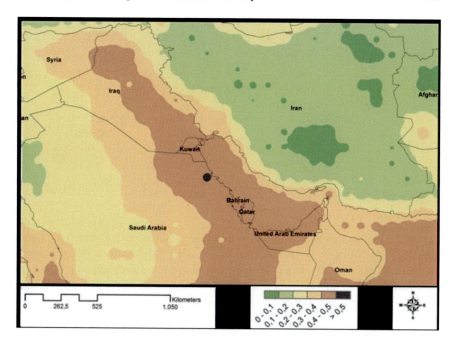

Fig. 7 Mean AOD in the middle east from 2003 to 2016 (Dundar et al. 2017)

increased with increasing AOD values. In general, the AOD Weighted trajectory analysis conducted for Kuwait, Bahrain, the UAE, and Saudi Arabia from April to August within the 2002–2012 period shows that Syria and other areas in Saudi Arabia on both its border with Iraq and its eastern province are the most extensive and intensive dust sources affecting these neighboring countries. Dundar et al. (2017) concluded that the annual mean AOD for the years from 2013 to 2016 and the frequency of SDS are indirectly related to other periods, whereas the frequency of SDS is at its maximum from 2008 to 2012 (Fig. 7).

1.5 The Influence of Climate Change in Iraq

Iraq is significantly impacted by climatic changes worldwide because of its location in an arid region (Fig. 8) (USAID 2017; Adamo et al. 2018; Oleiwi and Al-Dabbas 2021; Al-Ali and Al-Dabbas 2022a). Despite the negative effects of climate change on the environment, such as an increase in the frequency of drought and dust storm events as well as the wide variations in precipitation and temperature that have an impact on water discharge rates and salinity levels (Fig. 9) (Al-Dabbas et al. 2017, 2021; Al-Ali and Al-Dabbas 2022b; Oleiwi and Al-Dabbas 2022a, b). Therefore, water discharge into the Iraqi territory decreases with time, and the available water cannot

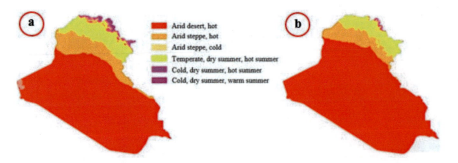

Fig. 8 Iraq's climate, according to the Koppen classification map, **a** for 1980 to 2026, **b** for 2071 to 2100 (USAID 2017)

support the rising population's demands in terms of quality and quantity (Al-Ansari et al. 2019; Sissakian et al. 2022). The lack of water has increased desertification, decreased annual rainfall, more frequent dust storms, and decreased wetlands and agricultural areas (Fig. 8).

The available historical climatological data such as the annual rainfall and temperature were plotted against time for Mosul, Baghdad, Kut, and, Amara meteorological stations for the Tigris River basin and for Haditha, Al-Hindiya, Samawa, and Nasitiya for the Euphrates River basin for 1987 to 2020 (Fig. 9) (Oleiwi and Al-Dabbas 2021; Al-Ali and Al-Dabbas 2022a, b; Muhammed and Al-Dabbas 2022, Al-Sekar and Al-Dabbas 2022). Climatic conditions in an area depend mostly on the geographic position of the earth's land; therefore, there are different climate zones in the Tigris and Euphrates river basins. The mean annual precipitation in Turkey reached 1000 mm, whereas it decreased to approximately 150 mm in Syria and 75 mm in Iraq's southern regions (Al-Ansari et al. 2019). In general, the Iraqi climate is well documented that are classified under the evidence of the dry climate characterized by high-temperature degrees during summer, and low rates of rainfall during winter (Oleiwi and Al-Dabbas 2021; Al-Ali and Al-Dabbas 2022a, b). Hot and dry climate conditions led to high averages of evaporation from the water surfaces, and the water loss from the Euphrates River was only about 5. 4 milliards m^3/ year, which exposes Iraq to certain losses of its water share (Oleiwi and Al-Dabbas 2021; Al-Ali and Al-Dabbas 2022a, b).

Climate change events have a positive effect on the water flow rates. The discharge of the Tigris and Euphrates Rivers decreased by more than 40% after the building of dams on the Rivers in Syria and Turkey, making Iraq a serious challenge as far as running the quantity of water is troubled if this problem will not be settled with shared countries, such as Turkey and Syria (Al-Ansari et al. 2019). Based on the circumstances mentioned above, the river's salinity increased downstream, and the average output of cations and anions downstream was relatively higher than that of the input upstream of the river. This was attributed to natural factors like increasing temperature degrees and decreasing rainfall in addition to the anthropogenic activities

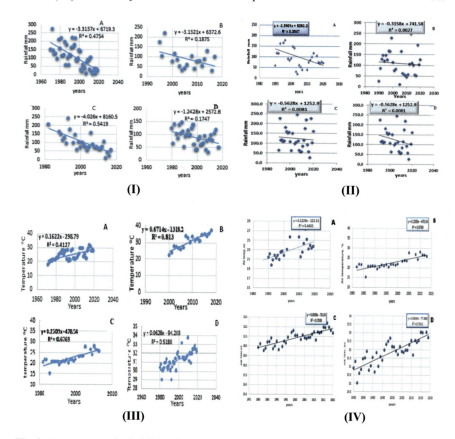

Fig. 9 Average annual rainfall (mm) and average annual temperature (°C) plots against time for the Tigris and Euphrates river basins for the years 1987–2000 (Oleiwi and Al-Dabbas 2022a, b; Al-Ali and Al-Dabbas 2022a, b; Iraqi Meteorological Organization 2022; Muhammed and Al-Dabbas 2022; Al-Sekar and Al-Dabbas 2022)

originating from agricultural activities and population communities along the river course (Oleiwi and Al-Dabbas 2022a, b; Al-Ali and Al-Dabbas 2022a, b).

(I) Average annual Rainfall (mm) plots against time for the Tigris River Basin (A) Mosul, (B) Baghdad, (C) Kut, and (D) the Amara Meteorological Station.
(II) Average annual rainfall (mm) plotted against time for the Euphrates River Basin (A) Haditha (B) Al-Hindiya, (C) Samawa, and (D) Nasitiya Meteorological Stations.
(III) Average Annual Temperature (°C) plotted against time for the Tigris River Basin (A) Mosul, (B) Baghdad, (C) Kut, and (D) Amara Meteorological Station.
(IV) Average Annual Temperature (°C) plotted against time for the Euphrates River Basin (A) Haditha (B) Al-Hindiya, (C) Samawa, and (D) Nasitiya Meteorological Stations.

1.6 The Storms Effect on Visibility and Their Classification

Storms cause visibility reduction, which deteriorates road and aircraft transportation, and increases the heat blanket effect. The calculation of the distance that a light or object is clearly discerned is defined as visibility. Visibility may deteriorate by particulates less than 2.5 µm, because of light absorption and scattering (Alonso-Pérez et al. 2013). The sand-dust storm classification according to visibility into four classes is as follows.

1.6.1 Severe Storm

It represents huge amounts of carrying loads that are distributed and blown by very high-speed winds, in which the visibilities are below 200 m.

1.6.2 Dust Storm

It represents visibilities from 1000 m to more than 200 m, reflecting strong winds.

1.6.3 Blowing Storm

This indicated the presence of aeolian dust. Blowing storms have visibilities ranging from 1 to 100 km.

1.6.4 Dust Haze

It represents very fine grains of silt and clay in suspension with a visibility of less than or equal to 10 km. According to Alonso et al. (2013), 17 stations in Iraq were used for visibility studies. Long-term measurements were used to analyze the available data on visibility values over Iraq.

2 Materials and Methods

2.1 Fieldwork

Fieldwork and activities included sampling storm-related loads (sand and dust) using portable field sampling and measuring equipment. The most crucial step in any sampling procedure is precise and exact site selection to produce the greatest amount of data for environmental evaluation, which can aid in determining the carrying loads, impacts, and connections of point sources for pollutants. The wind direction element was carefully considered when choosing the appropriate sampling locations because it has a notable impact on the spread of contaminants. Dust samples were collected from numerous Iraqi governorates between March 2007 and July 2010 (Fig. 1, Table 2). The samples were collected using either low-volume air samplers or large-volume plastic basins from the roofs of tall buildings (more than 3 m above the ground) (Sniffer L-30) (Fig. 10a), or by using the cyclone sampler air (Fig. 10b) by employing an electric motor to rotate the air inside the apparatus and pull and suction the air inside the apparatus to a height of 1.65 m or by using the pyramidal funnel trap (Fig. 10c). This device consists of a pyramidal funnel, 29 cm in height and width, stainless steel, and regular measurements of 91 cm (91 cm (square in shape). The surface of the funnel was covered with a wire net (with 20 mesh) to prevent foreign materials, plant leaves, and insects from entering the dust specimen. The funnel was connected to a 711 cm tube. The final collector of falling dust inside it was 29 cm high and 39 cm wide; therefore, the final height of the device was 791 cm. (Shehab Al-ddin and Aziz 2012). The collected samples ranged in weight from a few grams to >750 g. Depending on the duration of each sandstorm, the collection time ranged from less than 8 h to >24 h. The analysis was performed at the College of Science, University of Baghdad's laboratories, and the Ministry of Medicine, Medical City, Baghdad) for the collected dust samples.

Table 2 The investigated storms that blew on Iraq from 2007 to 2010

Storms number	Storms date	Number of storms	Storms date	Number of storms	Storms date
1	17-3-2007	17	7-6-2008	33	17-6-2009
2	18-4-2007	18	15-6-2008	34	28-6-2009
3	11-5-2007	19	28-6-2008	35	2–7-2009
4	16-5-2007	20	1-7-2008	36	4-7-2009
5	8-7-2007	21	7-7-2008	37	29-7-2009
6	16-7-2007	22	11-7-2008	38	30-7-2009
7	7-9-2007	23	27-7-2008	39	22-2-2010
8	19-2-2008	24	30-7-2008	40	3-4-2010
9	3-3-2008	25	15-9-2008	41	4-4-2010
10	15-3-2008	26	24-9-2008	42	13-5-2010
11	30-3-2008	27	16-10-2008	43	14-5-2010
12	4-4-2008	28	17-2-2009	44	6-6-2010
13	17-4-2008	29	20-2-2009	45	23-6-2010
14	27-4-2008	30	27-2-2009	46	24-6-2010
15	16-5-2008	31	9-3-2009	47	19-7-2010
16	25-5-2008	32	9-6-2009	48	20-7-2010

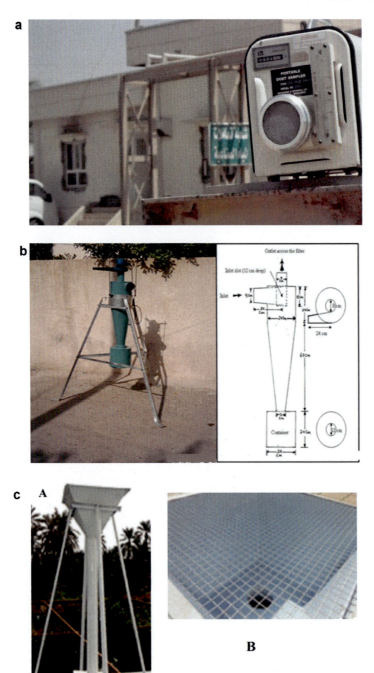

Fig. 10 a Low volume air sampler (Sniffer L-30) used for TSP sampling (Ali 2013). **b** One of the sites under study had its cyclone air sample taken (Al-Khafaji 2009). **c** The pyramidal funnel traps sampling at one of the studied sites (Shehab Al-ddin and Aziz 2012)

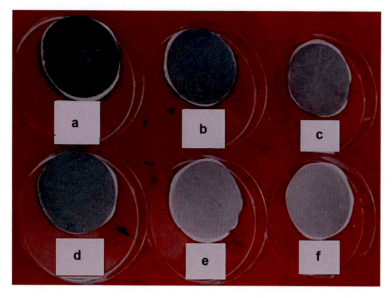

Fig. 11 The air filters (cellulose filters) for low volume sniffer devise type (L-30) that collects the TSP (Ali 2013)

2.2 Total Suspended Solids Sampling After a Dust Storm

The total suspended solid concentrations of the samples were determined using a low-volume sniffer device (Fig. 10a). The device measured air samples taken from the locations at one liter per minute. Rotheroe and Mitchell Ltd. outfit this device. The filter holder for this type of device (L-30) had a round shape and a diameter of 6 cm. Whitman cellulose candidate filters with a 6 cm diameter were used to capture the suspended particles in the airflow (Fig. 11). Additionally, a pump was used to remove and empty the air. A piece of clear plastic mounted in front of the gadget, which houses a small iron ball, measures the volume of airflow. The range of airflow employed during testing was between (40–70 L per minute).

2.3 Preparation of Dust Samples

Mineralogical, grain size and shape analyses were performed by applying the wet sieve method and utilizing the processes for hydropower (1953) and Folk (1974) to determine the grain size of the dust load. The mineralogical constituents of the dust particles were determined using Microscopic and XRD techniques (Carver 1971).

2.4 Trace Element in Dust

Trace elements Cu, Cd, Pb, Co, Zn, Fe, and Ni were detected by XRF and XRD. The geo-accumulation (I-geo), contamination factor (CF), and pollution load index were employed to analyze the heavy metals present in dust storms as indicators of pollution (PLI). Therefore, dust storm samples were collected during 2014 and 2016 from Basra City, Basra Governorate, Diwaniya City, Qadisiya Governorate, and Southern Iraq. Twenty-one samples of dust storms were gathered from seven sites in Basra city (1-Qurna, 2-Hartha, 3-Qarmmatt Ali, 4-Ashar, 5-Abu AlKhassib, 6-Fao, and 7-Umm Qasir sites). The same number of dust storm samples was collected from 12 sites in Diwaniya City (1-Hai Jamiaa, 2-Hathara, 3-Wahda, 4-Askan, 5-Karar, 6-Khadraa, 7-Sader, 8-Janoob, 9-Shurta, 10-Athari, 11-Jumhuria and 12-Sadiq sites). The samples were collected in dust containers. All samples were brought to Baghdad University's geochemistry lab, where they underwent drying procedures in an oven set to 60 °C. A few grams of the sample were then powdered, and trace elements were detected using XRF and XRD methods (Al-Jaberi 2014; Al-Dabbas et al. 2018).

2.5 Getting the Samples Ready

Low-volume air samplers, also referred to as sniffers because they capture airborne suspended particles, were used to collect the samples. The cellulose filters were weighed before being inserted into the filter holders and dried at 80 °C to remove any remaining moisture before the sampling collection process began (W_1). The Cellulose Filter is then inserted into the apparatus (in the filter holder), after which the airflow size is calculated. By calculating the volume difference between the flow of air that the iron ball read, represented as (V_1), and the flow of air reading after the device has run for an hour, represented as (V_2), it is possible to determine the most recent value (V_2). The Cellulose Filter was then removed and weighed once more, with the new weight represented as (W_2). Applying Eq. (1), we can determine the concentrations of suspended particles:

$$\text{TSP Concentration} = W_2 - W_1 / V_T \times 10^6 \tag{1}$$

where:

TSP   Concentration in micrograms per cubic meter.
W_2 Weight of the filter with TSP in grams.
W_1 Weight of the filter in grams.
V_T Volume of air in m^3.
10^6 Converting to microgram.

The following Eq. (2) is used to determine the air volume VT:

$$V_T = (V_1 + V_2/2) \times t/1000 \qquad (2)$$

where:

T   Per-minute sample time.
V_1 Air volume that the iron ball measured at the start of the measurement.
V_2 The iron ball ultimately measured the air volume after the completion of the measurement.

2.6 Samples of Air Filters that Have Been Digested

The following points were used to describe the digestive process:
1. Small bits of the filter paper are broken up and added to a Teflon beaker.
2. The sample is shredded and then 5 cc of nitric (HNO_3) acid is added.
3. The sample is heated with a heater until it is close to the drought's cure.
4. After being taken out of the heater, the samples are left to cool.
5. The cooled sample is heated again after being added to three milliliters of perchloric acid ($HClO_4$).
6. Before drying, the sample rises and then cools.
7. Hydrofluoric acid (HF) in the amount of two milliliters is added once again, and the sample is covered and left overnight until the solution is clear.
8. After that, the material is filtered in a volumetric flask with a 100 ml capacity.
9. The capacity is filled to 100 ml with deionized water, and the solutions are stored in plastic bottles until they are needed.

A (Blank Solution) must be prepared under the same circumstances as the samples for treatment while making those measurements, and this blank sample must then go through the automated components' analysis procedure. All measurements were performed in accordance with Eq. (3):

$$\text{Metal Conc.} (\mu g/m^3) = C * Vi / VT \qquad (3)$$

where:

C   The amount of each element in the sample that was collected in (ppm) units.
Vi   The sample's volume in milliliters.
VT   The sum of the air drawn, expressed in cubic meters.

2.7 Detection of Radioactivity

Radioactivity was detected for dust particles collected during the 2-4/7/2009 and 3-4/4/2010 storms that blew over the Ramadi and Baghdad districts, applying Mahdi et al. (2011) and Marouf.

2.8 Pollen Identification

The pollen types and concentrations were determined using the procedure described by Moore and Webb (1978).

2.9 Identification of Microorganisms

Microorganisms have been identified as fungi, algae, Viruses, and Bacteria using international procedures (Al-Khafaji 2009).

3 Results and Discussion

3.1 Dust and Sand Storms

As mentioned before, the number of days of dust storms indicated that the Basra, Qadisiya, and Muthana governorates had the worst dryness index, less annual rainfall, and higher mean annual temperature with a maximum number of days of dust storms (Fig. 1, Table 1) (Jassim et al. 2012; Sissakian et al. 2013). The average monthly falling dust weights in gm/m^2 that were dispersed throughout southern and middle Iraq show that the Muthana governorate had the maximum fallen dust that reached 250 gm/m^2 /month, followed by the Iraqi southern governorates as Basra and Qadisiya (Fig. 12) (Al-Dabbas et al. 2012).

3.2 The Shape and Texture

The results of the shape of quartz grains with 7 degrees of hardness indicate that the roundness ranges between sub-rounded of approximately 82% of the total samples, to approximately 18% of the total samples, indicating a long transportation distance (Fig. 13). The texture analysis results show the composition mainly of sand (min = 8%, max = 18%, mean = 13%), silt (min = 18%, max = 63%, mean = 32%), and clay (min = 20%, max = 71%, mean = 55%), with maximum sandy silty clay percentages (71.4%) and sandy clayey silt percentages (28.6%) (Ropmi 1985; Power 1953).

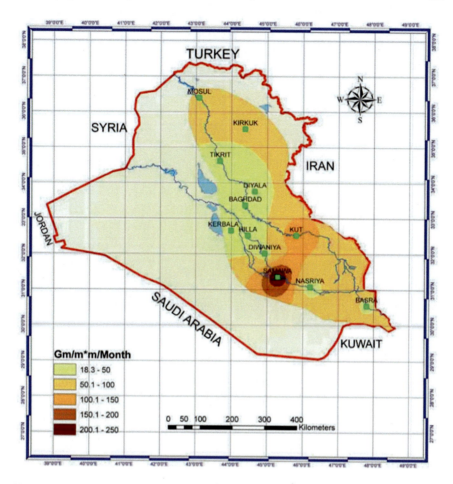

Fig. 12 The average monthly falling dust weights in gm/m^2 that were dispersed throughout southern and middle Iraq (Al-Dabbas et al. 2012)

3.3 The Dust Mineralogy

The results for Heavy and Light minerals indicated 98% were light minerals of approximately 52% quartz, 6% feldspar, 34.5% calcite and dolomite, 5.5% gypsum, and anhydrite, and approximately 2% heavy minerals composed of hornblende, chlorite, zircon, garnet, pyroxene, epidote, and opaque heavy minerals (Fig. 13). The same findings are noticed by applying XRD technique, which concurs with the studies of Al-Jannabi and Ali Jawad (1988); and Hamparsoum (2002). The recognition and identification results of clay minerals are indicated as Chlorite, Illite, Montmorillonite, Palygorskite, and Kaolinite (Grim 1968). These findings are in concordance

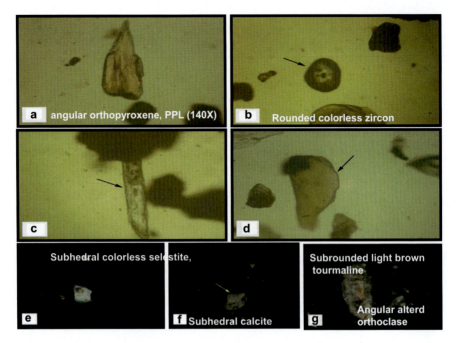

Fig. 13 Representative dust storm minerals and shape under the polarized microscope (Al-Khafaji 2009)

with the results of Al-Jannabi and Ali Jawad (1988), Dougramedji and Hamparsoum (1996), Darmoian (2000), Al-Ali (2003), Al-Ali et al. (2005).

3.4 The Effect of Heavy Metals

The heavy metals Zn, Cu, Pb, Fe, Cd, Ni, and Co have been identified by several scientists and institutes, such as the WHO (1996) and Poschl (2005). The results of heavy metal identification and measurements indicated that Fe (2937 ppm) was the highest, followed by Pb (432 ppm), Zn (374.7 ppm), Ni (154 ppm), Co (89.6 ppm), Cd (61 ppm), and finally the lowest values of Cu (56.5 ppm). These findings are in concordance with those of WHO (1996), Poschl (2005), and Abd El-Aal et al. (2023). The geo-accumulation (I-geo), contamination factor (CF), and pollution load index were employed to analyze the heavy metals present in dust storms as indicators of pollution (PLI). Therefore, dust storm samples were collected during 2014 and 2016 from Basrah City, Basrah Governorate, Diwaniyah City, Qadisiyah Governorate, and Southern Iraq. The PLI results ranged from class 0 (perfect of no pollution) to class 2 (deterioration of site quality), indicating local contamination at all tested sites. The results from Diwaniyah City were lower than those from Basrah City, with the exception of Fe and Cu readings, which were higher. While I-geo for Fe and Cu displays relative values of class 0, which indicates no pollution, I-geo for Co, Zn, Pb, and Ni in both Diwaniyah and Basrah city sites shows relative values of class 1, which indicates mild contamination. While the contamination factor for Fe and Cu was classified as class 1, which indicates little pollution, the Co, Zn, Pb, and Ni contamination factors were classified as class 2 in both Diwaniyah and Basrah cities.

3.5 Measurement of Radioactivity

Many radionuclides are present on Earth, such as Radium R226 with about 1620 years long life, which is an essential radionuclide in the U238 series and can precipitate in bones (Marouf et al. 1992, 1993).

The average absorbed dose of uranium was determined for storm particulates that blew over Ramadi and Baghdad during storms–2-4/7/2009 and 3-4/4/2010. The measurement unit was Bq/kg, in which 12 × 106 Bq/Kg of U238 was equal to 1 ppm. The findings revealed that the average uranium activity ranged from 5.43 to 9.56 Bg/kg for storms during 2-4/7/2009 and the average uranium activity ranged from 7.32 to 18.96 Bg/kg for storms during 3-4/4/2010, and they are within normal range and in agreement with Mahdi et al. (2011) conclusions.

3.6 Pollen Particle Identification

The pollen grains were identified and showed the following grains: Pine, Chenopodiaceae, Palmae, Graminea, Olea, Typha, and Artemisia with various Palynomorphs such as Algae, Cuticle, Fungi, Lycopodium spores, Sphagnum spores, Micro Spines, and unidentified spores (Table 3). The different pollen percentages were directly related to speed, wind direction, and pollen source. Such findings may provide strong evidence of local dust storms that originated in distant regions, as suggested by pine pollen with sacs to keep them afloat and transported over great distances by the wind, which may have come from northern Syria, Turkey, or neighboring countries with a similar climate to Iraq (Figs. 14, 15, and 16) (Willard et al. 2004; Elbert et al. 2007; Al-Dousari et al. 2012). These findings are in agreement with those of Al-Dousari et al. (2012) (Fig. 16).

3.7 Identification of Fungi, Bacteria and Viruses

The results of identified Fungi and Bacteria within the studied storms are about 40.6% of the gram-positive bacilli (*Bacillus* species), about 8.4% of *E. coli*, 7.4% of *S. pneumonia*, 5.8% of *E. cloacae*, 4.1% of *S. epidermidis*, less than 3.0% of both *P.*

Table 3 Pollens, Spores, Fungi, Algae, and microspines identified from storm loads blown over Iraqi Governorates (Al-Dabbas et al. 2012)

Pollen grains identified	Numbers of samples	Range %
Chenopodiaceae	23	Few to 83
Graminea	17	Few to 70
Pine	15	Ten to 65
Artemisia	6	Nil to 50
Palmae	5	Nil to 15
Olea	5	Nil to 10
Typha	4	Nil to 5
Cuticle	30	Ten to 80
Fungi	24	Few to 60
Algae	8	Nil to 20
Lycopodium	7	Nil to 75
Sphagnum	5	Nil to 8
Unidentified spores	3	Nil to 6
Micro spines	3	Nil to 20

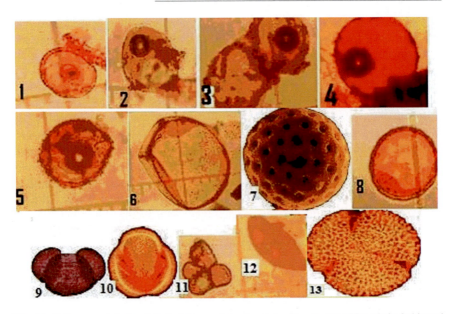

Fig. 14 Pollen counts in Iraqi dust of the most dominant families during 2007–2010: Cultivated Graminea (1–6), Chenopodea (7–8), Bisacate Pinus (9), Artemisia (10), Typha (11), Palmae (12), and Olea (13), (Al-Khafaji 2009)

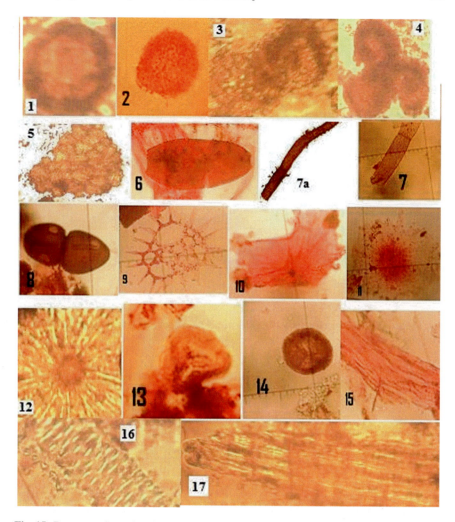

Fig. 15 Representative palynomorphs such as cuticle, micro spines in iraqi dust during 2007–2010: (1)—Sphagnum spores, (2)—Lycopodium spores, (4–8)—Fungi, (9–12)—Algae, (13–14)—Unnamed spores (15–17)—Cuticle (Al-Khafaji 2009)

aeruginosa and *S. aureus*, with less than 2.0% of both *E. aerogenes*, with *P. mirabilis*, *K. pneumoniae* and *P. vulgaris* is of 0.64%, (Table 4). The fungal particulate isolates also contained approximately 14.5% *Aspergillus* species and 7.7% *C. albicans*. The closest known pathogenic bacteria, fungi, and viruses that are distributed during storms and that affect human life are Bacillus circulans, Bacillus ticheniform, and *Staphylococcus aureus* (Wieser and Busse 2000; Kuske et al. 2006; Al-Dousari et al. 2012).

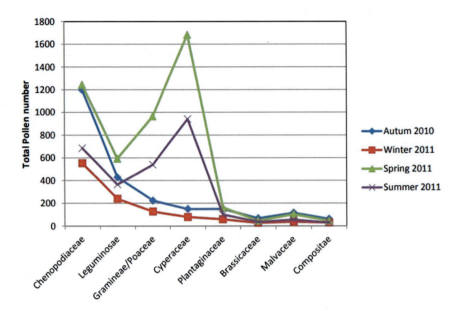

Fig. 16 Seasonal pollen counts in Kuwait dust of the most dominant families during 2010–2011, (Al-Dousari et al. 2012)

Table 4 Identification of fungi and bacteria in Iraqi storms (Al-Dabbas et al. 2012)

Microorganisms identification	Isolates counts	Percentages
Gram-positive cocci		
Staphylococcus aureus	8	2.6%
Staphylococcus epidermidis	13	4.2%
Staphylococcus pneumoniae	23	7.4%
Gram-positive rods		
Bacillus species	126	40.6%
Gram-negative cocci	0	Nil %
Gram-negative rods		
Escherichia coli	26	8.4%
Enterobacter cloacae	18	5.8%
Proteus mirabilis	5	1.6%
Pseudomonas aeruginosa	9	2.9%
Klebsiella pneumoniae	5	1.6%
Enterobacter aerogenes	6	1.9
Proteus vulgaris	2	0.6
Fungi		
Aspergillus species	45	14.5%
Candida albicans	24	7.7%
Total	310	100%

3.8 SDS Risk Assessment

Numerous wide-ranging environmental effects of wind erosion and atmospheric transport of sand and dust pose major risks to human populations in dry places (Middleton and Kang 2017). There are few analyses of regional patterns of vulnerability to SDS effects on human health and other types of economic activities. Knowledge of the distribution and composition of SDS sources can significantly improve the understanding of SDS risks, impacts, and dust cycle interactions with the environment and other climate system cycles. Sand and dust storm sources have a direct relationship with many environmental conditions, such as climate, weather, surface conditions, and human activities, which could negatively or positively impact SDS sources (Fig. 17). Knowledge of SDS sources provides information for SDS risk and impact assessment, SDS forecasting, mitigation planning, and the establishment of SDS early warning systems. This interaction can vary in time and space and may increase or decrease SDS generation. It was concluded that the Enhanced Plant Index (EVI) from the MODIS sensor and dust occurrences for 15 stations showed that inadequate vegetation cover (EVI 0.1) over roughly 60% of Iraq's land exists in the country's west and south, which is thought to be one of the key causes of dust events. EVI exhibits a strong inverse relationship with dust events, particularly in the north and center of the nation (Halos 2019).

Therefore, the best way to get rid of the effects of wind is to construct the shelter belts such as Green belts and Windbreaks by growing:

1. Woody trees
2. Shrubs in lines or circles as belts around the protected project.

Fig. 17 Drivers impact SDS source activities (Middleton and Kang 2017)

The purpose of windbreaks is to decrease the wind speed by 50% (Al-Dousari et al. 2019). Other solutions may be considered:

1. Stations were distributed well with sophisticated instruments and telecommunications to provide information about SDS before reaching the protected project.
2. Satellite images with ordinary visible light
3. Short wave radar.

4 Conclusions

1. The maximum grain size of the storm load is 71.4% sandy silty clay and 28.6% sandy clayey silt, depending on the wind speed.
2. The recognized light minerals are about 52, 6, 34.5%, and dolomite, 5.5% gypsum and anhydrite, and about 2% heavy minerals composed of hornblende, chlorite, zircon, garnet, pyroxene, epidote, and opaque heavy mineral.
3. The results of the clay mineral identification and recognition are shown. Kaolinite, Palygorskite, Montmorillonite, Illite, and chlorite.
4. Three indices, geo-accumulation (I-geo), contamination factor (CF), and pollution load index (PLI), were used to use heavy metals as pollution indicators. Dust storm samples were collected in the southern Iraqi cities of Diwaniyah and Basrah between 2014 and 2016. In contrast to Basrah City, Diwaniyah City's heavy metals resulting from dust storms showed lower levels of Co, Zn, Ni, and Pb, with the exception of greater Fe and Cu readings. The PLI values for dust storms in all areas under study did not indicate any local contamination. In both city sites, I-geo for Co, Zn, Pb, and Ni indicated some level of pollution, whereas I-geo for Fe and Cu indicated no pollution. According to the contamination factor, Co, Zn, Pb, and Ni showed moderate pollution in both cities, whereas Fe and Cu showed low contamination.
5. The findings revealed that the average uranium activity ranged from 5.43 to 9.56 Bg/kg for storms during 2-4/7/2009 and the average uranium activity ranged from 7.32 to 18.96 Bg/kg for storms during 3-4/4/2010 and they are within normal range and lower than critical dose level.
6. Pollen grains were identified and found to be Pine, Chenopodiaceae, Palmae, Graminea, Olea, Typha, and Artemisia with various Palynomorphs such as Algae, Cuticle, Fungi, Lycopodium spores, Sphagnum spores, Micro Spines, and Unnamed spores.
7. The results of Identified Fungi and Bacteria within the studied storms are about 40.6 % of the gram-positive bacilli (Bacillus species), about 8.4 % of E. coli, 7.4% of S. pneumonia, 5.8% of E. cloacae, 4.1% of S. epidermidis, less than 3.0% of both P. aeruginosa and S. aureus, with less than 2.0% of both E. aerogenes, with P. mirabilis, K. pneumoniae and P. vulgaris is of 0.64%.

8. The fungal particulate isolates also contained approximately 14.5% Aspergillus species and 7.7% Candida albicans.
9. Viral etiology analysis did not reveal any viral isolates.

5 Recommendations

1. Mapping of the SDS source areas is recommended. It is improved by discovering new source areas, distinguishing hotspot sources, and using sources of information from many different observations. Human interventions could reflect information from anthropogenic SDS sources, such as sectors of the economy, agriculture, water, forestry, energy, transport, etc., with indirect and direct impacts. Indirect factors that affect SDS source activity may be represented by fires, deforestation, mining, etc. in some regions, loss of soil structure, land cover changes, and disturbance of the topsoil due to agricultural practices, while hygienic needs, industry, and use of water for irrigation may represent direct impact factors that affect SDS source activity.
2. Sand and dust storm sources are directly related to many environmental conditions such as climate, weather, surface conditions, and human activities. Knowledge of the distribution and composition of SDS sources can significantly improve the understanding of SDS risks, impacts, and dust cycle interactions with the environment and other climate system cycles.
3. Knowledge of the distribution and composition of SDS sources is anticipated to be useful for developing guidelines to help people protect themselves from dust.
4. Improvements in management methods, considering the capabilities of the land, proper land-use plans, revisions to the policies governing the management of water resources, and the identification of efficient methods for fixing erosion sources are all successful recommendations for dust control.
5. The government must act quickly to make people aware of the situation. By passing the necessary rules and legislation, the Iraqi people's awareness will aid in regulating the reasons that humans cause.

Funding This research received no external funding.

Conflicts of Interest The author declares that there are no conflicts of interest regarding the publication of this manuscript.

References

Abd El-Aal A, Al-Awadhi J, Al-Dousari A (2023) The geology of Kuwait. Regional Geol Rev, Springer. https://doi.org/10.1007/978-3-031-16727-0
Abdullah D, Ahmed N (2021) A review of most recent lung cancer detection techniques using machine learning. Int J Sci Bus 5(3):159–173

Adamo N, Al-Ansari N, Sissakian V, Knutsson S, Lau J (2018) Climate change: consequences on Iraq's environment. J Earth Sci Geotech Eng 8(3):43–58. ISSN: 1792-9040

Al-Ali J (2003) Dust storms contribution of khor al-Zubair and khor Abdallah deposits. J Iraqi Geogr Soc 53:54–78 (In Arabic)

Al-Ali J, Abdul Jabbar C, Hassan H (2005) Effect of Marshlands drying process on regional dust fallout. Wade Al- Rafedian Magazine 16:261–272 (In Arabic)

Al-Ali I, Al-Dabbas M (2022a) The effect of variance discharge on the dissolved salts concentration in the Euphrates River upper reach, Iraq. Iraqi J Sci 63(9):3842–3853. https://doi.org/10.24996/ijs.2022.63.9.16

Al-Ali I, Al-Dabbas M (2022b) Assessment of some organic and inorganic pollution Indices/ Euphrates River/ Iraq. Int J Health Sci 6(S3):12395–12417. https://doi.org/10.53730/ijhs.v6nS3.9484

Al-Ansari N (2013) Management of water resources in Iraq perspectives and prognoses. Engineering 5:667–684

Al-Ansari N, Adamo N, Sissakian V (2019) Water quality and its environmental implications within Tigris and Euphrates Rivers. J Earth Sci Geotech Eng 9(4):57–108

Al-Dabbas M, Abbas M, Al- Khafaji R (2011) The mineralogical and micro-organisms effects of regional dust storms over middle east region. Int J Water Resour Arid Environ 1(2):129–141

Al-Dabbas M, Abbas M, Al-Khafaji R (2012) Dust storm load analyses- Iraq. Arab J Geosci 5(1):121–131

Al-Dabbas M, Al-Khafaji R (2012) Some geochemical, textural and radioactive characteristics of the sandstorms loads blown over Baghdad and Ramadi cities, Middle Iraq. In: Proceeding of the 1st conference on dust storms, pp 57–66

Al-Dabbas M (2016) Desertification and sand–dust storms in Iraq: Razzaza–Habbaria area. Lamb. Acad. Pub, Germany

Al-Dabbas M, Abbas M, Al-Khafaji R (2017) Sand Dust Storms Source Identification and the Mineralogical and Micro-Organisms Effects of Regional Dust Storms - Middle East Region. In: 5th international workshop on SDS, Istanbul

Al-Dabbas M, Mahdi K, Al-Khafaji R, Obayes K (2018) Heavy metals characteristics of settled particles of streets dust from Diwaniyah City- Qadisiyah Governorate—Southern Iraq. J Phys: Conf Series 1003:1–11

Al-Dabbas M (2018) Monitoring of SDS, ground observation with dust sample analysis. In: 6th international workshop on SDS, 12–15 Nov. 2018, Istanbul

Al-Dabbas M, Hussain T, Al-Kubaisi Q, Al-Dabbas H (2021) Impact of climate change on the hydrochemistry of Debaga unconfined aquifer, Kurdistan region, Iraq. IOP Conf Series: Earth and Environ Sci 779. https://doi.org/10.1088/1755-1315/779/1/012068

Al-Dousari A, Aba A, Misak R, Ramadan A, Ismail M, Ahmed M, Ismaeel A (2012) Monitoring and assessment of dust fallout and associated pollen within the state of Kuwait. Internal report, Kuwait Inst for Sci Res

Al-Dousari A, Domenico D, Modi A (2017) Types, indications and impact evaluation of sand and dust storms trajectories in the Arabian Gulf. Sustainability 9(9):1526. https://doi.org/10.3390/su9091526

Al-Dousari A, Ahmed M, Al-Dousari N, Al-Awadhi S (2019) Environmental and economic importance of native plants and green belts in controlling mobile sand and dust hazards. Int J Environ Sci Technol. https://doi.org/10.1007/s13762-018-1879-4

Al-Hemoud A, Al-Dousari A, Al-Shatti A, Al-Khayat A, Behbehani W, Malak M (2018) Health impact assessment associated with exposure to PM10 and dust storms in Kuwait. Atmosphere 9(6):87–104

Al-Sekar H, Al-Dabbas M (2022) The influence of the Tigris River discharge on the hydrochemistry with time, from Baghdad to Amara, Southern Iraq. Iraqi Geol J 55(2D):140–156

Al-Zubaidi E, Al-Sulttani A, Rabee F (2022) Sand dunes spectral index determination using machine learning model: case study of Baiji sand dunes field Northern Iraq. Iraqi Geol J 55(1F):102–121

Ali K, Shejiri S (2019) The radiological effects of dust storms in Baghdad-Ramadi area. J Sci 60(2):255–262

Ali L (2013) Environmental impact assessment of Kirkuk oil refinery. Ph.D. thesis, Univ. of Baghdad

Al-Jaberi M (2014) Heavy metals characteristics of settled particles during Dust storms in Basrah city, Iraq. Int J Sci Res 3(10):2277–8179

Al-Jannabi K, Jawad A (1988) Origin and nature of s and dunes in the alluvial plain of southern Iraq. J Arid Environ 14:27–34

Al-Jumaily K, Ibrahim M (2013) Analysis of synoptic situation for dust storms in Iraq. Int J Energy Environ 4(5):851–858

Al-Khafaji R (2009) Effects of dust storms on some Iraqi territories. Ph.D., thesis, College of Sci, Univ. of Baghdad

Alonso S, Al-Dabbas M, Ali A, Al-Waeli T (2013) Establishing a national program to combat sand and dust storms in Iraq. Spanish National Research Council (CSIC)/meteorological state agency of Spain (AEMET), Spain

Awadh S (2023) Impact of North African sand and dust storms on the middle east using Iraq as an example: causes, sources, and mitigation (review). Atmosphere 14(180):1–20. https://doi.org/10.3390/atmos14010180

Beg A, Al-Sulttani A (2020) Spatial Assessment of drought conditions over Iraq using the standardized precipitation index (SPI) and GIS techniques. Environmental Remote Sensing and GIS in Iraq. Springer

Boloorani A, Nabavi S, Azizi R, Kavosi M, Abasi E, Mirzapour F, Bahrami H (2012) The investigation of dust storms entering the west of Iran using remotely sensed data and synoptic analysis. Rec. Dust and dust Storms Conf, State of Kuwait

Carver R (1971) Procedures in sedimentary petrology. Willey, NY

Climatological Atlas of Iraq (2000) Iraqi meteorological organization. Unpublished internal report, Iraqi Meteorological Organization, Ministry of Transport

Darmoian S (2000) Sedimentary characters and accumulation of dust fallout southern Mesopotamian plain. Basrah, J Sci 18(1):141–156

Dougramedji J, Hamparsoum K (1996) Geomorphological and mineralogical study for some sand dunes near to Saddam River which covered surface soil strata. Iraqi J Sci Agric 27(2):9–17 (In Arabic)

Dundar C, Ayse G, Kahraman O, Gulen G (2017) Spatial and temporal analysis of sand and dust storms between the years 2003 and 2016 in the middle east. In: 5th international workshop on SDS, 23–25 Oct. 2017, Istanbul

Elbert W, Taylor P, Andreae M, Poschl U (2007) Contribution of fungi to primary biogenic aerosols in the atmosphere: wet and dry discharged spores, carbohydrates, and inorganic ions. Atmos Chem Phys 7:4569–4588

Folk R (1974) Petrology of sedimentary rocks. Hemphill, Austin

Goldman J (2003) Dust storms, sand storms and related NOAA activities in the middle east. NOAA magazine. Internet data

Griffin D, Kellogg C, Garrison V, Shinn E (2002) The global transport of dust. Am Sci 90:228–235

Griffin D, Kubilary N, Kocak M (2007) Airborne desert dust and aero microbiology over the Turkish Mediterranean Coastline. Atmos Environ 41:4050–4062

Grim R (1968) Clay mineralogy, 2nd edn. Mc Grow-Hill book, NY

Halos S (2019) Analytical study of the effect of vegetation cover on dust events over Iraq in spring season. In: 17th Iraqi geological conference Baghdad, Iraq

Hamparsoum A (2002) Geomorphologic and mineralogical study of sand in Al- MASSAB Al-AMM. J Sci Agric 33(2):55–79 (In Arabic)

IAEA (International Atomic Energy Agency) (2003) Environmental contamination by naturally occurring radioactive materials (NORM) and technological options for mitigation. Technical Report Series 419 Vienna

Iraqi Meteorological Department (2022) General Climatological Data of all Meteorological Stations. Unpub. Report, directorate of climatology and directorate of Hydro- and Agricultural Meteorology, Ministry of Transportation, Iraqi Meteorological Organization

Jacquelyn C (2009) Climate analysis and long-range forecasting of dust storms in Iraq. Unpub. M.Sc. thesis, United States Naval Academy, Naval postgraduate school, USA.

Jassim N, Habbib E, Hantosh T (2012) The climatic change effects on the precipitation and dust phenomenon in IRAQ. Unpub. Internal Report, Ministry of Transport, Preceding of the 1st Conference on Dust Storms and their environmental effects, pp 74–93

Kuske C, Barns S, Grow C, Merrill L, Dunbar M, Dunbar J (2006) Environmental survey for four pathogenic bacteria and closely related species using phylogenetic and functional genes. J Forensic Sci 51:458–558

Mahdi K, Nasif R, Yassin K (2011) Measurements of radioactivity pollution of dusty storms in the middle and western parts of Iraq in 2007 and 2008. Kufa Univ J Phys 2(3):38–55

Marouf B, Mohamad A, Taha J, Al-Haddad I (1992) Population doses from environmental gamma radiation in Iraq. Health Phys 62:443–444

Marouf B, Mohamad A, Taha J (1993) Assessment of exposure rate and effective dose equivalent in the city of Baghdad due to natural sources of radiation. Sci Tot Environ 13:133–137

Middleton N, Kang U (2017) Sand and dust storms: impact mitigation. Sustainability 9(6):1053. https://doi.org/10.3390/su9061053

Mohammed A, Khadum J, Al-Lami A, Al-Shamarti H, Al-Salihi A (2022) Dynamical transportation of aerosols for extreme dust storms over Al-Jazira region. Iraqi Geol J 55(2D):181–194

Moore P, Webb J (1978) An illustrated guide to pollen analysis. Hodder & Stoughton Educational Division. ISBN 10: 034021449X. ISBN 13: 9780340214497

Morton L, Evans C, Ester G (2002) Natural Uranium and thorium distributions in podzolized soils and native blueberry. J Environ Qual 31:155–162

Muhammed M, Al-Dabbas M (2022) Impact of climatic change and shortage of water flow on the Euphrates river water salinity between Al-Hindyia and Al-Nasiriya, Southern Iraq. Iraqi Geol J 55(2A):139–152. https://doi.org/10.46717/igj.55.2A.10Ms-2022-07-26

Nasif R (2011) Study the texture composition of some blowing dust storms on Ramadi City in 2010. Mustansiriya Univ., college of science. J Sci 22(7):89–94

Oleiwi A, Al-Dabbas M (2021) Hydrochemical evaluation of the Tigris River from Mosul to South of Baghdad Cities, Iraq. Int J Environ Climate Change 11(6):44–58

Oleiwi A, Al-Dabbas M (2022a) Relationship of annual flow with Hydrochemical analysis of the Tigris River and evaluation of water for drinking and irrigation uses. Eco Env & Cons 28 (January Suppl. Issue):S461–S472). https://doi.org/10.53550/EEC.2022.v28i01s.063

Oleiwi A, Al-Dabbas M (2022b) Assessment of contamination along the Tigris River from Tharthar-Tigris Canal to Azizziyah, middle of Iraq. Water 14:1194. https://doi.org/10.3390/w14081194

Poschl U (2005) Atmospheric aerosols: composition, transformation, climate and health effects. Angew Chem Int Edit 44:7520–7540

Power M (1953) A new roundness scale for sedimentary particles. J Sed Pet 23:117–119

Rasheed M, Al-Ramahi F (2021) Detection of the impact of climate change on desertification and sand dunes formation east of the Tigris River in Salah Al-Din governorate using remote sensing techniques. Iraqi Geol J 54:69–83

Ropmi (1985) Dust fallout in the northern part of Kuwait. ROPME Arid, Final report of the Regional organization for the protection of the Marine Environment

Shehab Al-ddin S, Aziz N (2012) Aromatic hydrocarbons multiple nuclei (PAHs) in dust falling in the province of Basra. In: Proceeding of the 1st conference on dust storms and their environmental effects

Shi Z, Shao L, Jones T, Lu S (2005) Microscopy and mineralogy of airborne particles collected during severe dust storm episodes in Beijing, China. J Geophys Res 110:D01303. https://doi.org/10.1029/2004JD005073

Sissakian V, Al-Ansari N, Knutsson S (2013) Sand and dust storm events in Iraq. Nat Sci 5(10):1084–1094

Sissakian V, Jassim H, Adamo N, Al-Ansari N (2022) Consequences of the climate change in Iraq. Global J Human-Soc Sci: B Geogr, Geo-Sci, Environ Sci Disaster Manage 22(2). Online ISSN: 2249-460x & Print ISSN: 0975-587X

Walker M (2005) Iraq: a full year climatology study. Air force combat climatology center (AFCCC), Asheville, North Carolina

Wahab Z (2007) Environmental analysis to Geographical factors influential the quantity and quality of Fallen air in DhiQar governorate. M.Sc. thesis in geography, College of Education University of Basrah

WHO (1996) Air quality guide lines for Europe. World Health Organization, Regional Office for Europe, Copenhagen

Wikipedia, the Free Encyclopedia (2009) Dust storms. Wikipedia, the free encyclopedia. Internet data. http://en.wikipedia.org/wiki/Dust-storm

Wieser M, Busse HJ (2000) Rapid identification of Staphylococcus epidermidis. Int J Syst Evol Microbiol 50:1087–1093

Willard D, Cooper S, Gomez D, Jensen J (2004) Atlas of pollen and spores of the Florida Everglades. Palynology 28:175–227

UNESCO (2013) Sand and dust storm fact sheet. http://reliefweb.int/report/iraq/sand-and-dust-storm-fact-sheet

UNEP, WMO, UNCCD (2016) Global assessment of sand and dust storms. United Nations Environment Program, Nairobi

USAID (2017) Climate risk profile: Iraq. https://www.climatelinks.org/resources/climate-change-risk-profile-Iraq

Monitoring Drifting Sand Using Spectral Index and Landsat TM/OLI Datasets in Bahr An-Najaf Area, Iraq

Ghadeer F. Al-Kasoob, Ahmed H. Al-Sulttani, Ayad M. Fadhil Al-Quraishi, and Ragad N. Hussein

Abstract This chapter analyzed the spatial and temporal characteristics of sand dunes, focusing on the rates of creep and movement of both dunes and sand. The study area comprised four types of sand features: Barchan dunes, Longitudinal dunes, Nabkha, and Sand sheets. Hazard analysis of the sand dunes was conducted using a combination of remote sensing techniques and field observations. Four Landsat TM/OLI images (22 Aug. 1988, 18 Aug. 1998, 13 Aug. 2008, and 9 Aug. 2018) have been used to achieve the objectives. The study area utilized the Normalized Difference Sand Dunes Index (NDSDI) to identify and map drifting sand. Notably, the size of the sand dunes underwent significant changes between 1988 and 2008 but remained relatively stable from 1978 to 1988. During these periods, the sand dunes expanded in area, but the valleys in the research region impeded their progress. This expansion and migration of sand dunes substantially threaten human activities in Al-Rahima, Eitha, Khiribah, and Um-Tharawi, with the risk increasing over time. Every decade, the encroachment rate of these sand dunes has been calculated, and it has been found that they approach areas and communities within a distance of 3000 m, indicating the proximity of the hazard. Among the dune fields, the third field demonstrated the fastest creep rate. Overall, the risk posed by these sand dunes has intensified, leading inhabitants in specific regions (southwest of Al-Rahima and

G. F. Al-Kasoob
Department of Administrative and Financial Affairs, Al-Qasim Green University, Babel, Iraq
e-mail: gadeerfaham@uoqasim.edu.iq

A. H. Al-Sulttani (✉)
Department of Environmental Planning, Faculty of Physical Planning, University of Kufa, Najaf, Iraq
e-mail: ahmedh.alsulttani@uokufa.edu.iq

A. M. F. Al-Quraishi
Petroleum and Mining Engineering Department, Faculty of Engineering, Tishk International University, Kurdistan Region, Erbil 44001, Iraq
e-mail: ayad.alquraishi@gmail.com; ayad.alquraishi@tiu.edu.iq

R. N. Hussein
Map Production Department, General Authority for Survey, Ministry of Water Resources, Baghdad, Iraq

© The Author(s), under exclusive license to Springer Nature Switzerland AG 2024
A. M. F. Al-Quraishi and Y. T. Mustafa (eds.), *Natural Resources Deterioration in MENA Region*, Earth and Environmental Sciences Library,
https://doi.org/10.1007/978-3-031-58315-5_12

Al-Ruhba) to abandon their houses and farmlands. Furthermore, industrial activities, including public services such as utility poles and roadways, are also affected.

Keywords Spectral index · NDSI · ENDSI · Sand sheets · Sand dunes · Creep rate · Bahr An-Najaf · Iraq

1 Introduction

Sand dunes are a dynamic landscape morphology formed when sand grains are accumulated by wind and the influence of gravity (McHarg 1972). Sand dunes have different forms and sizes based on their interaction with the wind, and respond to changes in their local environment (Moreno Casasola 1982; Fadhil 2002; Goudie 2013; Yekeen et al. 2023). It exhibits interesting forms associated with biological and ecological activities (Pye and Tsoar 2008; Dong et al. 2004; Chlachula 2021; Feng et al. 2019). Aeolian features can be divided into three types according to their dynamics (mobility), depending on climatic parameters such as (wind power and direction, precipitation, and potential evapotranspiration), which are (active, non-active, and partially active) (Al-Quraishi 1992; Knight et al. 2004).

Several types of sand features exist based on their morphology. Sand sheets and Nabkha are considered to have low altitudes and small sand features. Sand sheets are formed over flat and bare land, whereas Nabkha is formed when a specific obstacle faces the wind. For example, as a vegetative or rock obstacle, wind deposits the sand it carries (Tang et al. 2023; Bagnold 2012; Cooke et al. 1993; Danin 2012; Goudie 2013; Fadhil 2003). Barchan Sand, longitudinal, and star dunes are large features. Barchan has formed a crescent shape owing to the stable wind direction and has two wings that point to the opposite side of the wind direction. Longitudinal Dunes require more than one wind direction to form (Bagnold 2012; Cooke et al. 1993). The most crucial type of sand dunes is Barchans because of wind direction's relative stability, increased mobility, and risk (Wasson and Hyde 1983; Goudie 2013; Azzaoui et al. 2016; Hugenholtz et al. 2012; Yao et al. 2007).

The importance of studying aeolian sand features is to understand their morphology and temporal and spatial changes that indicate climate change and help in assessing the hazard (Kumar et al. 1993; Lam et al. 2011; Azzaoui et al. 2016; Hugenholtz et al. 2012; Sun et al. 2005; Bullard et al. 1997). Several methods have been used to study and obtain information about sand dunes. In addition to field studies, which are essential but require time and energy, remote sensing and satellite images are helpful for monitoring and assessing spatial and temporal changes in aeolian sand features (Wang et al. 2023; Al-Dabi et al. 1997; Yao et al. 2007; Azzaoui et al. 2016; Hugenholtz et al. 2012; Teodoro and Goncalves 2012; Lam et al. 2011; Bandeira et al. 2011; Zhang et al. 2008; Chen et al. 2006).

The dynamic nature and homogenous sand mineralogy make it an ideal object to monitor and extract using automatic and semi-automatic techniques from satellite images (Bandeira et al. 2011; Hugenholtz et al. 2012; Al-Quraishi 2009, 2013). The spectral index Machine Learning (ML) approach has become widely used to study, extract, and monitor these features (Al-Zubaidi et al. 2023; Zhao and Chen 2005; Chen et al. 2005; Zha et al. 2003; Tanser and Palmer 1999; Karnieli 1997). Azzaoui et al. (2016) studied and extracted barchan dunes from Mars surface using high-resolution satellite images (HiRISE). Chowdhury et al. (2010) the research used a trained multilayer perceptron (MLP) as a classifier to map the dunal landform of a part of the Rajasthan desert using multispectral optical data (LISS III). A semi-automatic approach for extracting sandy bodies in Portugal using IKONOS-2 data based on global thresholding using Otsu's method was further refined through detected edges (GThE) (Teodoro and Goncalves 2012).

Other studies also used object extraction-based classification techniques to extract barchan dune shapes in the northern part of the Laâyoune-Tarfaya Basin, Morocco, using panchromatic and multispectral bands of Landsat satellite imagery (Aydda et al. 2019; Aydda et al. 2017). Several studies have used spectral indices to extract bare land and sand (Al-Quraishi 2009, 2013; Chen et al. 2005; Karnieli 1997; Karnieli et al. 1999; Tripathy et al. 1996; Zhao and Chen 2005). Crust Index (CI) is much more sensitive to ground features than the original images. The absence, existence, and distribution of soil crust provides essential information for desertification and climate change studies (Karnieli 1997). Chen et al. (2005) also developed a new index named the Biological Soil Crust Index (BSCI). However, their experiments showed that the index is only suitable for cold deserts (Chen et al. 2005). Fadhil (2009, 2013) developed two sand dune indices using Landsat imagery: the Normalized Difference Sand Dune Index (NDSDI) and Normalized Difference Sand Index (NDSI). The NDSI functions on two specific bands: SWIR2 and Red. The NDSI effectively detects the presence of water, as it yields high values, whereas sand can be identified with values above zero.

The spectral characteristics of sand vary significantly to different degrees, which is fundamental for quantitative assessment because of the differences in the soil properties (Pinet et al. 2006; Udelhoven et al. 2003; Lagacherie et al. 2008). The primary objective of this chapter was to utilize a spectral index and Landsat (TM/ OLI) data to extract and analyze the spatiotemporal changes in sand features within the Bahr An-Najaf area. Additionally, this chapter aims to evaluate the impact of sand and its potential risks on various human activities, including population centers, agricultural practices, and industrial operations.

2 Materials and Methods

2.1 The Study Area

Bahr An-Najaf area is located in the Najaf governorate in southwest of Iraq. It is bordered to the north by Najaf Tar, east by Najaf City and Euphrates River, and south by Wadi Abu-Talah. These borders have been defined to include all human activities,

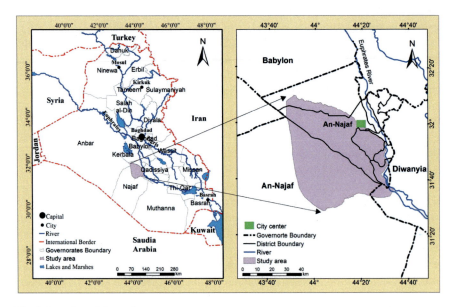

Fig. 1 Location of the study area

such as residential, industrial, and agricultural. The study area covers 3,603.72 km^2, which is located between Latitudes 31° 29′ 05.57" to 32° 10′ 05.87" N and Longitudes 43° 47′ 35.90" to 44° 33′ 40.90" E (Fig. 1). The climate is characterized by annual precipitation, which reaches up to 101 mm in total; the temperature is moderate in winter at 17 °C and increases in summer to 43 °C, while the average value of relative humidity is 49.7% (Beg and Al-Sulttani 2020; Mail et al. 2016). Geological deposits are represented mainly by quaternary sediments (Gypcrete, Aeolian, and Sebkha) and Paleocene-Miocene limestone and dolomitic limestone formations (Buday 1980; Jassim and Goff 2006; Al-Sulttani 2001).

Geomorphologically, the study area constitutes a portion of Iraq's western desert, and its prominent feature is the Bahr An-Najaf Depression. Six seasonal valleys flow southwest to northeast in this area, ultimately draining into the Bahr An-Najaf depression. These valleys originate from the western desert and extend towards the northeast, including Khur, Rhaimawi, Abo-Khamsat, Maleh, Hosoob, and Abo-Talah. The aeolian sand features in the study area are the Barchan, Longitudinal, Sand Sheet, and Nabkha (Fig. 2). All types, excluding the longitudinal dunes, were dynamic and changed. From the field study, the longitudinal direction is more stable because it is located in a marshland area to the south of the Bahr An-Najaf depression, which is formed from coarse sand grains mainly mixed with granules and pebbles of lime (Fig. 3).

Fig. 2 Barchan sand dunes with the direction of mobility SE (**a**), Longitudinal sand dunes (**b**), Sand sheet (**c**), and types of Nabkha accumulated after a human barrier (**d**)

2.2 The Remotely Sensed Datasets and Preprocessing

This chapter utilized Landsat satellite data (TM/OLI) with a spatial resolution of 30 × 30 m, encompassing four specific dates: August 22, 1988, August 18, 1998, August 13, 2008, and August 9, 2018. These data were used to examine spatiotemporal changes in the sand dunes, as illustrated in (Fig. 4). Before conducting the analysis, all images underwent geometric, radiometric, and atmospheric corrections using ENVI ver. 5.3. August was chosen for the image dataset in each study year to minimize the impact of cloud coverage on the analysis.

The NDSI was applied to differentiate the reflection between sand dunes and other phenomena in the study area. The index showed significant results in the study area. It was applied using the following equation (Al-Quraishi 2013) (Fig. 5):

$$NDSI = \frac{SWIR2 - \text{Red}}{SWIR2 + \text{Red}} \quad (1)$$

SWIR2: short infrared, representing Band-7 in Landsat (TM/OLI).

R: Red reflection within visible radiation represents Band-3 in Landsat TM and Band-4 in Landsat OLI.

Fig. 3 Section of the large longitudinal dune shows the bedding and moisture of **a** and **b** the grain sizes of sand and lime

The NDSI is sensitive to water, sand dunes, and drifting sand sheets. The index is considered efficient for determining aeolian sands. However, values close to (1) are water, so ArcGIS 10.7 has been used to determine the sand value thresholds because the index is sensitive to water. The sand values were in the range 0.101–0.3 (Fig. 6). A slight enhancement has been suggested to minimize the impact of water and exaggerate the brightness of aeolian sand, as explained by the Enhanced Normalized Difference Sand Dunes Index (ENDSI), as shown in Eq. 2.

$$ENDSI = 1 - \left(\frac{SWIR2 - Red}{SWIR2 + Red}\right) \qquad (2)$$

The ENDSI showed a higher sensitivity to aeolian sand than the NDSI. Karnieli (1997) used the same enhancement method on his Soil Crust Index (CI) but used normalized differences of *the Red* and *Blue* bands. The normalized differences of *the SWIR2* and *Red* bands show a higher sensitivity of aeolian sand than *the Red* and *Blue* bands. The pixels whose values were within the above-mentioned numerical

Monitoring Drifting Sand Using Spectral Index and Landsat TM/OLI ...

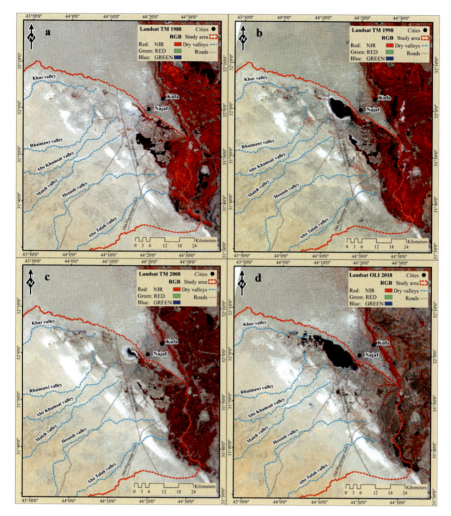

Fig. 4 Landsat images **a** Landsat 5 (TM) 22–8-1988, **b** Landsat 5 (TM) 18-8-1998, **c** Landsat 5 (TM) 13-8-2008, and **d** Landsat 8 (OLI) 9-8-2018

category for the four date images are shown in (Fig. 6). In addition, some statistical properties were calculated (the temporal and spatial variation of aeolian sand area, temporal and spatial variation of sand drifting, and the proximity of sand fields from settlements).

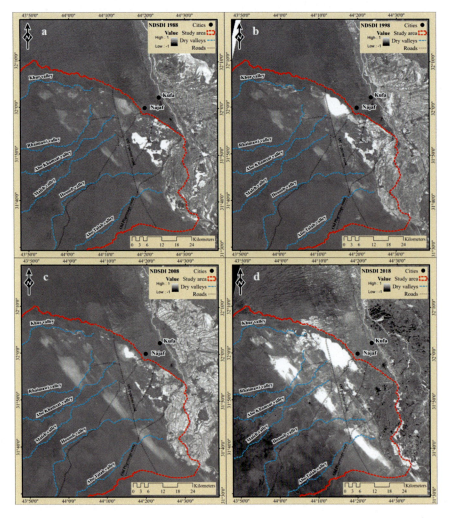

Fig. 5 Normalized Difference Sand Dune Index (NDSDI) **a** NDSDI 1988, **b** NDSDI 1998, **c** NDSDI 2008, and **d** NDSDI 2018

3 Results and Discussion

A spatiotemporal analysis of the dune area was conducted to examine the rates of encroachment and movement of these dunes on four different dates. The study revealed considerable variations in the sand dune area, particularly between 1988 and 2008, when these dunes expanded significantly owing to drought conditions. Notably, the area has experienced multiple periods of drought, with a prominent one occurring between 1998 and 2008. In 2008, the entire region, including all the Iraqi territories, endured an extended and severe drought period (Fadhil 2011; Mail et

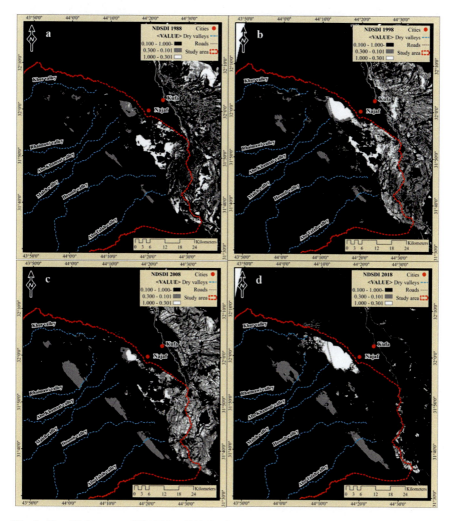

Fig. 6 Classified images of NDSDI **a** 1988, **b** 1998, **c** 2008, and **d** 2018

2016; Beg and Al-Sulttani 2020; Awadh et al. 2022). Figure 7 presents the variation in the sand dune area over the four years. The aeolian sand areas were determined using ArcGIS software. The results indicate that the aeolian sands' area expanded by 29.04 km^2 in 1988, 44.37 km^2 in 1998, 29.11 km^2 in 2008, and a substantial 143.64 km^2 in 2018 (refer to Table 1 and Fig. 8). The most notable change in dune areas occurred in 2008.

The study identified three primary sand dune fields in the area, all oriented in the NW–SE direction and aligned with the frequent wind direction. The first North Field (Field 1) is situated near the Arab Al-Rahima village and Al-Maurrat area, followed by the second Field (Field 2) located in Um-Tharawi, and finally, the south third

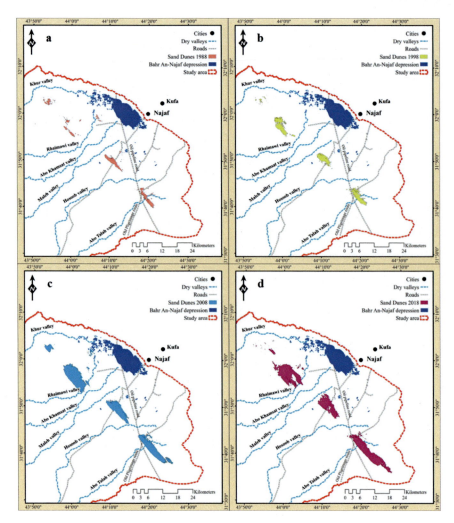

Fig. 7 Extracted sand dunes fields **a** 1988, **b** 1998, **c** 2008, and **d** 2018

Field (Field 3), starting from the Al-Ruhba and Al-Hassan springs (refer to Fig. 9). These fields were situated close to or along the direction of drifting sand. Analyzing the sand fields' areas in 1988, 1998, 2008, and 2018, it appeared that the first field expanded from 4.04 km^2 to 17.11 km^2, significantly increased to 51.76 km^2 in 2008, and slightly to 52.79 km^2 in 2018. The second field covered an area of 9.91 km^2 in 1988, increased to 11.90 km^2 in 1998, considerably expanded to 29.76 km^2 in 2008, and further expanded to 33.07 km^2 in 2018. As for the third field, its area was 10.28 km^2 in 1988, grew to 14.14 km^2 in 1998, substantially expanded to 38.77 km^2 in 2008, and even further to 55.13 km^2 in 2018 (see Table 2 and Fig. 10). The most

Table 1 Statistical properties of the extracted aeolian sand

Date	Area (km^2)	Change in area (km^2)[a]		
		1988–1998	1998–2008	2008–2018
1988-08-22	29.04	15.33	84.74	14.53
1998-08-18	44.37	Change percentage (%)[b]		
2008-08-13	129.11	1988–1998	1998–2008	2018–2008
2018-08-09	143.64	52.79	190.98	11.25
Total	346.16	Annual mean change[c]		
		1988–1998	1998–2008	2008–2018
		5.28	19.10	1.13

[a] Change in area (km^2) = Area in a Date—Area of the previous date
[b] Change percentage = (Area in a Date/Area in 1988) × 100
[c] Annual mean change = Change percentage/number of years

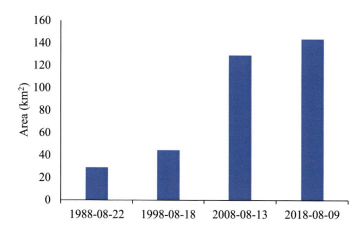

Fig. 8 Temporal changes in the aeolian sand area

significant expansion in sand fields occurred in 2008, primarily because of the severe drought period.

The area of the first field (Field 1) showed a significant percentage change at different time intervals. From 1988 to 1998, there was a remarkable increase of 323.65%, with an average annual change of 32.36%. Between 1998 and 2008, the area experienced substantial growth of 202.43%, with an average annual change of 20.24%. However, between 2008 and 2018, the change was minimal, with only a 2.00% increase and an average annual change of 0.20%. Field (2) witnessed a 20.16% change between 1988 and 1998, with an average annual change of 2.02%. Between 1998 and 2008, there was a notable percentage change of 149.99%, with an average annual change of 15.00%. From 2008 to 2018, the area experienced an 11.14% change, with an average annual change of 1.11%. Concerning Field (3), the

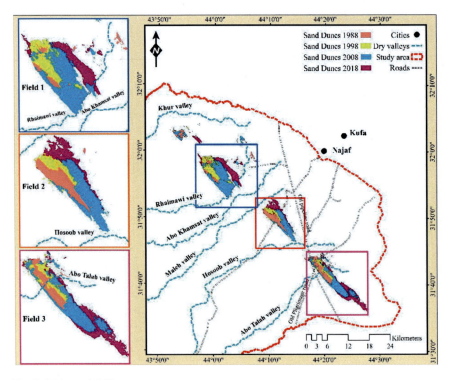

Fig. 9 Main sand field

Table 2 Area of the main sand fields

Date	Area (km^2)		
	Field 1	Field 2	Field 3
1988-08-22	4.04	9.91	10.28
1998-08-18	17.11	11.90	14.21
2008-08-13	51.76	29.76	38.77
2018-08-09	52.79	33.07	55.13

area underwent a 38.20% change between 1988 and 1998, with an average annual change of 3.82%. A substantial increase of 172.81% was observed between 1998 and 2008, with an average annual change of 17.28%. Finally, from 2008 to 2018, there was a percentage change of 43.21%, with an average annual change of 4.22% (Table 3 and Fig. 11).

Figure 11 illustrates a remarkable expansion in Field (1) between 1988 and 1998, while all sand field areas experienced significant growth from 1998 to 2008. The reduction in expansion in Field (1) can be attributed to the obstruction caused by the Rahimawi Valley. On the other hand, Field (2) also faced some obstruction in Um-Tharawi, primarily due to Hosoob Valley and circular agricultural farms (Fig. 12).

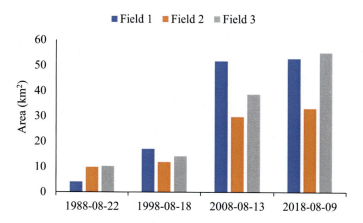

Fig. 10 Temporal variation in the main sand field area

Table 3 Statistical properties of the main sand fields

Sand field	1988–1998			1998–2008			2008–2018		
	Change in area (km^2)	Change (%)	Annual mean change	Change in area (km^2)	Change (%)	Annual mean change	Change in area (km^2)	Change (%)	Annual mean change
Field 1	13.07	323.65	32.36	34.64	202.43	20.24	1.03	2.00	0.20
Field 2	2.00	20.16	2.02	17.85	149.99	15.00	3.31	11.14	1.11
Field 3	3.93	38.20	3.82	24.56	172.81	17.28	16.36	42.21	4.22

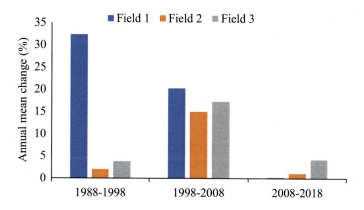

Fig. 11 Annual mean changes in the main sand fields

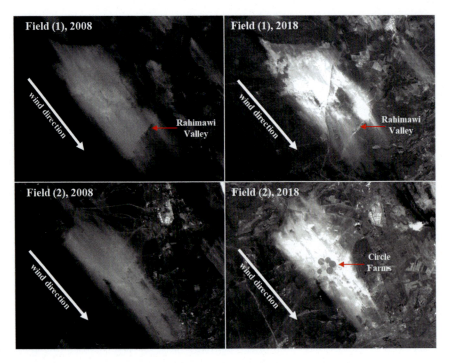

Fig. 12 ENDSI results for Fields (1 and 2) in 2008 and 2018 show obstructions

However, Field (3) showed no significant obstruction, except for the northern part of the Abo Talah valley, which is evident from Fig. 12.

Temporal encroachment and drifting variations of the dune fields were measured and analyzed. The endpoints of each field polygon were identified for each year, and the distances between these points were determined to calculate the differences (Table 4 and Fig. 13). For Field (1), the encroachment between 1988 and 1998 amounted to 1037.00 m, with an annual average of 103.70 m. Between 1998 and 2008, the encroachment reached 4718.59 m, with an annual average of 471.86 m. However, between 2008 and 2018, there was no movement as it reached 0.00 m. As for Field (2), the encroachment between 1988 and 1998 resulted in a distance of −2425.00 m (indicating drifting is negative), with an annual average of −242.50 m. Between 1998 and 2008, the drifting amounted to 5871.83 m, with an annual average of 587.18 m. Between 2008 and 2018, the drifting reached −2692.45 m, with an annual average of −269.25 m. These measurements provided valuable insights into the dynamics of dune fields and their changes over time.

The reduction in encroachment along the wind direction is attributed to obstructions that block the way of Field (1), causing its expansion to occur laterally. Between 1988 and 1998, Field (2) experienced a movement of −541.17 m (indicating drifting is negative), with an annual average of −54.1 m. In subsequent years, from 1998 to 2008, there was a significant drifting of 8517.86 m, with an annual average of

Table 4 Drifting distances of sand fields

Sand field	1988–1998		1998–2008		2008–2018	
	Drifting (m)	Annual mean (m)	Drifting (m)	Annual mean (m)	Drifting (m)	Annual mean (m)
Field 1	1037.00	103.70	4718.59	471.86	0.00	0.00
Field 2	−2425.00	−242.50	5871.83	587.18	−2692.45	−269.25
Field 3	−541.17	−54.1	8517.86	851.79	4664.89	466.49

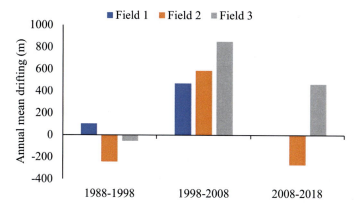

Fig. 13 Annual mean drifting distances of sand fields

851.79 m. Between 2008 and 2018, the encroachment reached 4664.89 m, with an annual average of 466.49 m. Field (3) faced obstructions from Abo Talah Valley and the city of Al-Ruhba, which were extensively invaded by sand, especially in the 19's years of the twentieth century. All residents of the Al-Rahba area were displaced and migrated to the city of Najaf because of the encroachment of sand (Plates 3a and 3b). These observations highlight how various factors can influence the movement and dynamics of dune fields, leading to changes in their shapes and expansion patterns over time.

Drifting of these fields every ten years causes the risk for villages, industrial activities, and public service places whose distances are approximately 3000 m if Field (3) drifting distance is 4213.86 m/10 years. Accordingly, to determine the levels of dune risks in 1988, the danger increases for Shuaib Al-Marat, Um-Tharawi, and Al-Ruhba as it approaches a distance of (125–3000) m from the dune's fields. At the same time, the danger increased in the Al-Hassan springs in 1998, with distances ranging between (101–3000) m (Fig. 14). Due to the increase in drifting in 2008 and its progress, the risk impact increased in the areas of (Eithha, Arab Al-Rahima, Shuaib Al-Murrat, Um-Tharawi, Al-Shajij springs, Al Ruhba, Khuraiba, and Imam Al-Hassan springs). In 2018, the dune fields impacted significant areas in Shuaib Al-Murrat, Um-Tharawi, and Imam Al-Hassan springs. Sand covers many of them,

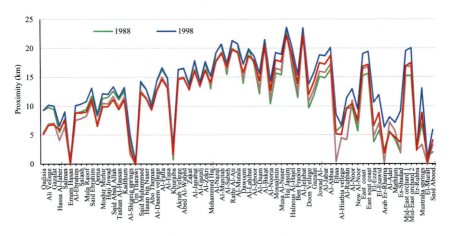

Fig. 14 Proximity of main sand fields to main settlements in the study area

threatening houses, farms, irrigation channels, roads, and limestone quarries (Figs. 15 and 16).

4 Conclusions

The spectral index (NDSI) has shown significant results in accurately extracting aeolian sand. The study area exhibited variations in the area of sand dunes on four different dates, totaling an area of (346.16) km^2. The dune fields demonstrated considerable variation, with an initial area of (29.04) km^2 on (22-8-1988), which expanded to (143.64) km^2 on (09-08-2018). In the study area, four types of sand features were identified: Barchan, Longitudinal, Sand sheets, and Nabkha. Additionally, the area has three distinct sand dune fields, each differing in size and expansion rate. Obstructions, particularly valleys, have hindered and delayed the expansion of specific fields. The southernmost field (3) exhibited the most significant expansion, owing to the absence of obstacles. The presence of these dune fields poses a threat to various locations and human activity in the study area. Field (1) threatens the Al-Ruhima and Al-Murrat areas as numerous houses and farms are covered by sand. Field (2) impacts agricultural lands in Um-Tharawi, whereas Field (3) affects the Al-Ruhba area. These findings emphasize the importance of understanding and managing the dynamics of these dunes to mitigate the potential risks to both the environment and human activities in the region.

Fig. 15 **a** and **b** Houses covered by sand in the Al-Ruhba area; **c** Houses covered by sand in the Al-Rahima area; **d** Irrigation channel filled with sand; **e** and **f** Farms surrounded by Nabkha and longitudinal dunes

Fig. 16 Barchan sand dunes drifting toward Al Al-Rahima road

5 Recommendations

In light of the results extracted through the spectral index, the sand dune area is steadily expanding due to drought and climate change. Maps and proximity analysis showed the encroachment of dunes in certain places towards agricultural lands and villages. Field studies have shown that some villages have been abandoned because the dunes have covered them. Developing effective plans, programs, and policies to confront the risks of desertification and its direct environmental and economic impact.

The development of artificial intelligence techniques in remote sensing, and according to the results of the current work, which showed the variation in the expansion of dune fields and the rates of encroachment, we recommend that a type of study be conducted in the field of classifying dunes according to their type and the relationship of this to the rates of encroachment and the degree of their danger. Such studies will significantly help develop treatments and solutions according to the type of dunes and their risks.

References

Al-Dabi H, Koch M, Al-Sarawi M, El-Baz F (1997) Evolution of sand dune patterns in space and time in north-western Kuwait using Landsat images. J Arid Environ 36(1):15–24

Al-Quraishi AMF (1992) Effects of sand dunes stabilization methods on some parameters of wind erosion in Baiji area, Iraq. Iraqi J Agric Sci 23:105–114

Al-Sulttani AH (2001) Bahar Al-Najaf a study of solution features in calcareous rocks. Mustansiriyah University, Baghdad, Iraq, Iraq, Geography Department

Al-Zubaidi EA, Rabee F, Al-Sulttani AH (2023) Calculating spectral index based on linear SVM methods for landsat OLI: Baiji sand dunes a case study, Iraq. Int J Comput Digital Syst 13(1):437–450

Awadh SM, Al-Sulttani AH, Yaseen ZM (2022) Temporal dynamic drought interpretation of Sawa Lake: case study located at the Southern Iraqi region. Nat Hazards 112(1):619–638. https://doi.org/10.1007/s11069-021-05198-3

Aydda A, Algouti A, Algouti A (2017) A qualitative assessment of desertification change in the Tarfaya basin (Morocco) using panchromatic data of Landsat ETM+ and oli: sand encroachment approach. In: MATEC web of conferences,. EDP sciences, p 09002

Aydda A, Althuwaynee OF, Algouti A, Algouti A (2019) Evolution of sand encroachment using supervised classification of Landsat data during the period 1987–2011 in a part of Laâyoune-Tarfaya basin of Morocco. Geocarto Int 34(13):1514–1529

Azzaoui M, Adnani M, El Belrhiti H, Chaouki I, Masmoudi C (2016) detection of barchan dunes in high resolution satellite images. International Archives of the Photogrammetry, Remote Sensing & Spatial Information Sciences 41

Bagnold RA (2012) The physics of blown sand and desert dunes. Courier Corporation

Bandeira L, Marques JS, Saraiva J, Pina P (2011) Automated detection of Martian dune fields. IEEE Geosci Remote Sens Lett 8(4):626–630

Beg AAF, Al-Sulttani AH (2020) Spatial assessment of drought conditions over Iraq using the standardized precipitation index (SPI) and GIS techniques. In: Environmental remote sensing and GIS in Iraq. Springer, pp 447–462

Buday T (1980) The regional geology of Iraq. Stratigraphy and Paleogeography, vol 1. Dar Al-Kutub, Mosul 445
Bullard J, Thomas D, Livingstone I, Wiggs G (1997) Dunefield activity and interactions with climatic variability in the southwest Kalahari Desert. Earth Surf Process Landforms: The J Brit Geomorphol Group 22(2):165–174
Chen J, Yuan Zhang M, Wang L, Shimazaki H, Tamura M (2005) A new index for mapping lichen-dominated biological soil crusts in desert areas. Remote Sens Environ 96(2):165–175. https://doi.org/10.1016/j.rse.2005.02.011
Chen X-L, Zhao H-M, Li P-X, Yin Z-Y (2006) Remote sensing image-based analysis of the relationship between urban heat island and land use/cover changes. Remote Sens Environ 104(2):133–146. https://doi.org/10.1016/j.rse.2005.11.016
Chlachula J (2021) Between sand dunes and hamadas: environmental sustainability of the Thar desert, West India. Sustainability 13(7):3602
Chowdhury PR, Deshmukh B, Goswami AK, Prasad SS (2010) Neural network based dunal landform mapping from multispectral images using texture features. IEEE J Selected Topics Appl Earth Observations and Remote Sensing 4(1):171–184
Cooke RU, Warren A, Goudie AS (1993) Desert geomorphology. CRC Press
Danin A (2012) Plants of desert dunes. Springer Science & Business Media
Dong Z, Chen G, He X, Han Z, Wang X (2004) Controlling blown sand along the highway crossing the Taklimakan desert. J Arid Environ 57(3):329–344
Fadhil AM (2002) Sand dunes fixation in Baiji district. J Earth Sci 13(1):67–72
Fadhil AM (2003) Evaluation of sand dunes stabilization techniques in Baiji district, Iraq. J Earth Sci 14(1):59–64
Fadhil AM (2009) Land degradation detection using geo-information technology for some sites in Iraq. Al-Nahrain J Sci 12(3):94–108
Fadhil AM (2011) Drought mapping using Geoinformation technology for some sites in the Iraqi Kurdistan region. Int J Digital Earth 4(3):239–257
Fadhil AM (2013) Sand dunes monitoring using remote sensing and GIS techniques for some sites in Iraq. In: PIAGENG: intelligent information, control, and communication technology for agricultural engineering. International Society for Optics and Photonics, p 876206
Feng X, Qu J, Fan Q, Tan L, An Z (2019) Characteristics of desertification and short-term effectiveness of differing treatments on shifting sand dune stabilization in an Alpine rangeland. Int J Environ Res Public Health 16(24):4968. https://doi.org/10.3390/ijerph16244968
Goudie AS (2013) Arid and semi-arid geomorphology. Cambridge University Press
Hugenholtz CH, Levin N, Barchyn TE, Baddock MC (2012) Remote sensing and spatial analysis of aeolian sand dunes: a review and outlook. Earth Sci Rev 111(3–4):319–334
Jassim SZ, Goff JC (2006) Geology of Iraq. Geological Society Publishing House
Karnieli A (1997) Development and implementation of spectral crust index over dune sands. Int J Remote Sens 18(6):1207–1220
Karnieli A, Kidron GJ, Glaesser C, Ben-Dor E (1999) Spectral characteristics of cyanobacteria soil crust in semiarid environments. Remote Sens Environ 69(1):67–75
Knight M, Thomas DS, Wiggs GF (2004) Challenges of calculating dunefield mobility over the 21st century. Geomorphology 59(1–4):197–213
Kumar M, Goossens E, Goossens R (1993) Assessment of sand dune change detection in Rajasthan (Thar) desert, India. Int J Remote Sens 14(9):1689–1703
Lagacherie P, Baret F, Feret JB, Madeira Netto J, Robbez-Masson JM (2008) Estimation of soil clay and calcium carbonate using laboratory, field and airborne hyperspectral measurements. Remote Sens Environ 112(3):825–835
Lam DK, Remmel TK, Drezner TD (2011) Tracking desertification in California using remote sensing: a sand dune encroachment approach. Remote Sens 3(1):1–13
Mail AA-SM, Somorowska U, Al-Sulttani AH (2016) Seasonal and inter-annual variation of precipitation in Iraq over the period 1992–2010. Prace i Studia Geograficzne 61(3):71–84
McHarg I (1972) Best shore protection-natures own dunes. Civ Eng 42(9):66–70

Moreno Casasola P (1982) Ecología de la vegetación de dunas costeras: estructura y composición en El Morro de La Mancha, Ver. I, vol 60

Pinet PC, Kaufmann C, Hill J (2006) Imaging spectroscopy of changing Earth's surface: a major step toward the quantitative monitoring of land degradation and desertification. CR Geosci 338(14):1042–1048

Pye K, Tsoar H (2008) Aeolian sand and sand dunes. Springer Science & Business Media

Sun D, Dawson R, Li H, Li B (2005) Modeling desertification change in Minqin County, China. Environ Monit Assess 108(1–3):169–188. https://doi.org/10.1007/s10661-005-4221-9

Tang Y, Wang Z, Jiang Y, Zhang T, Yang W (2023) An Auto-detection and classification algorithm for identification of sand dunes based on remote sensing images. Int J Appl Earth Obs Geoinf 125:103592

Tanser FC, Palmer AR (1999) The application of a remotely-sensed diversity index to monitor degradation patterns in a semi-arid, heterogeneous, South African landscape. J Arid Environ 43(4):477–484

Teodoro AC, Goncalves H (2012) A semi-automatic approach for the extraction of sandy bodies (sand spits) from IKONOS-2 data. IEEE J Selected Topics Appl Earth Observations Remote Sens 5(2):634–642

Tripathy G, Ghosh T, Shah S (1996) Monitoring of desertification process in Karnataka state of India using multi-temporal remote sensing and ancillary information using GIS. Int J Remote Sens 17(12):2243–2257

Udelhoven T, Emmerling C, Jarmer T (2003) Quantitative analysis of soil chemical properties with diffuse reflectance spectrometry and partial least-square regression: a feasibility study. Plant Soil 251(2):319–329

Wang Z, Shi Y, Zhang Y (2023) Review of desert mobility assessment and desertification monitoring based on remote sensing. Remote Sens 15(18):4412

Wasson R, Hyde R (1983) Factors determining desert dune type. Nature 304(5924):337–339

Yao Z, Wang T, Han Z, Zhang W, Zhao A (2007) Migration of sand dunes on the northern Alxa Plateau, Inner Mongolia, China. J Arid Environ 70(1):80–93

Yekeen ST, Balogun AL, Aina YA (2023) A review of the development in the remote sensing of sand dunes. Sand Dunes of the Northern Hemisphere: Distribution, Formation, Migration and Management 1:39

Zha Y, Gao J, Ni S (2003) Use of normalized difference built-up index in automatically mapping urban areas from TM imagery. Int J Remote Sens 24(3):583–594

Zhang Y, Chen Z, Zhu B, Luo X, Guan Y, Guo S, Nie Y (2008) Land desertification monitoring and assessment in Yulin of Northwest China using remote sensing and geographic information systems (GIS). Environ Monit Assess 147(1):327–337

Zhao H, Chen X Use of normalized difference bareness index in quickly mapping bare areas from TM/ETM+. In: Geoscience and remote sensing symposium, 2005. IGARSS'05. Proceedings. 2005 IEEE international, 2005. IEEE, pp 1666–1668l

Assessment of the Growth of Urban Heat Island in a Mediterranean Environment: A Pathway Toward a Sustainable City

Mohammed El Hafyani, Narjisse Essahlaoui, Ali Essahlaoui, Meriame Mohajane, Abdelali Khrabcha, and Anton Van Rompaey

Abstract Rapid urbanization can induce significant changes in the urban climate. It is a great threat to the urban environment and human life quality. Therefore, sustainable urban development needs to find a balance between urban expansion and the thermal environment. This chapter proposes a methodology for monitoring and quantifying the growth of urban heat island (UHI) in response to land-use land-cover (LULC) change impacts based on a time series of Landsat remote sensing-based images and geospatial tools. The work was conducted in the city of Meknes, located in the central part of Morocco, as a case study. First, Land Surface Temperature (LST) was calculated from 1990 to 2020 using the thermal bands of Landsat images. Second, LULC and several vegetation indices were assessed, and the MOD11A1

M. El Hafyani (✉)
National Institute for Scientific and Technological Research in Water, City of Innovation Souss Massa, Ibn Zohr University, 80000 Agadir, Morocco
e-mail: m.elhafyani@uiz.ac.ma

Applied Geology and Geo-Environment Laboratory, Faculty of Sciences, Ibn Zohr University, 80035 Agadir, Morocco

N. Essahlaoui · A. Essahlaoui · A. Khrabcha
Department of Geology, Laboratory of Geoengineering and Environment, Research Group "Water Sciences and Environment Engineering", Faculty of Sciences, Moulay Ismail University, Zitoune, BP11201 Meknès, Morocco
e-mail: narjisse.essahlaoui@edu.umi.ac.ma

A. Essahlaoui
e-mail: a.essahlaoui@fs-umi.ac.ma

A. Khrabcha
e-mail: khrabchaabdelali68@gmail.com

M. Mohajane
ITC-CNR, Construction Technologies Institute, National Research Council, 70124 Bari, Italy
e-mail: mohajane.meriame13@gmail.com

A. Van Rompaey
Department Earth and Environmental Science, Geography and Tourism Research Group, KU Leuven, Celestijnenlaan 200E, 3001 Heverlee, Leuven, Belgium
e-mail: anton.vanrompaey@kuleuven.be

© The Author(s), under exclusive license to Springer Nature Switzerland AG 2024
A. M. F. Al-Quraishi and Y. T. Mustafa (eds.), *Natural Resources Deterioration in MENA Region*, Earth and Environmental Sciences Library,
https://doi.org/10.1007/978-3-031-58315-5_13

MODIS product was used to develop the LST time series from 2000 to 2021 using the Google Earth engine platform. The results showed that the maximum LST in Meknes has increased by almost 10 °C during the past decades, and an overlay of the land surface temperatures with several land cover indicators shows that the increase in land surface temperatures is clearly correlated with the conversion of vegetation cover to built-up areas. The analysis of land use land cover changes (LULCC) showed an increase for the urban area and the arboriculture of about 12.6 and 6.07 km^2, respectively, in the same period, whereas the cereals and peri-urban areas were reduced by 16.18 and 2.52 km^2 respectively.

Keywords Urban heat island (UHI) · Land surface temperature (LST) · Land use land cover changes (LULCC) · Land cover indicators · MOD11A1 MODIS product · Landsat · Meknes · Morocco

1 Introduction

Mediterranean cities have grown rapidly in the past decades, in some cases above 5% per year (Debolini et al. 2015; Houpin 2011; Veron 2007), mainly because of ongoing rural–urban migration (El Hafyani et al. 2020). Consequently, the landscape surrounding the historical city centers has been modified intensively because of conversions from agrarian or natural land covers to built-up areas (Musakwa and Niekerk 2013). Many new city dwellers moved to the city for better livelihoods based on non-farming income. These economic goals were met to a certain extent, but the rapid urban growth rates also pose several challenges to the sustainability of quality of life in the new urban and suburban landscapes (Fung-Loy et al. 2019). In the view of climate change, a clear threat is the increasing temperature that is amplified because of the UHI island effect In the view of climate change, a clear threat is the increasing temperature that is amplified because of the UHI (Patz et al. 2005). UHI appear when the surface radiation balance is modified when due to radiation absorption by the building materials absorb more radiation during the day and only nighty energy gradually releases their energy during the night (Oke 1982).

UHI negatively affects the health of the urban population, especially in Mediterranean, subtropical and tropical climates where heat stress is already prevailing in normal circumstances without urbanization (Collier 2006). Several health studies (Gabriel and Endlicher 2011; Laaidi et al. 2012; Alqasemi et al. 2020; Son et al. 2012; Tan et al. 2010; Tong et al. 2014) pointed out that urban mortality increased during severe heat waves of the last decade. Secondly, some authors (Guhathakurta and Gober 2007) reported increasing water consumption levels during heat waves. The cities' growth of cities would then lead to an exponential increase in water consumption because of (i) the increasing number of people and (ii) the increasing water consumption per capita. This phenomenon is especially alarming in urban areas with seasonal water shortages. Urban planners and landscape architects (Imran et al. 2018; Li et al. 2014; Razzaghmanesh et al. 2016; Sharma et al. 2016) point out that the UHI effect can be remediated to a certain extent by providing urban green and surface water in the build environment. In traditional Mediterranean cities (Axhausen 2000; Parés-Franzi et al. 2006), these urban green and blue infrastructures were present in the form of parks with water surfaces, gardens and vertical green walls. The recent urban neighborhoods, especially those of the lower income groups, are often deprived from urban green and blue and are dominated by concrete structures and impervious surfaces, leading to environmental injustice (Fung-Loy et al. 2019).

A sound urban planning strategy aiming at reducing the UHI effect should be based on an accurate and spatially distributed inventory of the land surface temperatures in the cities and their evolution over time. However, assessing the total UHI effect remains difficult, especially in cities with limited temperature observations. Urban remote sensing seems to be a privileged tool for assessing the extent and configuration of UHI (Estoque and Murayama 2015; Divya et al. 2020; Sharma and Joshi 2016).

This paper explores the potential of satellite-based monitoring of UHI using the city of Meknes in Morocco as a case-study. Firstly, the spatio-temporal evolution of the lLST and the LULCC during the last 3 decades is mapped based on medium resolution satellite imagery. Secondly the changes in the LST are correlated with land cover indicators. The MOD11A1 MODIS product has been used to develop the times series of LST from 2000 to 2021 using the google earth engine platform. Finally, policy recommendations for the sustainable management of Mediterranean cities are formulated. It should be highlighted that this is the first study carried out in this region and thus, the authors believe that this research's outcomes can be a crucial reference to follow in any management activities and interventions in this area.

2 Materials and Methods

2.1 Study Area

The study area is in central of Morocco. Its latitude ranges from 33°50′22″N to 33°55′22″N, and longitude from 5°27′07″W to 5°36′22″W at an altitude of 576 m (Fig. 1). It includes 4 urban communes; Meknes municipality and the townships of Al Machouar Stinia, Toulal and Ouislane, with a population of 630 079 inhabitants according to the 2014 national census (RGPH 2014), and an area covering 110 km^2. The city of Meknes is one of the most important cities in Morocco, due to its geographical location, historical role and human capacities (El Hafyani et al. 2020). It is one of the four imperial cities of Morocco, founded in the eleventh century by the Almoravid dynasty during the reign of the Alaouite Sultan Ismail Ben Sherif. Since the foundation of this city, in addition to natural factors, historical depth has always been a key factor in the current city structure organization. Therefore, its structure, configuration, and the way its site has been developed result from the choices made during the Ismaili and colonial historical periods (Agence Urbaine 2009). Climatically, the annual average rainfall is about 500 with a dry period extending from June to October (El Hafyani et al. 2023) (Fig. 2).

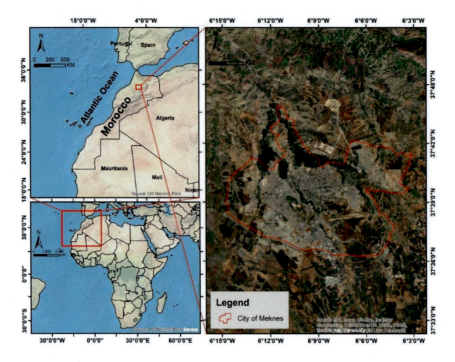

Fig. 1 Study area

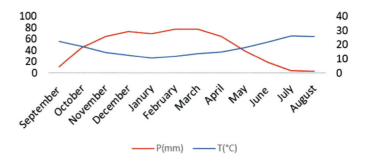

Fig. 2 Ombrothermic diagram of Meknes station (Period of 1998–2018)

2.2 Data

Currently, the data provided by Landsat is a special database for global change research to characterize processes on a human scale. They benefit the academic community and societal applications such as water resources and natural hazard management (Chen and Zhang 2017; El Ouali et al. 2021). In this study, sixteen satellite images downloaded from the United States Geological Survey (USGS) website, acquired in July and August from 1991 to 2021 with a step of 2 years. After having radiometrically calibrated and atmospherically corrected the satellite images, the LST has been calculated using the thermal bands, and the land cover indicators have been calculated using the bands of Visible and Near Infrared spectrum. While the land cover maps have been created using the support vector machine classification and evaluation of the different classifications was made based on the confusion matrices, the kappa index (Congalton 1991; Stehman 1996; Ijlil et al. 2022). While an analysis has been done for the LST and the different spectral indices. Then, the MOD11A1 MODIS daily per-pixel LST with 1 km (km) spatial resolution product has been used to develop the times series of LST from 2000 to 2021, using the google earth engine platform. At the end, the UHI was assessed for the study area (Fig. 3).

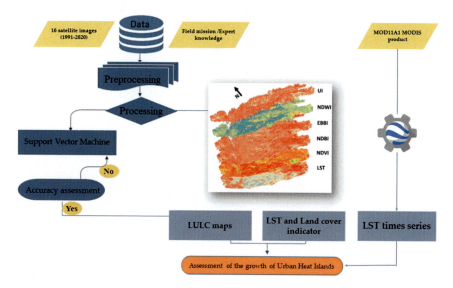

Fig. 3 Flowchart of the adopted methodology

2.3 Estimating Land Surface Temperature (LST) and the Spectral Index

Landsat 1–8 Level-1 data products consist of calibrated scaled Digital Numbers (DN) representing the multispectral image data (Zanter 2014). The LST is calculated following serial steps that include the radiance computation using the radiometric readjustment coefficient provided in the metadata file (MTL.txt). Landsat Level-1 data can be converted to TOA spectral radiance using the radiance rescaling factors in the MTL file:

Conversion to Radiance

$L\lambda = ML Qcal + AL$

where:

$L\lambda$ = TOA spectral radiance (Watts/ (m2 * srad * μm))

ML = Band-specific multiplicative rescaling factor from the metadata (RADIANCE_MULT_BAND_x, where x is the band number)

AL = Band-specific additive rescaling factor from the metadata (RADIANCE_ADD_BAND_x, where x is the band number)

Qcal = Quantized and calibrated standard product pixel values (DN)

Conversion to Top of Atmosphere Brightness Temperature

Thermal band data can be converted from spectral radiance to top-of-atmosphere brightness temperature using the thermal constants in the MTL file:

$$T = \frac{K_2}{\ln(\frac{K_1}{L_\lambda+1})}$$

where:

T = Top of atmosphere brightness temperature (K)where:

$L\lambda$ = TOA spectral radiance (Watts/(m2 * srad * μm))

K1 = Band-specific thermal conversion constant from the metadata (K1_CONSTANT_BAND_x, where x is the thermal band number)

K2 = Band-specific thermal conversion constant from the metadata (K2_CONSTANT_BAND_x, where x is the thermal band number)

Then, the surface temperature was calculated using the Artis and Carnahan (Artis and Carnahan 1982; Duan et al. 2019; Xiong et al. 2012) method, applying the following equation.

$$LST = \frac{T}{1 + \left(\lambda + \frac{T}{\rho}\right)\ln_\varepsilon}$$

where:

LST: is the LST in Kelvin (K)

T: is the Top of atmosphere brightness temperature in Kelvin

λ is the wavelength of the emitted radiance (for which the peak response and the average of the limiting wavelengths ($\lambda = 11.5$ μm)

$\rho = hc/\sigma = (1.438 \times 10 - 2$ m·K); σ is the Stefan-Boltzmann constant ($5.67 \times 10\text{--}8$ W·m $- 2$·K $- 4 = 1.38 \times 10 - 23$ J·K $- 1$); h is Planck's constant ($6.626 \times 10 - 34$ J·s); c is the velocity of light (2.998×108 m·s $- 1$);

ε is the land surface spectral emissivity of the Landsat TM/ETM + band 6, and OLI/TIRS band 10.

The derived LST unit has been converted to degrees Celsius using the equation:
LST (°Celsius degree) = LST (Kelvin)- 273,15

The spectral indices used in this study are normalized difference vegetation index (NDVI), which is used worldwide for the global observation of vegetation. Normalized difference water index (NDWI) permits the identification of water bodies. Normalized difference built-up index (NDBI), urban index (NDBI), and enhanced built-up and bareness index (EBBI) which allow to identify the agglomerations. These indices were necessary for the characterization of land cover types. The NDVI is calculated by the combination of red and near-infrared reflectance (Eq. 1). This index, proposed for the first time by (Rouse et al. 1973) It is relatively correlated with the vegetation cover growth (El Hafyani et al. 2021; Mohajane et al. 2018; Mohajane et al. 2017; Rondeaux et al. 1996). McFeeters, S. K. 1996 developed the standardized index of water (Eq. 2), making it possible to identify aquatic surfaces. The combination of green and near-infrared reflectance calculates the soil moisture. Another index used in this study that is sensitive to the built-up area is NDBI (Eq. 3) (Zha et al. 2003). It is a very important index for urban areas mapping. The last two indices used in this work are urban index (UI) (As-syakur et al. 2012) (Eq. 4), and enhanced built-up and bareness index (EBBI) (Eq. 5) (Kawamura 1996). They helped us to establish the relationship between the different components and the land surface temperature.

$$NDVI = \frac{NIR - Red}{NIR + Red} \quad (1)$$

$$NDWI = \frac{b3 - b5}{b3 + b5} \quad (2)$$

$$NDBI = \frac{b6 - b5}{b6 + b5} \quad (3)$$

$$UI = \frac{b7 - b5}{b7 + b5} \quad (4)$$

$$EBBI = \frac{b5 - B6}{\sqrt[10]{b6 + b10}} \quad (5)$$

3 Results

3.1 Spatio-Temporal Analysis of Land Surface Temperature (LST)

The LST was calculated using the radiance values of the Landsat sensor thermal bands (Band 6 for the TM and ETM + sensors, and Band 10 and 11 of the TIRS sensor for Landsat 8). The images were acquired in July and August during the period 1991–2020. Figure 4 shows the spatio-temporal variation of surface temperature for the city of Meknes during the last 30 years. The results showed an increase of the LST during this period by more than 10 °C from a maximum surface temperature of 37.62 °C in 1991 to 49.74 °C in 2020.

During the last decades, agricultural lands and greenery areas have been intensely degraded due to the rapid expansion of urban areas. According to the study carried out in the Meknes region by (El Hafyani et al. 2020) more than 1590 ha of agricultural areas have been transformed into urban ones during the period 1990–2018, with an increase of 183%, from 2577 ha in 1990 to 7314 ha in 2018. Overlaying the surface temperature maps with high-resolution Google Earth images shows that very high land surface temperature values characterize urban areas, while low values of this parameter characterize vegetation areas. Figure 5 shows the surface temperature in some areas of the study area. It shows the effect of the UHI on these variables. The difference in LST in the same area varies from 2 to 5 °C between urban and their surroundings.

3.2 LST with NDVI, NDWI, NDBI, UI, and EBBI

Pearson's correlation coefficient indicates that all land cover indicators are correlated with the LST. NDVI and NDBI have a stronger relationship with LST than other indices. NDVI is negatively correlated with LST, while NDBI, IU, EBBI have a positive effect on LST, with a correlation index of –0.61, 0.75, 0.70 and 0.75, respectively (Fig. 6). While LST has a random relationship with NDWI. The LST and NDBI scatterplot also clearly showed a "warm edge" and a "cold edge", located at the upper and lower edge of the scatterplot (Fig. 7).

3.3 MODIS Images for LST Assessment

In this section, the MOD11A1 MODIS product has been used to develop the times series of LST from 2000 to 2021, using the google earth engine platform. The MOD11A1 is a tile of daily LST product at 1 km spatial resolution on the sinusoidal projection. For this issue, various points were chosen for the urban and rural

Assessment of the Growth of Urban Heat Island in a Mediterranean ... 269

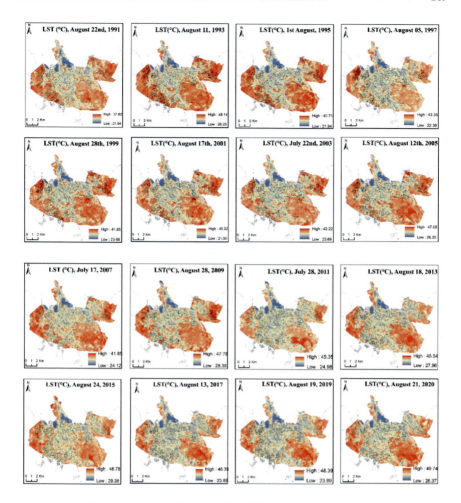

Fig. 4 Land surface temperatures (LST) from 1991–2020

areas to extract the times series of the LST (Fig. 8). The data visualization showed that the LST in the urban area is always higher than the LST in vegetation area. The recorded maximum temperature values for the urban and vegetation area are 54.43 and 50.27 °C respectively. Whereas the minimum temperature values are 4.91 and 3.09 °C respectively for urban and vegetation areas. For the average, the recorded values are 31.61 and 29.4 °C respectively for vegetation and rural areas with a difference of about 2 °C. The monthly average value of the temperature showed that the recorded values of LST in urban areas are always higher than the values of the vegetation areas, and sometimes by about 3 °C (Fig. 9).

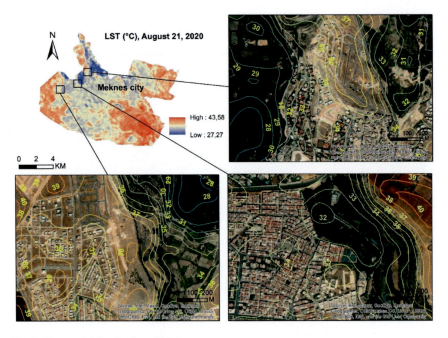

Fig. 5 The spatial distribution of LST

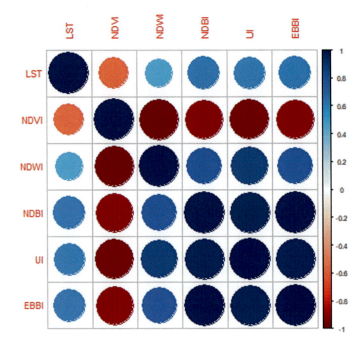

Fig. 6 Pearson's Correlation matrix

Assessment of the Growth of Urban Heat Island in a Mediterranean … 271

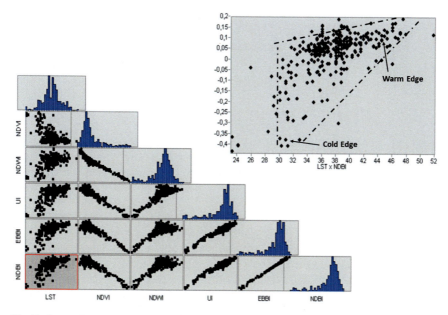

Fig. 7 Scatterplots of land surface temperature (LST, °C) versus spectral indices corresponding to 400 pixels (30 × 30 m each)

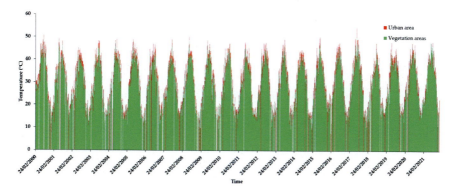

Fig. 8 LST times series for the period of (2000–2021)

3.4 Spatial Autocorrelation of LST and Land Cover Indicators

This section aims to measure the correlation of a variable with itself. A spatial autocorrelation is considered when the observations are considered with a spatial shift (Alitane et al. 2022). It is defined as the correlation, positive or negative, of a variable with itself due to the spatial location of the observations. The Global Moran's

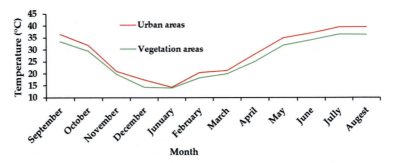

Fig. 9 The monthly average of the LST for the period 2000–2021 using MOD11A1 MODIS product

I Spatial Autocorrelation tool simultaneously measures spatial autocorrelation by location and feature values. The range of this index is calculated using a z-score and a p-value to evaluate the range of this index. Thus, autocorrelation was applied for the LST variable and the land cover indicators (NDBI, NDVI, and NDWI). The Moran's index was calculated for the different varieties presented in Fig. 10. The results show a positive autocorrelation for all variables with Moran's index of 0.56, 0.62, 0.57, and 0.52 for LST, NDBI, NDVI, and NDWI respectively. This shows that similar values of the variables studied are grouped geographically, and consequently the presence of spatial autocorrelation for all variables. This was confirmed with the p-value values for all variables (-value lower than 0.01). In addition, the z-score for all variables is greater than 2.58, demonstrating that the trend of spatial aggregation of the different variables to the overall distribution is significant.

3.5 Overall Accuracy and Classification Performance

The assessment of the accuracy of the LULC classification was performed by ground-truthing points, including survey field data, and high-resolution Google Earth imagery, and using the confusion matrix. The classification showed perfect agreement for the last three times (2009, 2015, and 2021) with an overall accuracy of 81.11%, 89.16%, 87.65% respectively. While those for 1991, 1997, and 2003 showed moderate agreement with an overall accuracy of 77.21, 75.30, and 76.82%, respectively (Tables 1, 2, and 3). While the kappa index showed a good agreement for the two times of 2015 and 2021 with a value of 0.83 and 0.81, respectively. For the other classification, the kappa index showed a moderate agreement with a value of 0.68, 0.66, 0.67, 0.73 for 1991, 1997, 2003, and 2009 respectively (Tables 1, 2, and 3). The classification validation using the confusion matrix showed some confusion between some classes, such as: Arboriculture, cereals. This is especially due to the similarity of the spectral properties of the different categories.

Assessment of the Growth of Urban Heat Island in a Mediterranean ... 273

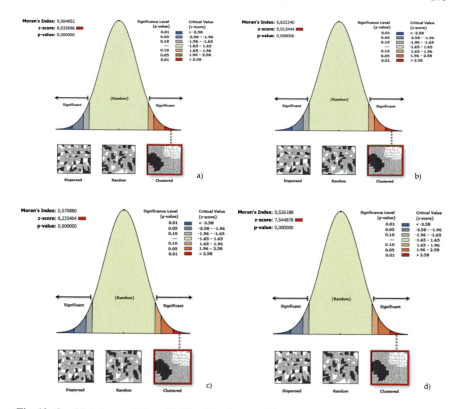

Fig. 10 Spatial Autocorrelation of LST and land cover indicators

Table 1 Overall accuracy and kappa coefficient

	1991	1997	2003	2009	2015	2021
Overall accuracy	77.21	75.30	76.82	81.11	89.16	87.65
Kappa coefficient	0.68	0.66	0.67	0.73	0.83	0.81

3.6 LULCC

This section aims to understand the evolution of LULCC in the study area. Therefore, six different times have been included: 1991, 1997, 2003, 2009, 2015 and 2021. Figure 11 shows the different maps that have been made for the six times. The analysis of LULCC showed an increase for the urban area and the arboriculture of about 35 and 87%, respectively in the same period, whereas the areas of the cereals and the peri-urban agricultural were reduced by 30 and 16%, respectively (Figs. 11 and 12).

Table 2 Confusion matrix of the classifications 1991, 1997, and 2003

Class	Ground truth (%)															
	A			C			P. A			B. A			W		
Year	1991	1997	2003	1991	1997	2003	1991	1997	2003	1991	1997	2003	1991	1997	2003
A	20.59	34.31	47.06	7.20	21.60	23.20	7.79	3.25	3.25	1.05	0.00	1.63	0.00	0.00	25.00
C	54.90	49.02	38.24	84.00	63.20	59.20	1.95	5.19	7.79	4.21	4.74	6.12	0.00	0.00	0.00
P. A	21.57	15.69	11.76	3.20	10.40	8.80	90.26	90.91	87.66	1.05	1.05	0.00	0.00	0.00	0.00
B. A	2.94	0.98	2.94	5.60	4.80	8.80	0.00	0.65	1.30	93.68	94.21	92.24	75.00	100.00	50.00
W	00.0	0.00	0.00	0.00	0.00	0.00	0.00	0.00	0.00	0.00	0.00	0.00	25.00	0.00	25.00

Table 3 Confusion matrix of the classifications 2009, 2015, and 2021

Class	Ground truth (%)												
	A			C			P. A			B. A			W
Year	2009	2015	2021	2009	2015	2021	2009	2015	2021	2009	2015	2021	2009	2015	2021
AA	30.39	59.26	59.26	6.40	8.00	10.40	0.00	7.01	6.37	0.00	0.05	0.50	0.00	0.00	0.00
C	53.92	24.07	24.07	77.60	85.60	76.00	6.49	2.55	5.37	2.45	1.25	0.50	0.00	0.00	0.00
P. A	12.75	14.81	14.081	10.40	4.00	4.80	93.51	90.45	87.90	0.00	0.00	0.00	0.00	0.00	0.00
B. A	2.94	1.85	1.85	5.60	2.40	8.80	0.00	0.00	0.00	97.55	98.25	99.00	100.00	50.00	25.00
W	0.00	0.00	0.00	0.00	0.00	0.00	0.00	0.00	0.00	0.00	0.00	0.00	0.00	50.00	75.00

Arboriculture (A); Cereals (C); Peri-urban agriculture (P.A); Built-up areas (B.A); Water (W)

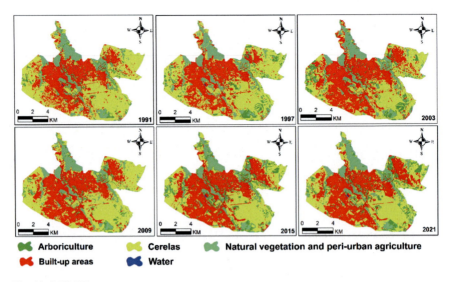

Fig. 11 LULCC maps

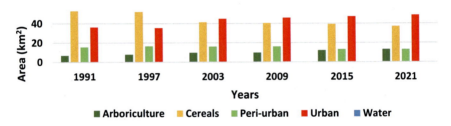

Fig. 12 Area changes in different LULCC categories from 1991 to 2021

4 Discussion

The LST of a given place is an indicator of its microclimate status and radiation exchange with the atmosphere. The UHI affects the local climate mainly in urban areas where the environment is much warmer than the surrounding rural areas due to human activities. It can be a direct indication of global warming. Worldwide, several works related to this topic have been carried out (Alexander 2020; Azmi et al. 2021; Bokaie et al. 2016; Guha et al. 2021; Guo et al. 2015; Sahani 2021; Zhang et al. 2009; Burnett and Chen 2021; Lu et al. 2022).

We cite for example (Kikon et al. 2016), who conducted research based on integrating Landsat thermal data for the years 2000 and 2013, and other field data to assess temporal changes in the UHI in the city of Noida, India. In another context, (Addas et al. 2020), carried out a study in Saudi Arabia. The authors used the LST as an indicator for evaluating the UHI, based on Landsat-8 satellite data. Also, the relationship between LST and land cover indicators has been studied by calculating several spectral indices, such as, NDBI, NDVI, and UI. The study's results indicated the importance of satellite data for evaluating this phenomenon and managing and planning green urban areas facing climate change. In Morocco, (Fathi et al. 2019), carried out a study with the main objective of evaluating the heat island throughout Morocco. In line with our results, this study showed that most cities have experienced an increase in the magnitude of the urban UHI. Indeed, on average, town centers are warmer than the rural peripheries by 1.51 C during the day.

In 2022, a study by (Gourfi et al. 2022) aims to investigate, in the daytime, the relationship between green surfaces, built-up areas, and the surface urban heat island (SUHI) in Marrakesh, Morocco. The results showed a difference of 3.98 °C in the ground surface between the different areas of the city. And the correlation between the vegetation index and SUHI decreases over time. This is due to significant changes in the region's urban planning policy and urban growth. In the same context, the LULCC maps showed that urbanization has known a huge expansion of about 35% from 1991 to 2021, this can be explained by several parameters, such as: the launch of the "cities without slums program" in Morocco in 2004 and improvement of infrastructure, transportation, and mobility networks. This increase has influenced the LST and consequently, the UHI has increased within the city of Meknes. As green spaces disappear and roads, buildings and other constructions replace natural areas, LST increases leading to a subsequent increase in UHI effect. Thus, decision-makers are recommended to reorient the city's development plans, by encouraging green building technology, condensing the green areas, and stimulating reforestation to confront this issue.

5 Conclusions

This study aimed to assess the spatio-temporal variation of UHI in Meknes city using thermal data from the Landsat sensor. Therefore, the LST has been calculated and evaluated for different land use types through various spectral indices. The results showed a significant UHI increase related to the dynamics that Meknes has experienced during the last 30 years, triggered by population growth and improvement of infrastructure, transport, and mobility networks. The results of this study invite decision-makers to reorient urban development plans by the condensation of green zones, stimulation of planting and reforestation, and the promotion of green building technologies.

Acknowledgements The authors thank Thematic Project 4, Integrated Water Resources Management of the institutional university cooperation, and VLIR-UOS for the financial support, equipment, and mission at KU Leuven, Belgium.

References

Addas A, Goldblatt R, Rubinyi S (2020) Utilizing remotely sensed observations to estimate the urban heat Island effect at a local scale: case study of a university campus. Land 9(6):191

Agence Urbaine (2009) Agence Urbaine 2009, Projet de Plan d'action de Développement Local Meknès

Alexander C (2020) Normalised difference spectral Indices and Urban land cover as indicators of land surface temperature (LST). Int J Appl Earth Obs Geoinf 86:102013

Alitane Abdennabi, Ali Essahlaoui, Mohammed El Hafyani, et al. (2022) Water erosion monitoring and prediction in response to the effects of climate change using RUSLE and SWAT equations: case of R'Dom watershed in Morocco. Land 11(1). Multidisciplinary Digital Publishing Institute: 93

Alqasemi AS, Hereher ME, Al-Quraishi AMF (2020) Retrieval of monthly maximum and minimum air temperature using MODIS aqua land surface temperature data over the United Arab Emirates. Geocarto Int 37:2996–3013

Artis DA, Carnahan WH (1982) Survey of emissivity variability in thermography of urban areas. Remote Sens Environ 12(4):313–329

As-syakur Abd Rahman, Wayan Sandi Adnyana I, Wayan Arthana I, Wayan Nuarsa I (2012) Enhanced built-up and bareness index (EBBI) for mapping built-up and bare land in an urban area. Remote Sens 4(10). Molecular Diversity Preservation International: 2957–2970

Axhausen KW (2000) Geographies of somewhere: a review of urban Literature. Urban Stud 37(10):1849–1864

Azmi R, Koumetio CST, Diop EB, Chenal J (2021) Exploring the relationship between urban form and land surface temperature (LST) in a Semi-Arid region case study of Ben Guerir City - Morocco. Environ Chall 5:100229

Bokaie M, Zarkesh MK, Arasteh PD, Hosseini A (2016) Assessment of urban heat island based on the relationship between land surface temperature and land use/ Land Cover in Tehran. Sustain Cities Soc 23:94–104

Burnett Michael, Dongmei Chen (2021) The impact of seasonality and land cover on the consistency of relationship between air temperature and LST derived from Landsat 7 and MODIS at a local scale: a case study in Southern Ontario. Land 10(7). Multidisciplinary Digital Publishing Institute: 672

Xu, Chen, Zhang Y (2017) Impacts of urban surface characteristics on spatiotemporal pattern of land surface temperature in Kunming of China. Sustain Cities Soc 32:87–99

Collier CG (2006) The impact of urban areas on weather. Q J R Meteorol Soc 132(614):1–25

Congalton RG (1991) A review of assessing the accuracy of classifications of remotely sensed data. Remote Sens Environ 37(1):35–46

Debolini M, Valette E, François M, Chéry J-P (2015) Mapping land use competition in the rural-urban fringe and future perspectives on land policies: a case study of Meknès (Morocco). Land Use Policy 47:373–381

Divya Y, Gopinathan P, Jayachandran K, Al-Quraishi AMF (2020) Color slices analysis of land use changes due to urbanization in a city environment of Miami Area, South Florida, USA. Model Earth Syst Environ 7:537–546

Duan S-B, Li Z-L, Wang C et al (2019) Land-surface temperature retrieval from Landsat 8 single-channel thermal infrared data in combination with NCEP reanalysis data and ASTER GED product. Int J Remote Sens 40(5–6):1763–1778

El, Hafyani, Mohammed AE, Fung-Loy K, Hubbart JA, Van Rompaey A (2021) Assessment of agricultural water requirements for semi-arid areas: a case study of the Boufakrane river watershed (Morocco). Appl Sci 11(21):10379

El, Hafyani, Mohammed AE, Van Rompaey A et al (2020) Assessing regional scale water balances through remote sensing techniques: a case study of Boufakrane river watershed, Meknes Region, Morocco. Water 12(2):320

Hafyani, Mohammed EL, Ali Essahlaoui, Narjisse Essahlaoui, Ali et al. (2023) Generation of climate change scenarios for rainfall and temperature using SDSM in a Mediterranean environment: a case study of Boufakrane river watershed, Morocco. J Umm Al-Qura Univer Appl Sci, 1–13

El, Ouali, Anas ME, Hafyani AR et al (2021) Modeling and spatiotemporal mapping of water quality through remote sensing techniques: a case study of the Hassan Addakhil Dam. Appl Sci 11(19):9297

Estoque RC, Murayama Y (2015) Classification and change detection of built-up lands from Landsat-7 ETM+ and Landsat-8 OLI/TIRS imageries: a comparative assessment of various spectral Indices. Ecol Ind 56:205–217

Fathi N, Bounoua L, Messouli M (2019) A satellite assessment of the urban heat island in Morocco. Can J Remote Sens 45(1):26–41

Fung-Loy K, Van Rompaey A, Hemerijckx L-M (2019) Detection and simulation of urban expansion and socioeconomic segregation in the greater Paramaribo region, Suriname. Tijdschr Econ Soc Geogr 110(3):339–358

Gabriel Katharina MA, Wilfried R Endlicher (2011) Urban and rural mortality rates during heat waves in Berlin and Brandenburg, Germany. Environ Pollut 159(8–9):2044–2050

Gourfi Abdelali, Aude Nuscia Taïbi, Salima Salhi, Mustapha El Hannani, Said Boujrouf (2022) The surface urban heat island and key mitigation factors in arid climate cities, case of Marrakesh, Morocco. Remote Sens 14(16):3935

Guha Subhanil, Himanshu Govil, Ajay Kumar Taloor, Neetu Gill, Anindita Dey (2021) Land surface temperature and spectral indices: a seasonal study of Raipur city. Geodesy Geodyn. https://linkinghub.elsevier.com/retrieve/pii/S1674984721000793. Accessed January 12, 2022

Guhathakurta S, Gober P (2007) The impact of the phoenix urban heat island on residential water use. J Am Plann Assoc 73(3):317–329

Guo G, Wu, Zhifeng, Xiao R et al (2015) Impacts of urban biophysical composition on land surface temperature in urban heat island clusters. Landsc Urban Plan 135:1–10

RGPH (2014) Population Légale d'après Les Résultats Du RGPH 2014 Sur Le Bulletin Officiel

Houpin S (2011) Urban mobility and sustainable development in the Mediter- Ranean: regional diagnostic outlook, blue plan papers

Ijlil Safae, Ali Essahlaoui, Meriame Mohajane et al. (2022) Machine learning algorithms for modeling and mapping of groundwater pollution risk: a study to reach water security and sustainable development (Sdg) Goals in a Mediterranean Aquifer System. Remote Sens 14(10). Multidisciplinary Digital Publishing Institute: 2379

Imran HM, Kala J, Ng AWM, Muthukumaran S (2018) Effectiveness of green and cool roofs in mitigating urban heat island effects during a heatwave event in the city of Melbourne in southeast Australia. J Clean Prod 197:393–405

Kawamura M, Jayamana S, Tsujiko Y (1996) Relation between social and environmental conditions in Colombo Sri Lanka and the Urban index estimated by satellite remote sensing data. Int Arch Photogramm Remote Sens 31:321–326

Kikon N, Singh P, Singh SK, Vyas A (2016) Assessment of urban heat islands (UHI) of noida city, India using multi-temporal satellite data. Sustain Cities Soc 22:19–28

Laaidi K, Zeghnoun A, Dousset B et al (2012) The impact of heat islands on mortality in Paris during the August 2003 heat wave. Environ Health Perspect 120(2):254–259

Li D, Bou-Zeid E, Oppenheimer M (2014) The effectiveness of cool and green roofs as urban heat island mitigation strategies. Environ Res Lett 9(5):055002

Lu Yuting, Penghai Wu, Kaijian Xu (2022) Multi-time scale analysis of urbanization in urban thermal environment in major function-oriented zones at landsat-scale: a case study of Hefei City, China. Land 11(5). Multidisciplinary Digital Publishing Institute: 711

McFeeters SK (1996) The use of the normalized difference water index (NDWI) in the Delineation of Open Water Features 17(7):1425–1432

Mohajane Meriame, Ali Essahlaoui, Fatiha Oudija, et al. (2018) Land Use/Land Cover (LULC) using landsat data series (MSS, TM, ETM+ and OLI) in Azrou forest, in the central middle Atlas of Morocco. Environments 5(12). Multidisciplinary Digital Publishing Institute: 131

Mohajane Meriame, Ali Essahlaoui, Fatiha Oudija, Mohammed El Hafyani, Ana Cláudia Teodoro (2017) Mapping forest species in the central middle Atlas of Morocco (Azrou Forest) through Remote Sensing Techniques. ISPRS Intern J Geo-Inform 6(9). Multidisciplinary Digital Publishing Institute: 275

Musakwa W, Van Niekerk A (2013) Implications of land use change for the sustainability of urban areas: a case study of Stellenbosch, South Africa. Cities 32:143–156

Oke TR (1982) The energetic basis of the urban heat island. Quart J Royal Meteorol Soc 108(455):1–24

Parés-Franzi M, Saurí-Pujol D, Domene E (2006) Evaluating the environmental performance of urban parks in Mediterranean cities: an example from the Barcelona metropolitan region. Environ Manage 38(5):750–759

Patz JA, Campbell-Lendrum D, Holloway T, Foley JA (2005) Impact of regional climate change on human health. Nature 438(7066):310–317

Razzaghmanesh Mostafa, Simon Beecham, Telma Salemi (2016) The role of green roofs in mitigating urban heat island effects in the metropolitan area of Adelaide, South Australia. Urban Forest Urban Green 15:89–102

Rondeaux G, Steven M, Baret F (1996) Optimization of Soil-Adjusted Vegetation Indices 55(2):95–107

Rouse et al. (1973) Monitoring the vernal advancement and retrogradation of natural vegetation. NASA/GSFCT Type II Report, Greenbelt, MD, USA

Sahani N (2021) Assessment of spatio-temporal changes of land surface temperature (LST) in Kanchenjunga biosphere reserve (KBR), India using landsat satellite image and single channel algorithm. Remote Sens Applic Soc Environ 24:100659

Sharma A, Conry P, Fernando HJS et al (2016) Green and cool roofs to mitigate urban heat island effects in the Chicago metropolitan area: evaluation with a regional climate model. Environ Res Lett 11(6):064004

Sharma R, Joshi PK (2016) Mapping environmental impacts of rapid urbanization in the national capital region of India using remote sensing inputs. Urban Climate 15:70–82

Son Ji-Young, Jong-Tae Lee, Brooke Anderson G, Michelle L Bell (2012) The impact of heat waves on mortality in seven major cities in Korea. Environ Health Perspect 120(4):566–571

Stehman, Stephen V (1996) Use of auxiliary data to improve the precision of estimators of thematic map accuracy. Remote Sens Environ 58(2):169–176

Tan Jianguo, Youfei Zheng, Xu Tang, et al. (2010) The Urban heat island and its impact on heat waves and human health in Shanghai. Intern J Biomet 54(1):75–84

Tong S, Wang XY, Yu, Weiwei, Chen D, Wang X (2014) The impact of heatwaves on mortality in Australia: a multicity study. BMJ Open 4(2):e003579

Veron J (2007) La Moitié De La Population Mondiale Vit En Ville 435:1–4

Xiong Y, Huang S, Chen F et al (2012) The impacts of rapid urbanization on the thermal environment: a remote sensing study of Guangzhou, South China. Remote Sens 4(7):2033–2056

Zanter K (2016) Landsat 8 (L8) Data Users Handbook. LSDS-1574 Version 2.0, vol 1. Department of the Interior US Geological Survey

Zha Y, Gao J, Ni S (2003) Use of normalized difference built-up index in automatically mapping urban areas from TM imagery. Int J Remote Sens 24(3):583–594

Zhang Y, Odeh IOA, Han C (2009) Bi-temporal characterization of land surface temperature in relation to impervious surface area, NDVI and NDBI, using a sub-pixel image analysis. Int J Appl Earth Obs Geoinf 11(4):256–264

Environmental Challenges, The Impacts of Climate Change in North Africa Region: A Review

Afeez Alabi Salami and Olushola Razak Babatunde

Abstract Global climate change effects, referred to as "environmental stresses," in North Africa (NA) may be anthropocentrically caused or naturally occurring. This region is one of the regions most sensitive to climate change because of its location and dry environment. Even though nations in the NA area have distinct political and social frameworks, access to natural resources, and wealth levels, the region is bound by major environmental issues and international disputes. This chapter examines the regional impacts of global climate change in North Africa by focusing on the most pressing environmental issues and potential remedies. This chapter clarifies that climate change is mostly to blame the environmental problems in the North African region, which threatens the long-term stability of the area, particularly in terms of poverty, inequality, and underdevelopment. However, attempts to address these issues have been hampered by genuine ignorance of their causes and potential solutions. As one of the most water-scarce regions in the world with a heavy reliance on climate-sensitive agriculture, North Africa's economic and social situation will likely worsen in the future. Mitigating climate change will remove resources from initiatives to combat poverty, unemployment, and subpar living circumstances, endangering the viability of the development process. North African nations must consider environmental conservation in the context of regional sustainable development and propose responsibilities that the government and outside aid should take.

Keywords Climate change · Environmental challenges · North African region

---

A. A. Salami (✉)
Department of Geography and Environmental Management, University of Ilorin, Ilorin, P.M.B. 1515, Nigeria
e-mail: 14-68mn023pg@students.unilorin.edu.ng; afeezsalami@bisnigeria.org

O. R. Babatunde
Department of Urban and Regional Planning, Federal University of Technology Owerri, P.M.B. 1526, Owerri, Nigeria
e-mail: olushola.babatunde@futo.edu.ng

1 Introduction

Climate change, an international phenomenon, is currently the biggest major environmental threat to our planet. This is true and affects our lives, especially in Africa (Salami and Tilakasiri 2020). The environment is immediately affected by climate change, which causes several catastrophes and strains that directly influence people. Environmental problems, such as deforestation, soil erosion, water scarcity, food hunger, danger, and hazardous living conditions, encourage people to leave safer locales (Fadhil 2009, 2011; Salami and Fenta 2022; Salami and Tilakasiri 2020; Tilakasiri et al. 2017). The NA area, which is one of the most sensitive areas to climate change, suffers terrible and catastrophic effects from climate change. The natural variety in North Africa is astounding, comprising enormous deserts, rough mountains, windswept plateaus, surrounding seas, and rivers that provide life to arid lands.

Desertification, droughts, insecurity, floods, tropical cyclones, erosion, and other environmental hazards caused by climate change are already known to influence North Africa (Aw-Hassan et al. 2014; De Châtel 2014; Salami and Tilakasiri 2017). Despite having varied geographies, political and social systems, natural resources, and income levels, North African countries are connected by the region's major environmental challenges (Table 1) and cross-border conflicts (Abumoghli and Goncalves 2020).

These problems threaten the stability of the area over the long term. Rising temperatures caused by climate change are predicted to diminish land in North Africa, which is suitable for farming, shorter growing seasons, and lower agricultural yields. These consequences will be worsened by the decline in annual precipitation that is anticipated for Northern Africa in the twenty-first century, especially in semi-arid and

Table 1 Major environmental Challenges, population and land areas of North African countries

Country	2022 population	Density (/km^2)	Environmental concerns
Algeria	44,903,225	18.9	Pollution of rivers and coastal waters by industries, desertification, soil erosion
Egypt	110,990,103	111.5	Air pollution which is causing a haze over Cairo, ten to one hundred times the acceptable world standards
Libya	6,812,341	3.9	Desertification, limited natural fresh water resources
Morocco	37,457,971	83.9	Desertification, soil erosion (from farming, overgrazing), destruction of vegetation, oil pollution of coastal waters
Sudan	46,874,204	25.1	Deforestation, land degradation, loss of biodiversity and habitat, pollution of air, land, and water, conflicts over diminishing natural resources, food insecurity and poor waste and sanitation services in urban areas
Tunisia	12,356,117	79.5	Deforestation, soil erosion (overgrazing), toxic and hazardous waste disposal, water pollution (from raw sewage)

desert regions that depend on irrigation for agricultural development. Water scarcity, the disappearance of arable land, air and water pollution, biodiversity loss, the depletion of marine resources, the degradation of coastal ecosystems, desertification, soil erosion, and inadequate waste management are some of the major environmental problems in North Africa (Table 1). In addition, an often-mentioned and sensationalized impact of climate change in the NA area is the risk of fresh violent conflict that is either directly caused or exacerbated by it. However, the reality is more nuanced than some pessimistic predictions made by so-called climate determinists. Most academics agree that climate change is the only factor among several that directly contributed to earlier conflicts and political turmoil in the NA area, including pervasive corruption, authoritarianism, economic marginalization, population pressures, and other governance ills.

Most of the environmental issues in North Africa are caused by desertification and soil erosion. Given the scarcity of water in arid regions, pollution is an important issue. The most water-stressed region on Earth's surface is North Africa because of its increasing water shortages. Additionally, it is a unique hotspot for the effects of adaptation.

Climate change and urbanization, which have resulted in rising population and urban growth rates, are the main contributors to these environmental risks (Salami and Fenta 2022). In the NA region, increasing temperatures, increased rainfall unpredictability, rising sea levels, and conflicts caused by drought are additional effects of climate change. The four main types of hazards caused by climate change in the NA are environmental, biological, radiation, and hazardous chemicals. Both natural and man-made threats fall under this category. According to the United Nations Environment Programme (UNEP) (2017), environmental hazards can be divided into episodic and persistent types. While episodic risks might be viewed as ongoing/sudden-onset threats, perennial hazards are known as slow-onset threats. The dangers associated with the first type are primarily those caused by geology, hydrometeorology, and accidents (Salami and Tilakasiri 2020).

The characteristics of the atmosphere are connected to enduring threats. Desertification, droughts, insecurity, floods, tropical cyclones, erosion, and other environmental hazards brought on by climate change are already known to influence the region of North Africa (Aw-Hassan et al. 2014; De Châtel 2014). The environmental issues brought on by climate change in North Africa are highlighted in this chapter.

2 Description of the North Africa Region

North Africa is a region of Africa that is between the Sahara Desert and the Mediterranean coastline region. The Sahara Desert covers much of Northern Africa. The largest of the five sub-regions in Africa by land area, NA is the northernmost region of Africa. Six nations make up North Africa: Egypt, Algeria, Libya, Morocco, Tunisia, and Sudan (Fig. 1). These countries are divided into three regions: the Arab Least Developed Country, which includes Sudan; the Maghreb area, which includes

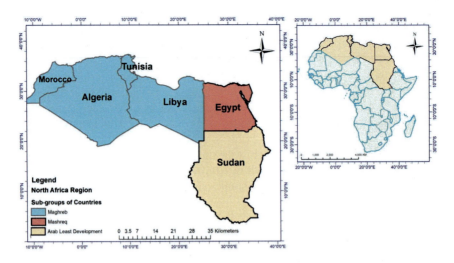

Fig. 1 North Africa region with the three sub-groups of countries

Algeria, Libya, Morocco, and Tunisia; and the Mashreq region, which includes Egypt. The North Africa (NA) area is most vulnerable to climate change, particularly rising temperatures, food and water scarcity, and economic crises due to its geographic location and dry meteorological conditions. Table 1 lists the nations of North Africa along with their densities, populations, and environmental issues. This study was conducted based on earlier research and published materials on the NA region's environmental problems and climate change. The material and data utilised in this investigation were based on what was already publicly available.

3 Environmental Challenges Induced by Climate Change in North Africa

Existing people may be displaced by the effects of climate change and natural disasters, such as water shortages or losses, rising sea levels along the coast, extended droughts, or the acceleration of desertification, and those who stay may face challenges. Numerous environmental assaults have been directed towards the human environment, which has resulted in losses for the environment, people, and finances. These losses depend on how tough and capable the human population is to put up with risks.

3.1 Desertification

One of the significant issues in the NA region is desertification, where sand covers hundreds of square hectares of land yearly (Tilakasiri et al. 2017). Even though climate change is not the only cause of this and other types of soil degradation and ecological depletion, which have been characteristics of the NA region since well before the industrial age, rising temperatures and dwindling water supplies will quicken the rate of desertification in the region. In addition to creating more water shortages, climate change will exacerbate aridity in several nations in sub-Saharan Africa and the NA region during the next century, reducing arable land and upsetting agricultural practices.

Additionally, existing dry land will continue to get dryer, and desert dust will continue to build up in the atmosphere and cause more sandstorms, particularly in Egypt, leading to desertification in the previous 20 years. Desertification may occur in two ways: through dune extension and soil compaction, which are often caused by over usage and overfarming.

Due to the loss of both livelihoods and livable places, desertification is a driving force behind migration (Tilakasiri et al. 2017). A specialist at the Desert Research Centre claimed that several villages, including Ganah and Moschee, whose residents were forced to relocate, have completely vanished as a result of desertification in Egypt's Western Desert. A similar thing happened in Egypt's Eastern Desert, where an Egyptian NGO is studying a tribe known as the "Ababda" who were driven from their homes by expanding dunes and droughts (Tilakasiri et al. 2017; Warner 2008).

Desertification negatively impacts ecosystem dynamics, air quality, human health, and land production. Governments, businesses, and international organizations have responded by implementing a variety of initiatives in the region, including reforestation, afforestation, and new agricultural projects, to stop desertification and lower atmospheric carbon levels. However, some academics have expressed doubt about these initiatives since they believe they are ineffective and environmentally harmful.

3.2 Water Scarcity, Stress and Shortages

It is impossible to overstate water's importance because it is utilised for many things, including home, industrial, and agricultural uses. On Earth's terrestrial surface, precipitation totals 110,000 km^3 every year (Guinness and Nagle 2021). This would be more than enough to meet the demands of the whole world's population, but a large portion is inaccessible and the remainder is divided quite unevenly. For instance, the world's dry areas (which include NA) account for 40% of the planet's geographical area yet only get 2% of the world's precipitation (Guinness and Nagle 2021).

The concept of water scarcity refers to when water supply falls below 1000 m^3 per person per year—the country faces water scarcity for all or part of that year, while water stress is when water supply in a country is below 1700 m^3 per person per year (Guinness and Nagle 2021). In other words, water scarcity has to do with the availability of potable water while water stress is where a country or area cannot meet its demand for fresh water.

Global water stress and shortage are on the rise, particularly in the Middle East and North Africa. These areas have extraordinarily high levels of water stress, making them among the driest places on Earth (Food and Agricultural Organisation 2018; United Nations Water, 2007). Since water supplies are being used more quickly than they are being restored by precipitation, the NA area is predicted to be among the first in the globe to "essentially run out of water." An already water-scarce region will likely see water availability continuing to decline, both overall and per capita, to critically low levels due to climate change and the stresses currently put on the water supply by a growing population. The degree of water stress is determined by comparing the amount of water being removed (pumped out) to the available amount.

While much of sub-Saharan Africa is affected by economic water scarcity, which occurs when a population lacks the necessary financial resources to utilise an adequate source of water, the NA region is most often associated with physical water scarcity, which refers to limited physical access to water. Because there isn't enough water for local food production, Egypt is an excellent illustration of a nation with physical water scarcity because it imports more than half of its food. Furthermore, a 1-m sea level rise in Egypt is predicted to impact around 6 million residents of the Nile delta basin, the majority of whom are 'poor'. This weakening in the Nile Delta, which is extensively populated and utilised for agriculture, would affect millions of people. According to recent research, the Nile River's flow would decline by 40–60%, increasing the frequency and severity of droughts, particularly in North African nations, which will have a significant negative socioeconomic and political impact on the region. The fact that 80% of the water that comes into Egypt is produced in Ethiopia, whose precipitation patterns have changed, is also underlined as a reason why Egypt would need to reform its water strategy.

Local militias in Libya have employed similar tactics to gain an advantage over competitors or the government, particularly when it comes to the Great Man-Made River. Residents of Tripoli's capital were forced to dig through poured concrete for water when a militia cut off the supply to compel a rival militia into freeing a commander who was being held captive. These sub-state conflict dynamics might become more pronounced due to climate change.

Temperatures, evapotranspiration rates, and precipitation levels are all relatively high in the NA region. Due to its geographic location, the NA area experiences extremely unpredictable rainfall, and most of its nations significantly rely on seawater desalination. Due to the growing population, rising freshwater demand, higher frequency of droughts due to climate change, fast urbanisation (Abumoghli and Goncalves 2020), rising living standards, and pumping more water than the rain can replenish, water stress is escalating quickly in the NA area. These and other causes are surpassing the region's supply capacity. These pose major hazards to lives, livelihoods, biodiversity, economic stability, and human security by increasing demand for already-scarce water supplies. Overuse has caused the quality of the readily available water sources to degrade. In NA nations, a larger portion (86%) of annually freshwater withdrawals go to the agricultural sector (FAO 2016).

Additionally, the region's water supply will continue to fall while demand for water rises. As the "sleeping tiger" of environmental issues, the water shortage has been portrayed as having the potential to endanger the global food supply, impede social and economic advancement, and lead to severe disputes between nations sharing drainage basins.

3.3 Rising Sea Levels and Temperatures

The NA region is also among the most vulnerable to the effects of climate change-related sea level rise; some estimates predict that by the end of the century, the average worldwide sea level would have risen by 30–122 cm (1 to 4 feet). There are several effects. Seawater may seep into coastal aquifers and wells when sea levels rise, salinizing the water and wreaking havoc on agricultural settlements along the shore. Additionally, the flooding of coastal areas might render several shoreline cities and villages in the NA region uninhabitable, adding to the already burdensome governments' economic and migration-related strains. The cities of Algiers, Benghazi, and Alexandria in particular will be vulnerable to even minor rises in sea level by 2050, making the nations of North Africa, particularly in danger in terms of the overall population affected by sea level rise.

The NA region is expected to experience temperatures much above this estimate, even if global warming is only projected to increase temperatures by 2 °C. The desert warming amplification phenomena, which precludes natural cooling effects and produces a feedback loop that further intensifies heat, would lead most of the region to experience global warming more rapidly. If greenhouse gas emissions continue to climb at their current rate, the region's temperatures are predicted to rise by at least 4 °C by 2050, and the frequency of heat waves is predicted to increase tenfold. Future temperatures in the area are predicted to rise over the point at which people can adapt, resulting in greater death rates for young people and the elderly. Similar to other climate change phenomena, global warming and its perceived effects in the region continue to widen socioeconomic gaps and put vulnerable populations at risk.

3.4 Food Security and Insecurity

The lack of water available for irrigation poses a major danger to the safety of the food supply in the NA area. Food security is seen as a vital problem since climate change affects not only people's livelihoods but also the expansion of the national economy and social stability. Food security is described as having sufficient access to food in both quality and quantity, whereas food insecurity is defined as having restricted access to food at the level of people or families owing to a lack of money and/or other resources (FAO 2018; Salami and Tilakasiri 2020). Agricultural production may decline due to climate change, adding to the already existing food shortage. Drops in food production brought on by climatic changes might cause a food crisis, food riots, and economic collapse in the world's least developed nations, particularly in the region of North Africa.

3.5 Water Pollution, Coastal and Marine Ecosystems

The NA region is home to some of the most distinctive coastal and marine environments in the entire world, with high levels of species endemism in the remote Mediterranean and the Red Sea (International Union for Conservation of Nature 2020) and remarkable habitats like mangroves, seagrass beds, coral reefs, and mudflats. These ecosystems provide a range of goods and services, such as freshwater through desalination, economic gains from tourism, and food and sources of income for those living along the shore (United Nations-Habitat 2011). Unrestrained coastal development spurred on by urbanisation and population growth has altered the coastal zones. The most urbanised coastline is that of the Maghreb, which comprises Algeria, Libya, Morocco, and Tunisia.

In the NA region, water contamination is a significant problem. Water contamination, which can affect humans, plants, and animals, is the contamination of water bodies brought on by the direct or indirect discharge of pollutants. The NA region's ecosystems, which provide food and a nursery for marine creatures, are disturbed and polluted by human activity. Particularly in Egypt, garbage, oil spills, and discharge from desalination facilities considerably impact coastal habitats. Other critical environmental challenges in the NA region include bycatch, overfishing, and sea level rise brought on by climate change. Low-lying nations, smaller islands, and river delta regions in the Mashreq (Egypt) and Maghreb (Tunisia and Libya) subregions are particularly susceptible to floods and salinization. Due to the loss of mangroves and coral reefs, coastal ecosystem degradation will compound these negative effects by increasing shoreline wave exposure (Abumoghli and Goncalves 2020).

The fisheries in Egypt's northern lakes, Manzala, Mary out, Edkou, and Borollos, which are contaminated by agricultural waste and saltwater, were researched by UNDP Cairo (Tilakasiri et al. 2017; Warner 2008). As a result, fish's quantity and quality have drastically declined; only those that can withstand water contamination are left. Since these fish do not command high market prices, the fishermen's income declines, and they are compelled to look for alternative sources of support in places like Eritrea, Somalia, Ethiopia, and Sudan. Others sailed across the Mediterranean to Cairo and Europe.

3.6 Conflict

In addition to posing a security risk, climate change can lead to disputes between pastoral and agricultural populations, sparked by water unpredictability and droughts made worse by the changing climate. These pressures may cause individuals to feel the need to migrate, raising the cost of cattle and other items in the process. They may also heighten tensions and violence across communities. In some circumstances, this may also contribute to a rise in disease outbreaks and malnutrition, harming food security. The Nuba tribe in Southern Sudan, for instance, has warned that they would renew the battle because Arab nomads who have been forced south into their land by drought are chopping down trees to feed their camels, according to Julian Borger, Diplomatic Editor of The Guardian (Tilakasiri et al. 2017). In other words, violence erupted between the Arabs (cattle and camel herders who were pushed south of Darfur, Sudan by repeated droughts and desertification) and Fur (a non-Arab, "black" tribe in western Sudan) over grazing areas and water resources.

According to research linking climate change to war in Sudan, the Sahel, which includes sections of southern Sudan, might experience a 70% reduction in grain production due to climate change (Tilakasiri et al. 2017). Many Sudanese fled to Khartoum, Kordofan, and other locations due to environmental difficulties in their home country, which in turn caused intense conflict in the migration destinations. Twenty-seven Arab tribes banded together between 1987 and 1989 to form the nomad militia Janjawid (armed men on horseback), which they used to fight the Fur and other non-Arab tribes. The armed Janjawid militia targeted civilian populations in a campaign of violence, including mass killings, rape, and the displacement of communities. Estimates of the death toll in Darfur vary, but it is generally agreed that hundreds of thousands of people lost their lives due to violence, disease, and starvation. The conflict also led to the displacement of millions, with many seeking refuge in internally displaced persons (IDP) camps or fleeing to neighbouring countries like Chad.

3.7 Air Quality

Due to the accumulation of particulate matter in the region of North Africa, air pollution is a significant environmental concern (Abumoghli and Goncalves 2020). This has a detrimental effect on people's health, way of life, and ecosystem. In 2015, Cairo (Egypt) recorded a mean annual PM10 concentration of 284 g/m^3 (World Health Organisation 2018). According to estimates, thousands of premature deaths occurred in the area in 2013 as a result of air pollution (World Bank 2016). Air pollution is naturally produced in the NA region since it is a part of the "dust belt," where most dust storms start (Furman 2003). They convey fine sand and dust over great distances, impeding several operations, including aircraft, traffic on the roads, supply chains, and agricultural cultures, and aggravating respiratory conditions (United Nations

Environmental Programme-UNEP 2017). Both natural and man-made forces cause the NA region's air pollution. Natural reasons include desertification, drought, and climate change, especially in the most vulnerable nations, while anthropogenic causes include industrial emissions fueled by human activity (UNEP 2017).

3.8 Rising Temperature

Greater than normal global warming is causing temperatures to rise. Extreme and protracted heat waves will alter both rural and urban regions, increasing the heat island effect in cities and reducing agricultural production while putting major pressure on already limited water supplies. Climate extremes are expected to get harsher over the next several decades. The region's population is anticipated to quadruple by 2050, compounding these consequences and placing a tremendous demand on the region's resources.

3.8.1 Land Resources Challenges

The desertification and land degradation that the NA region has been suffering over the past several decades is the consequence of contemporary phenomena, including climate change, political upheaval, and modifications in land-use patterns and practices spurred on by growing urbanization (Abumoghli and Goncalves 2020; Abumoghli 2019). Desertification is accelerated by indications of climate change, such as decreased precipitation rates and higher surface temperatures, through soil erosion, salinization, and more frequent dust storms. Reduced production, worsened biodiversity and ecosystem health, increased water shortages, and dwindling arable lands are all consequences of land degradation (FAO and Organisation for Economic Co-operation and Development- OECD 2018).

3.8.2 Biodiversity

The NA region's terrestrial, coastal, and marine environments support rich biodiversity. Many wild descendants or relatives of essential crops, such as cereals, pulses, oil- and fibre-yielding plants, vegetables, and fruits can be found in the diverse habitats of Mashreq country in Egypt, which includes mountain upland plateaus, inland, riverine and coastal plains, sand deserts, and wetlands. Development, climate change (increasing sea temperatures cause coral bleaching), conflict, and ineffective regulations like limited protected area coverage or conservation and restoration programs are the leading causes of species loss and habitat reduction in NA regions. The region needs to develop over time. Development is something that does - or does not -happen over time and across territories (Salami et al. 2017).

4 Solutions to Address the Impacts of Climate Change and Environmental Issues

Many NA governments have pledged to reduce global warming by switching to renewable energy and green technology. Some have started moving in that direction in light of urgent threats associated with climate consequences in the area. These risks from climate change and other environmental issues may be reduced by a range of political, institutional, and societal factors working together with adaptation strategies and investments. As a result, global solutions should be adopted to solve climate issues. Given their differences in income and development, North African countries can attain regional governance and cooperation, especially in policy coordination and agenda setting, research and knowledge and information sharing, technical assistance and capacity building, and financial leverage.

The nations of North Africa are increasingly aware of these climate stresses and committed to finding solutions, particularly by building dams and increasing the use of water desalination. While beneficial, such local solutions may unintentionally enhance dangers and vulnerabilities nationally and worldwide. Additionally, governments in the NA region are working to alter the incentive frameworks for agricultural water use, to develop licensing and regulation systems for groundwater management, and occasionally to decentralize water resources management to the local level while enhancing community responsibility for water efficiency (Ghazanfar et al. 2019). These methods have been successful in Jordan and Egypt.

An integrated approach that focuses solutions on water and industries that depend on it is required to address the problem of water shortage. This includes giving priority to adaptation action in transboundary water basins given that more than two-thirds of the region's freshwater resources cross one or more international boundaries and that climate change impacts on those resources have significant implications for climate security, water, energy, and food security, rural livelihoods, and economic development. Development is all about improving the life conditions faced by the majority, specifically this means reducing existing levels of poverty and inequality (Salami et al. 2017). The mainstreaming of adaptation into planning at the national and sectoral levels must optimize cross-sector synergies and promote coherence at all scales. This necessitates the integration of horizontal and vertical policies and synchronizing regional and national plans, water-related agenda, and national objectives and targets at the international level.

Better strategies to prevent a serious water crisis in the NA area include systematically reusing discarded or "grey" water, recycling and reclamation systems, shifting or transferring, desalination, reducing water losses, and cloud seeding. Israel ranks among the top countries in the world for water recycling with more than 87% of its wastewater being treated and utilised for agriculture (Marin et al. 2017). Similarly, China aims to move water from the River Yangtze in the South to the North through "South-North Water Transfer Project, also known as the Great Canal" in 2002, with an estimated cost of over $79 billion. The rationale of water transfer in China is that the south of China gets plenty of rain, the north is dry, but it has much of China's agriculture. The institutional, statutory, and technological components that support this achievement serve as a model for other countries that face severe water stress and scarcity. Governments can take use of their shared culture, language, and religion to find cooperative solutions to local climatic and environmental concerns because socioeconomic situations and natural resource endowments differ. Some of these decisions consist of:

i. these threats could be mediated by a range of political, institutional, and societal factors, alongside adaptation policies and investments;
ii. North African countries should give urgent attention to address the issue without wasting time;
iii. actions such as litigation and adaptation, which the regional government already establishes should be linked to policies and regulations needed to achieve them;
iv. local innovations and technologies should be enhanced and used by not relying on foreign technical know-how which is not designed and suitable for the region;
v. collaboration among North African countries in addressing the impacts of climate change and other environmental challenges should be intensified. Since the countries in North Africa share similar geographical characteristics and natural resources, the impacts of climate change are likely to be felt across borders. This needs to be properly managed as it could lead to frequent conflicts or insecurity; and
vi. understanding the structural reality of climate change in its wide dimensions and with its medium- and long-term effects, which must be integrated into the political and economic decisions made by these nations. The relationship between this phenomenon and other socioeconomic issues that are more integrated at the policy level has to be made explicit and direct.

5 Conclusions

Due to its predominately semi-arid and dry nature, the NA area has traditionally been one of the most subject to natural stresses and is presently one of the most vulnerable to climate change. Over the past 20 years, significant changes like population growth, economic development, accelerated urbanization, especially in coastal areas, and geopolitical conflicts in some countries have made the region's most pressing environmental challenges—water scarcity, arable land depletion, ecosystem degradation, biodiversity loss, waste management, and air pollution—worse. These issues are interconnected and their consequences on the socioeconomic system are inextricably intertwined. As one of the most water-scarce regions in the world with a heavy reliance on climate-sensitive agriculture, North Africa's economic and social situation will likely worsen in the future. While significant climate change events like droughts and floods will likely delay economic growth for many years, adaption strategies are typically not in line with development concerns.

A key source of employment and revenue in Morocco and Tunisia, tourism may be negatively impacted by sea level rise, leading to significant population shifts.

Morocco, Egypt, and Algeria's ambitious national plans to transition to renewable energy are evidence that the region is moving closer to fulfilling the Sustainable Development Goals (Goals 7 and 13). New water-related technologies are also crucial for the area's sustainable growth, as shown by Israel's inventive achievements in water desalination, recycling, and reuse. Experts warn that efforts are still far from being sufficient to appropriately address environmental concerns, despite several case studies showing that NA governments are paying greater attention to them. Policy improvements encounter difficulties when it comes to effective enforcement and implementation. Experts generally agree that for integrated environmental management to be accomplished in the NA nations, institutional and regulatory reforms, long-term investments, efficient governance, regional coordination for concerted efforts, and the engagement of all stakeholders are necessary.

References

Abumoghli I (2019) Environmental outlook for the West Asia region. In Environmental challenges in the MENA region: the long road from conflict to cooperation. In Pouran H, Hakimian H (eds) Gingko library. Chapter 2, pp 10–30

Abumoghli I, Goncalves A (2020) The environmental challenges in the Middle East and North Africa Region Paper. United Nations Environment Programme and Faith for Earth Initiative

Aw-Hassan A, Rida F, Telleria R, Bruggeman A (2014) The impact of food and agricultural policies on groundwater use in Syria. J Hydrol 513:204–215. https://doi.org/10.1016/j.jhydrol.2014.03.043

De Châtel F (2014) The role of drought and climate change in the Syrian uprising: untangling the triggers of the revolution. Middle East Stud 50(4):521–535. https://doi.org/10.1080/00263206.2013.850076

Fadhil AM (2009) Land degradation detection using geo-information technology for some sites in Iraq. J Al-Nahrain Univers Sci 12:94–108. https://doi.org/10.22401/jnus.12.3.13

Fadhil AM (2011) Drought mapping using geoinformation technology for some sites in the Iraqi Kurdistan region. Intern J Dig Earth 4:239–257. https://doi.org/10.1080/17538947.2010.489971

Food and Agriculture Organization, AQUASTAT Data (2016) Annual freshwater withdrawals, agriculture (% of total freshwater withdrawal). https://data.worldbank.org/indicator/ER.H2O.FWAG.ZS. Accessed 13 December 2022

Food and Agriculture Organization of the United Nations (FAO) (2018) Water management in fragile systems. Building resilience to shocks and protracted crises in the Middle East and North Africa. FAO and World Bank. Rome and Washington DC

Furman HKH (2003) Dust storms in the Middle East: sources of origin and their temporal characteristics. Indoor Built Environ 12(6):419–426. https://doi.org/10.1177/1420326X03037110

Ghazanfar SA, Böer B, Khulaidi AWA, El-Keblawy A, Alateeq S (2019) Plants of sabkha ecosystems of the Arabian Peninsula. In: Gul B et al. (eds) Sabkha ecosystems. Springer, pp 55–80

Guinness P, Nagle G (2021) IGCSE Cambridge and O level geography, 3rd edn. UK, Hodder Education

International Union for Conservation of Nature - IUCN (2020) The IUCN red list of threatened species. Version 2020–1. https://www.iucnredlist.org, Accessed 19th Dec 2020

Marin P, Shimon T, Joshua Y, Klas R (2017) Water management in Israel: key innovations and lessons learned for water-scarce countries. World Bank, Washington, DC

Organization for Economic Co-operation and Development (OECD) and the Food and Agriculture Organization (FAO), 2018Organization for Economic Co-operation and Development (OECD) and the Food and Agriculture Organization (FAO) (2018) OECD-FAO Agricultural Outlook 2018-2027, OCDE editions, Paris/FAO editions, Rome. https://doi.org/10.1787/agr_outlook-2018-fr

Salami AA, Fenta AA (2022) Spatio-temporal evaluation of open access precipitation products with rain gauge observations in Nigeria. Arab J Geosci 15:1785. https://doi.org/10.1007/s12517-022-11071-9

Salami AA, Tilakasiri SL (2020) Impact of climate change on refugees and food shortage crises in Africa. Intern J Multidisc Stud (IJMS) 6(2):58–78

Salami AA, Tilakasiri SL, Ahamed YA (2017) The indicators and indices of development. In Tilakasiri SL (ed) Geography in development: issues and perspectives (pp 77–102). Sri Lanka: Stamford Lake (Pvt) Ltd

Tilakasiri SL, Olanrewaju RM, Salami AA (2017) Climate change induced migration and conflict in African States. In: De Silva S (ed) Development dynamics: transforming societies for sustainable futures. Saga University, Japan, pp 337–335

United Nations Environment Programme (2016) GEO-6 Regional Assessment for West Asia. Nairobi

United Nations Environment Programme (2017) The North Africa Middle East Air Quality Regional Report. Nairobi

United Nations-Water, Food and Agriculture Organization (2007) Coping with water scarcity. Challenge of the twenty-first century

United Nations-Habitat (2011) Lebanon urban profile: a desk review report. United Nations Human Settlements Programme. http://unhabitat.org/lebanon-urban-profile-a-desk-review-report-october-2011/

Warner K (2008) Human security, climate change and environmentally induced migration. UN University, Environment and Human Security. www.ccir.ciecin.columbia.edu/nyc

World Bank (2016) The cost of air pollution: strengthening the economic case for action (English). World Bank Group, Washington, D.C.

Soil Salinization Impacts on Land Degradation and Desertification Phenomenon in An-Najaf Governorate, Iraq

Sa'ad R. Yousif

Abstract The danger of Desertification is one of the most important global problems facing large areas of the world, the affected ones by the extreme conditions of aridity that maliciously change all environmental, economic and social levels. The problem has been greatly exacerbated in Iraq after 1992, for many reasons including: climate change, low rates of rainfall, and urbanization at the expense of agricultural land without regard to rules and regulations, unsustainable farming practices and poor irrigation. The increasing decline in surface water levels of Iraq's rivers has been further aggravated by this phenomenon, resulting in its expansion to regions that were previously considered among the most fertile agricultural areas in the world. Soil salinization results from high temperatures, especially in the summer, which may reach 50 °C and sometimes more. It represents the most common features that leads to land degradation, which increases the rate of evaporation in surface water, increases the deposition of salts, and affecting the growth and production of the plant. The result is increasing soil erosion and degradation, which is widely regarded as a a serious indicator of Desertification. Remote sensing data provide great potential for assessing, monitoring and controlling the criteria of Desertification to large areas of the world, because of the ability of remote sensing to provide a broad and frequent view of the Earth's surface, and thus play a major role in the continuous monitoring of the Earth and its various resources.

Keywords Desertification · Soil salinization · An-Najaf province · Remote sensing

S. R. Yousif (✉)
Department of Geology, University of Kufa, An-Najaf, Iraq
e-mail: saad.aboghnaim@uokufa.edu.iq

1 Introduction

Desertification is a prominent humanitarian issue that poses a significant environmental challenge. It leads to the decline of agricultural land and productivity in regions characterized by arid and semi-arid climates, including Iraq. Iraq is currently confronted with a pressing issue that poses a significant threat to its food security. This issue stems from the low productivity of its arable land, which exacerbates problems such as soil salinization, waterlogging, vegetation cover degradation, and sand dunes encroachment. These challenges are further compounded by the occurrence of dust storms, which adversely affect agricultural, irrigation, railway, and urban development projects. Additionally, this phenomenon contributes to environmental pollution and various forms of pollution.

The objective of this study is to provide a comprehensive analysis of the issue of Desertification on a global, regional, and local scale. This will be achieved by examining the concept of Desertification and the international standards that have been established to address this problem. Additionally, the study will explore the various factors contributing to Desertification, including natural and human-induced causes. Furthermore, the impacts of Desertification on the biosphere and the environment will be investigated. Lastly, the study will assess the role of remote sensing in detecting the presence of Desertification and its associated effects. The research's emphasis lies in examining this issue within the context of global developments, taking into account the United Nations' and international organizations' concern regarding the gravity of this problem. A distinction exists between the concepts of desert and Desertification. The former refers to regions characterized by harsh environmental conditions that are inhospitable to human habitation. In contrast, Desertification pertains to the process by which desert areas expand, encroaching upon previously stable regions and suitable for human activities. The process of expansion being discussed is not solely attributed to natural causes. Desertification can manifest within desert regions rather than solely in their peripheral areas. Despite the challenging environmental conditions that deserts exhibit, they frequently exhibit a delicate equilibrium comprising agricultural regions, natural grazing lands, diverse fauna, and water reservoirs. The utilization of these resources is carried out by the human population, which has successfully adjusted to the challenges posed by the environment's harsh natural conditions.

The notion has been reiterated since the inaugural United Nations-sponsored international conference on Desertification, which took place in Nairobi, Kenya, in 1977, until the conclusion of the second United Nations Conference on Environment and Development, held in Rio de Janeiro, Brazil in 1992. This includes: According to Johnson (1977), Desertification refers to the decline in arid regions' productivity due to natural and anthropogenic factors. However, it is worth noting that this concept not only pertains to reduced productivity in dry environments, but also encompasses the issue of declining productivity in sub-humid and even humid environments. This decline is primarily attributed to human activities and their impact on natural resources. According to Menshing (1977), Desertification refers to expanding desert-like conditions into regions previously characterized by wetland and sub-humid environments. Based on Menshing's definition, this concept is limited to regions with high humidity and moderate humidity, disregarding arid and semi-arid regions that are most affected by desertification characteristics. In these areas, the level of aridity increases significantly, transforming highly arid regions into arid desert areas. The definition of Desertification was established during the UNESCO World Conference in Nairobi in 1977. It refers to the widespread degradation of ecosystems caused by both climatic conditions and excessive human resource exploitation. This degradation reduces the ability of land-based biological production and overall land degradation (Al-Hadadin 1996). Desertification refers to the process by which arid and semi-arid regions expand their boundaries, transforming once fertile areas into barren wastelands. This phenomenon has been extensively studied by scholars such as Al-Rihani (1986) and Khawlie (1990). The concept of Desertification has been delineated by various authoritative bodies, including the United Nations Environment Programme (UNEP) in 1990, the United Nations Conference on Environment and Development in 1992, and the United Nations Conference on combating Desertification in 1995. According to these sources, Desertification refers to the degradation of land in regions characterized by dry, semi-arid, arid, and sub-humid conditions, which occurs as a consequence of both climatic shifts and human interventions. The phenomenon of Desertification encompasses the decline in the environmental bioenergy, leading to a diminished capacity to sustain agricultural and pastoral activities on the affected land. Additionally, it results in the degradation of essential ecosystem functions (Ghunaimi 1997). The concept of Desertification has been delineated by various authoritative bodies, including the United Nations Environment Programme (UNEP) in 1990, the United Nations Conference on Environment and Development in 1992, and the United Nations Conference on combating Desertification in 1995. According to these sources, Desertification refers to the degradation of land in regions characterized by dry, semi-arid, arid, and sub-humid conditions, which occurs as a consequence of both climatic shifts and human interventions. The phenomenon of Desertification encompasses the decline in the environmental bioenergy, leading to a diminished capacity to sustain agricultural and pastoral activities on the affected land. Additionally, it results in the degradation of essential ecosystem functions (Ghunaimi 1997). According to the United Nations Convention, Desertification refers to the decline in land productivity in arid regions due to natural factors such as frequent droughts, irregular rainfall patterns, and processes like aeolian and fluvial erosion. These factors contribute to the loss of fertile surface layers in agricultural land. Additionally, local factors such as the migration of sand dunes from deserts towards inhabited areas or the conversion of agricultural land for urban development further exacerbate the problem. In both cases, the common outcome is a reduction or destruction of the land's biological capacity, resulting in the prevalence of semi-desert conditions. This is primarily influenced by changes and fluctuations in rainfall, as well as increased erosion activity. According to Lehman et al. (2017), the outcome is a decline in both the quantity and quality of terrestrial ecosystems, as well as the depletion or loss of bio-energy resources, which may ultimately result in the emergence of desert-like conditions. Desertification is a dynamic process characterized by three distinct phases. The initial phase involves an environmental imbalance resulting from the accumulation of salts in the soil or a reduction in vegetation cover, encompassing both natural and agricultural plant species. The second stage of environmental degradation is characterized by a notable decline in the quality of environmental elements, particularly in terms of reduced agricultural or pastoral land productivity, decreased soil cohesion, and the presence of sand dunes. The aforementioned characteristics become more apparent during the third phase, wherein Earth's capacity to sustain its soil and vegetation diminishes, resulting in the transformation of these areas into arid deserts that are incapable of supporting plant growth (Al-Rihani 1986).

1.1 Desertification Situations

Views on desertification situations and degrees were identified by the Conference on Desertification held in Nairobi 1977 in four main ways that could be included as follows:

Slight Desertification: It is represented by minor damage or destruction of vegetation and soil, accompanied by the destruction of parts of agricultural land to be of desert properties.

Moderate Desertification: It represents the imbalance and damage to the environment to a moderate degree in its vegetation, the formation of small sand dunes, the formation of a number of protrusions, and the apparent salinization of the soil, which reduces the return of production by 10–50%. It reaches 35.6% of the arid lands area in the world.

Severe Desertification: The effects of this type is mainly emerge through changes in plant life, especially the spread of unpalatable grass and shrubs at the expense of economically beneficial plants. In addition, there is a clear activity of erosion (aeolian and fluvial), a shortage of production, and a rise in salinity to the extent that the land use is difficulty followed in such areas, and extreme Desertification accounts for 28.3% of the world's drylands, and 28.9% of the area of Africa's alone.

Very Severe Desertification: This type of Desertification is represented by a high degree of severity. This form of Desertification is characterized by the transformation of arable land into a state of complete unproductivity. Remedying this situation typically requires substantial financial investment and economic resources, focused on a limited area of land affected by Desertification. However, it is worth noting that the effectiveness of such interventions is often limited.

1.2 Desertification in the World and the Arab World in Particular

The majority of regions susceptible to Desertification are situated within developing nations in Africa, Asia, Latin America, and the Caribbean. It is important to acknowledge that a significant number of lands that are either free from Desertification or at risk of Desertification are situated in the Arab world. Statistical data reveals that approximately 357,000 square kilometers of agricultural or arable land, which accounts for approximately 18% of the total land area of 1.98 million square kilometers, is affected by the issue of Desertification. The issue of Desertification significantly affects agricultural land in the Arab world, as these regions are particularly susceptible to this problem due to their location within arid and semi-arid climatic zones. Consequently, the twelfth regional conference of the Food and Agriculture Organization (FAO) of the Middle East, convened in Jordan in 1974, put forth a recommendation for the creation of a regional institute dedicated to the examination of the escalating issue of Desertification. According to studies conducted by the United Nations, approximately 357,000 square kilometers of agricultural or arable lands in the Arab world experience annual degradation. Furthermore, it is observed that 80% of the total land area, which amounts to 1980,000 square kilometers, is affected by Desertification. The area of the Arab World, which is 14 million km^2, represents about 10.2% of the world's area. Because of its extension within several latitudes showing various environmental and climatic zones, including arid and semi-arid areas with precipitation of less than 300 mm per year. This amount is considered a little useful compared with the high evaporation rate that equals 15% of the total rainfall, and falls on an area of 66.6% of the Arab World, which means that about 11.5 million km^2 of the Arab World area are arid and semi-arid areas. This is about 23.5% of the total area of arid and semi- arid areas on Earth's surface. The area of agricultural land in the Arab world is estimated at 198 million hectares (1.98 million km^2). Of which about 40 million hectares (20%) are used, and 80% of the remaining agricultural area depends on rain, despite the huge water resources in the Arab world, and the forests cover an area of 268 million hectares. Moreover, those who depend on these two aspects (agricultural and pastoral) reach about 20 million persons, representing 43% of the total labor force, or about 30% of the total population (Khawlie 1990).

1.3 Soil Salinization as an Aspect of Desertification

The phenomenon of Desertification is often characterized by a significant increase in soil salinity, particularly in the topsoil. This salinity has a detrimental impact on soil characteristics, leading to reduced productivity and subsequent neglect of the land. In some cases, it even results in abandonment and migration away from affected areas. This phenomenon is observed in extensive geographical areas across the globe, particularly in regions characterized by longstanding agricultural practices that rely on traditional techniques and irrigation methods. Additionally, the limited efficiency of irrigation systems, coupled with the neglect of irrigation and drainage projects, collectively contribute to an initial increase in soil salinity and subsequent waterlogging.Soil salinity, in contrast to salinization, is a naturally occurring characteristic, while soil salinization is mainly a human-induced process.The salinization represents an increase in the elevation of salt concentration in the root zone caused by the buildup of salt in the uppermost layer of the soil,which impedes plant growth. The increase in salinity results from the use of high salinity water in low permeable soil, or the amount of irrigation water is insufficient to wash salt from the soil resulting in an increase in salinization and this is helped by high evaporation and salinity accumulation on the soil surface (Al-Ashraf 1994). This phenomenon is one of the manifestations of Desertification in the arable areas that irrigated by-products or near the beaches, or areas of presence of the halite minerals, a problem that arises in central and southern Iraq where the presence of chloride salts and the appearance of saline-bearing plants is considered. Dougrameji and Clor (1977) pointed out that it is not easy to control salinization in the irrigated areas of southern Iraq, owing to the soft soil tissue and its slow permeability, as well as the topography of the flat area, which makes it difficult to provide efficient water-disposal and salinity collection. One feature of Desertification in Kuwait is the soil salinization phenomenon of white, which reflects the accumulation of salt and its accumulation on the soil surface as a result of the activity of the poetic characteristic, especially when the deaf is close to the soil surface. Lal and singh,1998 indicated that the salinization process is caused by inadequate irrigation management, and salinity in agricultural land is the cause of chemical degradation resulting in reduced productivity.

Agriculture, which has been going on for more than 4,200 years in the soil of the Mesopotamian Plain in Iraq, accompanied by the use of inefficient irrigation, has gradually increased the level of groundwater, as well as the weakness and inefficiency of drainage systems in some areas of the Mesopotamian Plain, all of which have contributed to the aggravation of the problem of salinity, alkalinity and soil waterlogging. Many agricultural land in the Mesopotamian Plain was subsequently converted into desertified areas (Al-Rihani 1986).Waterlogging represents soil saturation with moisture when the groundwater level in the root zone rises. This condition leads to the deprivation of oxygen for the roots, causing asphyxiation and hindering their ability to respire effectively (Al-Ashraf 1994). This is caused by wasteful irrigation water beyond the capacity of the field-based drainage, which is helped by the presence of water-less layers of dehydration near the surface of the earth, turning the area into unproductive marshes and sabkhas. The lack of good drainage and poor efficiency of the process exacerbates this phenomenon. The cane and papyrus plants exhibits a notable disparity when compared to other crops, and this problem seems very clear in marshes, swamps and low areas in Iraq. In addition to the use of the fallow land system, i.e. leaving the agricultural land fallow, especially in the long hot season, which increases soil salinity in its surface layer as a result of the rise of saline groundwater due to the capillary property and evaporation, leaving salts accumulated on the upper surface of the soil (Al-Rubaie 1986).The factors of Desertification in

Iraq reflect environmental, economic and social impacts. For the former, they represent the deterioration of plant and animal life, the degradation of soil and pasture land, the reduction of agricultural land area, the shortage of water resources and the deterioration of its quality, especially the high salinity; This is due to inefficient use of water sources and misuse of old irrigation methods.

The economic effects are summarized in the United Nations Environmental Survey 1972–1992 in which it states: Land degradation and Desertification affect the capacity of countries to produce food and the accompanying decline in regional and global food production capabilities. They also cause food shortages in areas threatened with Desertification and their effects on food needs and the world food trade. This is reflected in the report, which confirmed that the drying of marshes had a clear effect on the decline in livestock numbers, with the number of buffaloes having decreased from 148,000 in 1990 to less than 65,000 in 2001. Fishing declined from 31,500 metric tons in 1990 to less than 22,500 metric tons in 1996. The social consequences of Desertification have been to overcome social problems, primarily the migration of rural populations and pastoralists to cities for work and better life, and this migration will have an impact on increasing the population's pressure on the limited potential of cities, accompanied by increasing demand for services from housing, schools and other services. This increase in migrants also creates social problems with the special habits, social values and lifestyles of migrants, and the attendant social conflicts that threaten the social stability of these cities.

2 Geology of the Study Area

The study area represents An-Najaf Province, which extends in the western central part of the Republic of Iraq, 160 km southwest the Capital Baghdad within the geographical coordinates latitudes (29° 50'-32° 21') N and longitudes (42° 50'-45°44') E. It is bordered to the north by the Province of Babylon, to the northwest by the Province of Karbala, and to the west by Al-Anbar Province, but from the south by the Kingdom of Saudi Arabia, by the southeast by Muthanna Province, and from the east by the Qadisiyah province, it consists of three districts (An-Najaf, Al-Kufa, Al- Manadhira) and seven subdistricts (Al-Haidariya, Al-Shabaka, Al-Abbasiyah, Al-Hurriyah, Al-Heerah, Al-Mishkhab, and Al-Qadisiyah), (Fig. 1).

The area of the Province is (28,824 km^2)which accounts for approximately 6.6% of Iraq's area of (434128 km^2).

2.1 Geological Setting

The study area is influenced by natural factors that shape it with a variety of terrain that is the product of both the rock structure and the forces that formed the rocks during geological times, which has an important impact on the identification of the

Fig. 1 Location of An-Najaf Province (Al-Maliki et al. 2018)

characteristics of rocks as to their quality and composition, i.e. the identification of the type of mineral minutes from which the soil has formed and its dominant physical and chemical characteristics, which can be determined by the knowledge of the geological fabric by which the characteristics of topography and soil are determined. Tectonically, the study area is situated according to the divisions of Foad 2015 within the stable shelf, As-Salman Zone. This zone is notable for its simple structures, (Fig. 2.) The area is marked by a NNW-SSE trending subsurface folds. These folds are evident on the surface as a series of elongated convex linear features associated with faults(Al-Atiya 2006).

Located to the western side of An-Najaf City, specifically in the Ar-Ruhban Area, there exists a region characterized by unidentified fires and rising vapors. These phenomena are potentially linked to the NW–SE Abu-Jir fault zone, which is accompanied by a corresponding arrangement of springs and sinkholes in Madhlum and Hiadhiyah, among other locations. It is plausible that these springs and sinkholes

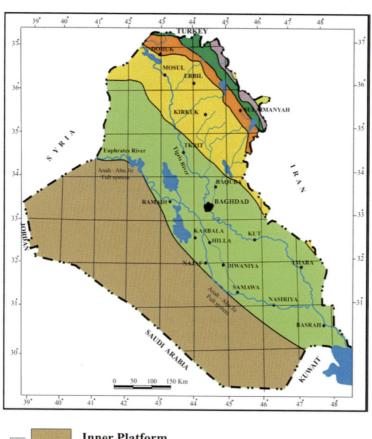
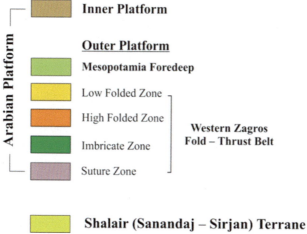

Fig. 2 Main structural zones of Iraq (Foad 2015)

are connected to the aforementioned fractures and faults. The Abu-Jir fault is classified as a normal fault, characterized by the eastern side undergoing grabbening. Furthermore, there are NE-SW transversal faults, such as the Schbicha-Najaf-Badra and Ruhaimawi-Hilla faults, as documented by Sessakian and Deikran in 1998. According to geological experts, it is widely accepted that the formation of the Abu-Jir fault occurred during the Triassic period. This fault is believed to have originated from the associated rifts that emerged during the opening of the Neo-Tethys along the passive margin of the Arabian Plate. The fault itself is characterized as a narrow grabben with relatively minor vertical displacement. During the Upper Cretaceous period, there was evident reactivation of the fault, followed by subsequent cessation. The occurrence of a dextral strike-slip along a fault has been attributed to the collision between the Arabian Plate and the Iranian and Anatolian Plates during the Alpine Movement (Dawood 2000).

2.2 Surface Features

The study area is part of the Mesopotamian Plain in Iraq, which occupies an area estimated at (1400 km^2) at a rate of (4.8%) of the total area of the Province, and it is characterized by its flattening and slope that has significant effects on surface properties, where an equal height line (20 m) passes in its northern sides, while the elevation line (15 m) passes in its southern sides, meaning that the basin of the Mesopotamian Plain decreases and is still continuing to subside due to the weight of the accumulated sediments and due to tectonic orogenies, (Fig. 3) This extension affects the poor surface water discharge, the rise of ground water levels, and the dominance of soft- texture clay soils that help to highlight the capillary characteristic. They are factors that contribute effectively to the spread of the desertification problem in the study area. In spite of the flattening of the Mesopotamian Plain, it contains accurate topographical features represented in the river levees and the adjacent slopes towards the low lands represented by the river basins, which are natural phenomena of the rivers that the river works to sediment the rocky fragments of big sizes close to its course, compared to the surrounding lowlands, where the smaller and lighter minutes are depositea. The areas of river levees in the study area extend along the Euphrates River and its tributaries of Al-Kufa and Al-Abassiya and their branches, and their height ranges between (2–3 m) from the level of the neighboring lands of river levees, this reflected its poor drainage and waterlogging soil due to the high level of ground water and the inefficiency of drainage projects. The study area is also part of the Western Desert of Iraq and occupies a large area in An-Najaf Province, as it extends from the western edge of the Mesopotamian Plain to the southwestern corner of the study area. The surface of the Western Desert is characterized by a gradual slope from the southwest towards the northeast, and its general slope rate (1 m per 2 km). Its surface is generally characterized by flatness, despite the high altitudes shown by wind weathering factors. In its eastern sides, it looks over Bahr An- Najaf, which is represents a sharp rocky cliff locally called the Tar An-Najaf. As

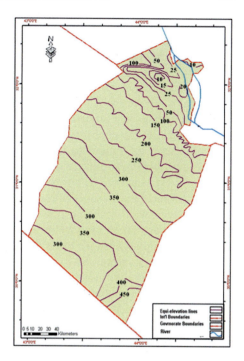

Fig. 3 Relief map of the study area (after the General Survey Authority 2011)

for the northern extension of this area towards Karbala, it is called Tar As-Sayyid. The sediments that appear An-Najaf Plateau represent aeolian Deposits in the form of Sand Sheet of 0,5–1 m thick in addition to the emergence of a number of sand dunes sand dunes of different sizes and elevations of Barchan type.

2.3 Stratigraphic Sequence

The Cenozoic Era, which spans a duration of 2–3 million years, represents the most recent geological period that Earth has undergone. This era, also known as the modern era of life, is further divided into two distinct epochs, namely the Pleistocene and Holocene epochs. The division of these epochs is based on the stratigraphic divisions.

within the geologic column. In the region of Iraq, specifically within the Mesopotamian Plain, which covers a significant portion of the country's surface area, the Pleistocene Deposits have played a prominent role. These deposits have formed an alluvial fan situated between Karbala and An-Najaf, which were deposited by the wadis of Al-Ubaiyedh and Al-Khir. The alluvial fan extends over an elevation range of 150–200 m. The Pleistocene deposits in the vicinity of the Euphrates River are characterized by peneplain sediments, specifically composed of silty clayey sand. These sediments are found in two tributaries of the river, namely Al-Kufa and Al-Abbasiya Rivers. Additionally, there are several sand dunes that extend towards the south and southwest of An-Najaf, in close proximity to the escarpment of Bahr An-Najaf Depression. According to Makki (2006), the thickness measures approximately between 4 to 7m. The Holocene epoch is commonly referred to as the age characterized by the retreat of ice. During the current era, the climate exhibits continental characteristics, which contribute to the occurrence of aeolian erosion in the desert plains and Mesopotamian Plain. This erosion is primarily driven by factors associated with aeolian processes, resulting in the formation of contemporary sand dunes. The Holocene deposits are found in a linear arrangement, running parallel to the course of the Euphrates River, starting from An-Najaf City and extending southwestward towards the city of Nasiriyah. Research has demonstrated that the source of these sands can be attributed to flood sands that were transported during the rainy period of the Pleistocene era. These sands were subsequently deposited and carried away by wind, resulting in the formation of sand dunes. Notably, these sand dunes encompass the Mesopotamian Plain and valleys basins sediments, with their thickness varying between 0.6 to 1.5 m depending on the specific location. Deposits of alluvial clay or mixed soil, commonly referred to as "playa," are often found alongside salt deposits. These depressions serve as collection points for rainwater and torrents. These phenomena occur when they are subjected to climatic conditions, resulting in the evaporation of water and the subsequent formation of salt clusters and sabkhas. Additionally, there are prominent depressions with significant expanses, such as An-Najaf, Salman, and Shabaka depressions. The depth of these depressions can reach several meters (Al-zamili 2001).The Stratigraphic column of the exposed formations of the Tertiary and deposits of the Quaternary periods is illustrated in: (Table 1).

2.4 The Climate

The long hot dry summer and the short cold winter with a wide range of temperature fluctuations from day to night and limited rainfall are the prevailing characteristics of the study area climate, as the climate governs in the dispersion and distribution of vegetation and the abundance of water resources including their physical and chemical properties. Besides climate parameters influence on soil and sediments formation, and governs on their depth, properties and transportation of mineral deposits due to erosion processes. The major climatic data of Iraq are summarized in Fig. 4

The valuable climatic parameters on the tapis, are those affect on the groundwater and the aquifer characteristics. The marked climatic elements were recorded in An-Najaf Meteorological Station (lies on longitude 44° 19′ 00" easterly, and latitude 33° 57′ 00" northerly) (Iraqi Meteorological Organization 2005), Fig. 5.

The two variables that define weathering are the Mean Annual Temperature and the Mean Annual Precipitation. Together, these two variables can be used to define a regions weathering type. Peltier (1950) defined seven graphs using these two variables that described different types of weathering phenomena, they are Chemical

Table 1 The stratigraphic sequence of the study area

Era	Period	Epoch	Age	Formatio/sediments	Symbol	Lithology
Cenozoic	Quaternry	Miocene Pliocene Pleistocene Holocne		Sand sheet	Qa	Loose sand
				Valley Fill	Qv	Pebbles, sand, silt, and clay
				Flood plain	Qf	Sand, silt, and clay
				Sabkha	Qi	
				Slope dep	Qs	Mud, silt, and clay
				Gypcrete	Qg	Sand, silt, clay, and rock fragments
				Valley ter	Qt	Secondary gypsum, and clastics
	Tertiary			Dibdibba	Dib	Sandy graveks
				Zahra	Zhr	S.st., claystone and L.st
			Upper	Injana	Inj	S.st., claystone, and siltstone
			Middle	Nfayil	Nfl	Green marl and L.st. (with Oyster)
			Lower	Euphrates	Euph	Conglomerate, L.st., dolostone, and marl
		Eocene		Dammam	Dam	Marl, fossiliforous and dolomitic L.st

Weathering, Frost Action, Weathering Regions, Pluvial Erosion, Mass Movement, Wind Action and Morphogenetic Regions.

Figures 6 and 7 show the actual graphs from Peltier's paper and contain a brief description of each of the graphs to better understand the role each graph plays in the weathering process as it is described in Peltier's paper.

Figure 8 shows Peltier's classification of the morphogenetic regions, which are 9 regions.

According to the climatic parameters and Peltier's graphs presented above, the study area properties can be summarized as follows: The study area lies within the arid region, very slight weathering, absent or insignificant frost action, minimum fluvial action and mass movement, and maximum wind action.

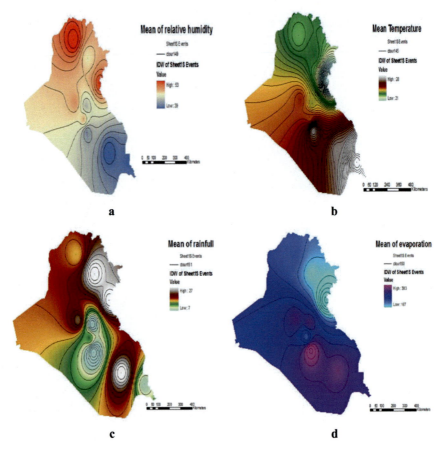

Fig. 4 The major climatic data of Iraq **a** Mean Temperature **b** Mean of Evapotation **c** Mean of Relative Humidity **d** Mean of Rainfall, (Al-Azawi 2013)

3 Concluding Remarks

3.1 Climatic –Water Balance of An-Najaf Province

Climatic water balance is one of the important criteria in determining the special water needs in arid and semi-arid regions that suffer from erratic rainfall and its species, as well as its fluctuation. The water balance is an expression of the quantitative relationship between precipitation and evapotraspiration when the amount of precipitation (p) is greater than the amount of evaporation/transpiration (E) there is a water surplus and vice versa when precipitation is less than evaporation/transpiration resulting in water deficit which refers to the amount and duration of the need for irrigation water, and it represents one of the indicators of Desertification in the study area.

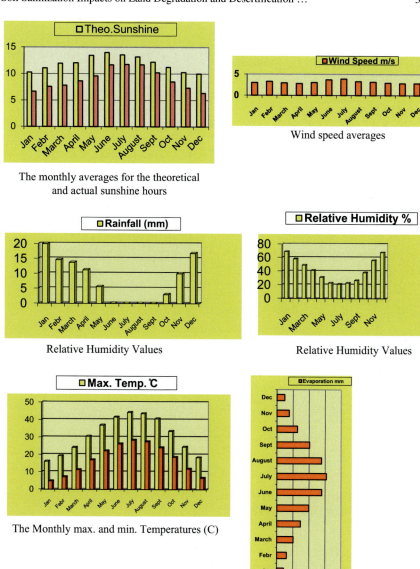

Fig. 5 The valuable climatic parameters in An-Najaf Meteorological Station for the period (1962–2002)

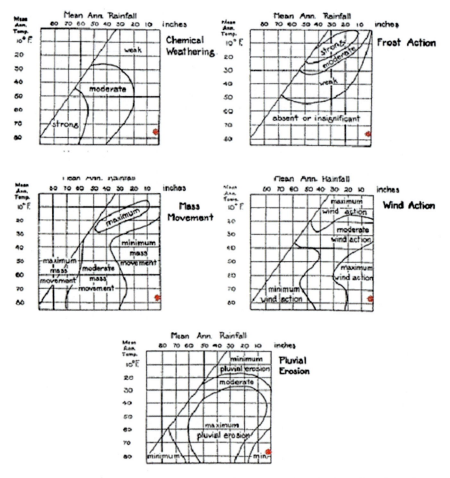

Fig. 6 Peltier's graphs including the different environments and weathering, where the study area is marked as a red point (after Peltier 1950)

The concept of water balance determines the amount of water entering the soil, what it evaporates from, the amount of moisture content of the soil, and the water available in the roots of the plants. As the Water Balance is one of the important criteria in determining the drought and its severity, the water balance shows the water deficit or surplus in what the region gets from the rain water and it is lost through evaporation/transpiration "Evapotranspiration". The most favourable method applied in calculating the water balance based on the rates of calculating the evaporation values E and thus calculating the amount of water loss from rain water is Ivanov equation which is

$$E = 0.00018(T + 25)2(100 - Hr)$$

Fig. 7 Peltier's graph of chemical weathering types and degrees, where the study area is marked as a red point (after Peltier 1950)

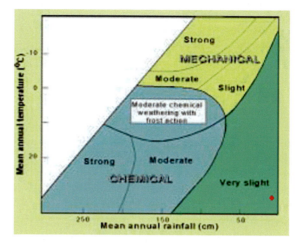

Fig. 8 Climatic boundaries of the morphogenetic regions, where the study area is marked as a red point (after Peltier 1950)

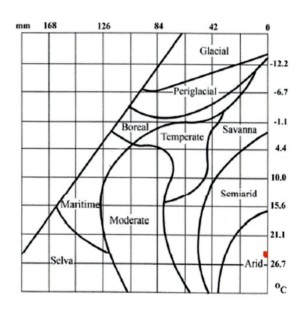

vsWhere:

3.1.1 E: Evaporation (mm)

3.1.2 T: Average of Annual Temperature (°C)

3.1.3 Hr: Relative Humidity

The characteristics of the water balance indicate that An-Najaf Province suffers from a large water deficit, as the values of evaporation/transpiration exceed the amount of rain in all months of the year, which leads to a lack of availability of water quantities that plants need for their growth due to the climate that is characterized by drought and reflects negatively on many matters, the most important of them are low discharge rates and low water levels for river networks and streams represented by surface water, well water and springs as well as a large lack of soil moisture content because plants depend on that stock to compensate for their water shortages, as well as a shortage of water needs that must be met to irrigate the cultivated areas. This results in a decrease in the amount of irrigation water reaching the agricultural lands compared to the amount of their water needs, especially in the hot season of the year, which leads to leaving areas suitable for cultivation to become fallow or wastelands it was also characterized by the appearance of drought properties throughout the year, as the relative humidity coefficient according to the study of Ivanov for the rainy months in the study area represented by months (January, February, March, April, May, October, November, and December). This indicates that the study area may witness continuous drought which leads to a sharp decrease in humidity in it.

Humidity Coefficient $= \frac{P(mm)}{E(mm)}$

Accordingly, if humidity coefficient *is* less than 0.12 Very arid desert.

0.13 – 0.29 Arid
0.30 – 0.59 Semi-Arid
0.60 – 0.99 Sub-Humid
1.0 – 1.49 Humid
Greater Than 1.50 Very Humid

Al- Khazaali (2013) depended on the classification of desertification features by identifying the causes that contributed to their existence and exacerbation based on the classification that were identified at the United Nations Conference on Desertification in Kenya (Nairobi) in 1977 and adopted on the basis of two indicators in agricultural land productivity, natural plant quality and soil erosion into four degrees, the aforementioned degrees are: Slight, Moderate, Severe, and Very Severe Desertification. So, the study area were classified due to soil salinity and the effect of Aeolian erosion, Figs. 9 and 10.

Accordingly, the spatial distribution of desertification stages in An-Najaf Province is illustrated in Fig. 11.

Fig. 9 Spatial Distribution of Soil Salinity Phenomena in An-Najaf Province (Al-Khazaali 2013)

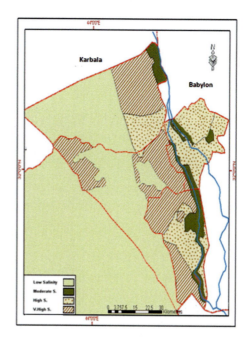

Fig. 10 Spatial Distribution of Aeolian Erosion in An-Najaf Province (Khazaali 2013)

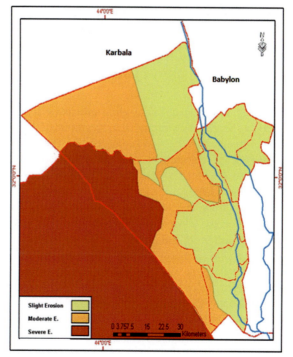

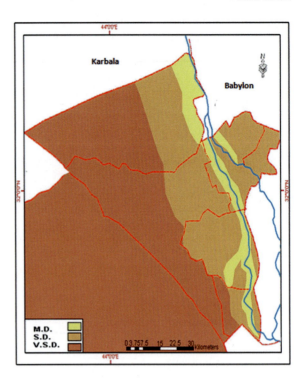

Fig. 11 Spatial Distribution of Desertified Areas due to Desertification Stages in An-Najaf Province (Khazaali 2013)

3.2 The Role of Remote Sensing Techniques in the Assessing the Soil Salinity of the Study Area

Remotely sensed data has a great efficiency for controlling dynamic processes, including Salinization through the integration of Remotely sensed data and Geoinformatics in research related to geomorphology agriculture, water, and marine sciences. It enables the researchers to overcome the difficulties of investigations, assessments and may lead to more understanding of sustainable development (Robbins and Wiegand 1990).

The Arab Organization for Agricultural Development, 1999 indicated that it is possible to take advantage of the data on remote sensing and satellite images in various processes of soil degradation, areas of spread, and its intensity depending on the interaction of energy with the objectives set out to the targets, as salts increase in soil, the brightness of this soil increases, depending on the type of salt found in the area. This helps to distinguish them when analyzing satellite images and digital data, especially those made in black and white, as saline soils appear in white, which increases in brightness as salts increase in soil (Al-A'dhami 2001).

Hussin 1979 concluded that the use of satellite data for multiple seasons of the year has given better results and significant differences when used for a single season, especially the data taken in the spring and summer seasons, giving the best results for distinguishing land use (forests, field crops, orchards, and agricultural land).

Remote sensing was successfully confirmed in detecting salinity phenomenon using Landsat imageries along with ancillary and field data to diagnose the spectral classes for salt-prone areas (Table 2).

Al-Khakani et.al. (2018) determined soil salinity indices (SSIs) using Landsat 8 Operational Land Imager (OLI) sensor and field data of electrical conductivity (EC) samples for a part of An-Najaf Province to make thematic maps of soil salinity, where there is a strong relationship between SSIs and ECm. The result of the study is making various soil salinity maps based on soil salinity indices, and deduce the predicted EC mapping due to the measured EC values as the formula explains:

Predicted (EC) = -7696.1 + 0.596 *SI5 + 0.265*B4

Al-Khakani and Yousif (2019) studied a part of An-Najaf Province to assess and detect soil salinity changes and vegetation cover for the period 2001 to 2015 by calculating Spectral vegetation indices (VIs) from multitemporal Landsat images of March 2001, 2009, and 2015 as keys to assess and monitor natural resource that cause salinity in soils via the controlling of change detection in vegetative cover, beside image differencing technique to establish the difference image accompanied to special indices that demonstrates the changes in salinity in the study area.

$$NDVI = \frac{NIR - R}{NIR + R}$$

$$DNDVI = NDVI(t1) - NDVI(t2)$$

$$SI = \frac{(SWIR1 - SWIR2)}{(SWIR1 + SWIR2)}$$

Table 2 The main indices suitable for determining and predicting salinity

References	Spectral Functions	Salinity Indices
Khan et al. (2005)	$NDSI = \frac{R-NIR}{R+NIR}$	Normalized Differential Salinity Index
Khan et al. (2005)	$BI = \sqrt{R^2 + NIR^2}$	Brightness Index
Douaoui et al. 2006()	$SI\,1 = \sqrt{B*R}$	Salinity Index 1
Douaoui et al. (2006)	$SI\,2 = \sqrt{G^2 + R^2 + NIR^2}$	Salinity Index 2
Douaoui et al. (2006)	$SI\,3 = \frac{SWIR1*SWIR2 - SWIR2*SWIR2}{SWIR1}$	Salinity Index 3
Abbas and Khan (2007)	$SI\,4 = \frac{R*NIR}{G}$	Salinity Index 4
Bannari et al. (2008)	$SI\,5 = \frac{G*R}{B}$	Salinity Index 5

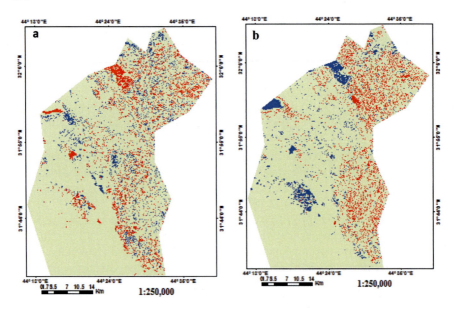

Fig. 12 Change/no-change maps of DNDVI, **a** for (2001–2009) periods, **b** for (2009–2015) periods

$$DSI = SI\,(t1) - SI\,(t2)$$

The results were summarized in Table 3 that demonstrates the accuracy of the comparison of difference images of Normalized Difference Vegetation Index and Salinity Index, and Figs. 12 and 13.

4 Conclusions

Numerous researchers have studied the issue of desertification and its effects on the An-Najaf Governorate. The resulting deterioration of environmental systems has been a cause for concern, particularly in relation to the soil, which plays a crucial role in various economic activities, including agriculture. The decrease in productivity caused by soil degradation has become a significant issue. There are many factors contributing to desertification and its environmental degradation, encompassing economic, social, and political aspects. Natural factors such as location, surface features, climatic characteristics, water resources, and natural vegetation, as well as human factors such as wrong agricultural methods, overgrazing, and population expansion, all contribute to erosion, salinization, and waterlogging.

The study area's location and geographical extension from southwest to northeast, as well as its position in relation to latitude circles, have contributed to the dryness of the soil and lack of vegetation cover. This location allows for an increase in incoming solar radiation and heating of the Earth's surface, resulting in a rise in

Soil Salinization Impacts on Land Degradation and Desertification ...

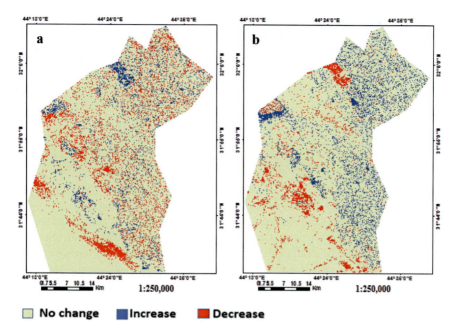

Fig. 13 Change/no-change maps of DSI, **a** for (2001–2009) periods, **a** for (2009- 2015) periods

Table 3 Change no-change classification accuracy of DNDVI and DSI for the two change periods

2009–2015				2001–2009				Change class
DSI		DNDVI		DSI		DNDVI		
UA%	PA%	UA%	PA%	UA%	PA%	PA%	UA%	
93.75	93.75	86.6	92.8	92.3	96	91.66	95.65	Change increase
85.75	100	92.8	86.6	95.45	95.45	90	96.42	Change decrease
94.44	85	93.7	83.3	96	88.88	93.10	84.37	No change
91.66		91.11		94.54		91.56		Overall accuracy
0.87		0.86		0.91		0.87		Kappa coefficient

UA = User's Accuracy PA = Producer's Accuracy

temperatures and increased evaporation values. The surface characteristics of the area are predominantly flat, with a lack of slope from the north and south, which affects the poor drainage of surface water and the high groundwater level. The dominance of soft clay soils and the lack of natural drainage make the area vulnerable to wind erosion.

Various studies have found that soil texture can vary significantly in different areas. River banks tend to have medium-textured mixed soils, while basin soils are typically fine-textured and heavy in clay. Sand dunes lands, on the other hand, have coarse sandy soils. Abandoned lands can also have different textures, ranging from clayey to sandy to mixed clay soils. Unfortunately, these abandoned lands are often prone to waterlogging, which leads to soil salinity and an increased risk of desertification. Remote sensing technology has been helpful in analyzing the effects of desertification, including the identification of saline areas.

5 Recommendations

The following recommendations could be beneficial:

1. Enhance the scientific level by providing ample opportunities for research and study and encourage scientific research both morally and materially.
2. Protect the environment from potential threats and restore its natural balance.
3. Restore and rehabilitate lands that have been affected by desertification, minimize their degradation, and safeguard them from salinity risk within the environmental limits of the region.
4. Protect and regulate the presence of environmental and water resources in the study area by preventing pollution.
5. Balance economic activities such as industry, agriculture, and urban activities to ensure the integration of land investments in the region and prevent expansion towards agricultural lands.
6. Increase awareness and agricultural extension among farmers and the general population. Offer training courses on environmental protection and its significance.
7. Encourage science students and researchers to study field and applied science. Establish specialized laboratories for scientific research and development purposes.

References

Abbas A, Khan S (2007) Using remote sensing techniques for appraisal of irrigated soil salinity. In Oxley L, Kulasiri D (eds) International congress on modelling and simulation (MODSIM) modelling and simulation society of Australia and New Zealand, Brighton, pp 2632–2638

Al- Khazaali HMK (2013) Deserting features in Al-Najaf province and its reflections on the reality and the future of agricultural status, (Unpub.) M.Sc. thesis, Univ. of Kufa, Iraq, p 299 (in Arabic)

Al-A'dhami RAM (2001) Using remote sensing methods to study movement Sand dunes in the middle of the Iraqi Mesopotamian Plain, (Unpub.) M.Sc. thesis, University of Baghdad, Iraq

Al-Ashraf M (1994) Protecting the pastoral environment from Desertification in the Kingdom of Saudi Arabia, Desert Studies Symposium, Kuwait (in Arabic)

Al-Atia MJ (2006) An-Najaf land: the history, geologic heritage, and natural wealth's, an-nubras foundation, An-Najaf, p 159 (in Arabic)

Al-Azawi FW (2013) Utilizing of remote sensing and GIS for desertification monitoring in the agricultural areas, Part of Iraq, (Unpub.) Ph.D. thesis, Univ. of Baghdad, Iraq, p 136

Al-Hadadin WJI (1996) Desertification in Madaba Province , (Unpublished) M.Sc. Thesis, College of Higher Studies, The Jordanian University (in Arabic)

Al-Khakani ET, Yousif SR (2019) J Phys: Conf Ser 1234:012023

AL-Khakani ET, Al-Janabi WF, Yousif SR, Al-Kazaali HM (2018) Using landsat 8 OLI data to predict and mapping soil salinity for part of An-Najaf Province. Ecol Environ Conserv 24(2):572-578

Al-Maliki LJ, Al-Mamoori SK, El-Tawel K, Hussain HM, Al-Ansari N, Al-Ali MJ (2018) Bearing capacity map for An-Najaf and Kufa Cities Using GIS, Engineering, vol 10, no 5

Al-Rihani AMN (1986) The phenomenon of desertification in Iraq and its effects on investing natural resources, (Unpub.) Ph.D. Thesis, University of Baghdad (in Arabic)

Al-Rubaie TAH (1986) The effect of agriculture, irrigation and leaving fallow on the land's Salinization, (Unpub.) M. Sc. Thesis, Univ. of Baghdad, Iraq (in Arabic)

Al-Zamili AJ (2001) A geographical analysis of the variation of surface shapes in An-Najaf province, (Unpub.) M.Sc. thesis, University of Kufa, Iraq (in Arabic)

Bannari A, Guedona AM, El-Hartib A, Cherkaoui Z, El-Ghmari A (2008) Characterization of slightly and moderately saline and sodic soils in irrigated agricultural land using simulated data of advanced land imaging (EO-1) Sensor. Commun Soil Sci Plant Analy 39(19-20):2795-2811

Dawood RM (2000) Mineralogy, origin of celestite and the factors controlling its distribution, Tar Al-Najaf, Najaf Plateau, (Unpub.) M.Sc. thesis, Univ. of Baghdad, Iraq (in Arabic)

Douaoui AEK, Nicolas H, Walter C (2006) Detecting salinity hazards within a semi-arid context by means of combining soil and remote sensing data. Geoderma 134(1-2):217-230

Dougrameji JS, Clor MA (1977) Case study on desertification Greater Mussayeb project. Iraq United Nation Conference on Desertification. A CONF. 74/10

Foad SFA (2015) Tectonic map of Iraq scale 1:1000000, 3rd ed., Iraqi Bulletin of Geology and Mining, vol 11, no 1

Ghunaimi ZA (1997) Environment and human: study of human problems with environment, knowledge facility, Alexandria, Egypt, p 350 (in Arabic)

http://www.jstor.org/stable/2561059

Hussin YA (1979) Regional inventory of forest and land use Classification in north Iraq using aerial and satellite data, (Unpub.)M.Sc. thesis, Univ. of Mosul, Iraq (in Arabic)

Iraqi Meteorological Organization (2005) Climatic Elements Data Recorded in Al-Najaf Station for the Period (1962–2005)

Johnson D (1977) The human dimensions of desertification. Econ Geogr Clark Univ 53:317–321. https://doi.org/10.2307/142968

Khan NM, Rastoskuev V, Sato Y, Shioza-wa S (2005) Assessment of hydrosaline land degradation by using a simple approach of remote sensing indicators. Agricult Water Manag 77(1–3):96–109

Khawlie MR (1990) Desertification in Arab homeland, Arabic unity studies center, Beyrouth, Lebanon, p 213 (in Arabic)

Lal R, Singh BR (1998) Effect of soil degradation on crop productivity in East Africa. J Sus Agric 13(1):15–35

Lehman RM, Cambardella CA, Stott DF, Acosta-Martinez V, Manter DK, Buyer JS, Maul JE, Smith JL, Collins HP, Halverson JJ, Kremer RJ, Lundegren JG, Ducey TF, Jin VL, Karlen DL (2017) Understanding and enhancing soil biological health: the solution for reversing soil degradation. In Sustainability, Special Volume, Karlen DL, Rice CW (eds) Enhancing soil health to mitigate soil degradation, MDPI, Switzerland

Makki MM (2006) Geographical characteristics of the middle euphrates region and its spatial relationship to regional specialization, (Unpub.) M.Sc. thesis, University of Kufa, Iraq (in Arabic)

Menshing H (1977) The problem of Desertification in and around Arid Land, Applied Science and Development, vol 10, London

Peltier LC (1950) The geographic cycle in periglacial regions as it is related to climatic geomorphology. Ann Assoc Am Geogr 40(3):214–236

Robbins CW, Wiegand CL (1990) Field and laboratory measurements. American society of civil engineers, Agricultural Salinity Assessment and Management

Sissakian VK, Derikan DB (1998) Neotectonic Map of Iraq, Scale 1:1000000, Geosurv, Baghdad, Iraq

Printed in the United States
by Baker & Taylor Publisher Services